Contemporary Irish Drama

Contemporary Irish Drama

Second Edition

Anthony Roche

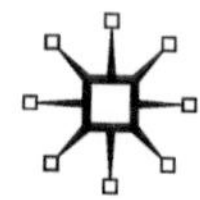

First edition published 1994 by Gill & Macmillan Ltd

Second edition published 2009 by
PALGRAVE MACMILLAN

Palgrave Macmillan in the UK is an imprint of Macmillan Publishers Limited, registered in England, company number 785998, of Houndmills, Basingstoke, Hampshire RG21 6XS.

Palgrave Macmillan in the US is a division of St Martin's Press LLC, 175 Fifth Avenue, New York, NY 10010.

Palgrave Macmillan is the global academic imprint of the above companies and has companies and representatives throughout the world.

Palgrave® and Macmillan® are registered trademarks in the United States, the United Kingdom, Europe and other countries.

ISBN-13: 978-0-230-21978-6 hardback
ISBN-10: 0-230-21978-0 hardback
ISBN-13: 978-0-230-21979-3 paperback
ISBN-10: 0-230-21979-9 paperback

This book is printed on paper suitable for recycling and made from fully managed and sustained forest sources. Logging, pulping and manufacturing processes are expected to conform to the environmental regulations of the country of origin.

A catalogue record for this book is available from the British Library.

A catalog record for this book is available from the Library of Congress.

10 9 8 7 6 5 4 3 2 1
18 17 16 15 14 13 12 11 10 09

Printed and bound in Great Britain by
CPI Antony Rowe, Chippenham and Eastbourne

To Katy, Merlin and Louis

HAMM: [*Anguished.*] What's happening, what's happening?
CLOV: Something is taking its course.

Samuel Beckett, *Endgame*

DIDO: What happened? Everything happened, nothing happened, whatever you want to believe, I suppose.

Frank McGuinness, *Carthaginians*

Contents

Acknowledgements viii

Introduction 1

1 Beckett and Behan: Waiting for Your Man 13

2 Friel's Drama: Leaving and Coming Home 42

3 Murphy's Drama: Tragedy and After 84

4 Kilroy's Doubles 130

5 Northern Irish Drama: Imagining Alternatives 158

6 The 1990s and Beyond: McPherson, Barry, McDonagh, Carr 220

Notes 261

Select Bibliography 278

Index 286

Acknowledgements

My first debt of gratitude is to Richard Pine who encouraged this book since its inception; his friendship and views on contemporary Irish drama were an unfailing resource. I would like to thank Terence Brown for commissioning this study for the series, Gill's Studies in Irish Literature, and for the guidance and patience with which he saw it through. I am grateful to my colleagues in the School of English, Drama and Film, University College Dublin, for their support, especially Ron Callan, Anne Fogarty, Declan Kiberd, the late Gus Martin, Frank McGuinness, Gerardine Meaney and Christopher Murray. I owe a particular debt of thanks to Bernard Loughlin and Pat Donlon, directors of the Tyrone Guthrie Centre, Annaghmakerrig, Co. Monaghan, where some of these chapters were written and revised.

This revised and updated second edition owes a great deal to the support and encouragement of Christabel Scaife, my editor at Palgrave Macmillan. In the first edition, I particularly thanked Elizabeth Doyle, Katy Hayes, Eileen Kearney and Conor McPherson. The last named is now a subject of the book and Katy Hayes married me the month the book was published. She remains my inspiration and sustainer. I would also like to thank Nicholas Grene, Anna McMullan, Paul Murphy, Mark Phelan and Melissa Sihra for their helpful comments and suggestions.

I would like to thank the following publishers for permission to quote from the plays: Faber and Faber, for Samuel Beckett, *Waiting for Godot, Endgame, Krapp's Last Tape, Happy Days, Play, Not I, Footfalls, Rockaby*; Brian Friel, *Philadelphia, Here I Come!, Faith Healer, Translations, Dancing at Lughnasa, Wonderful Tennessee*; Anne Devlin, *Ourselves Alone*; Tom Paulin, *The Riot Act*; Frank McGuinness, *Observe the Sons of Ulster Marching Towards the Somme, Carthaginians, Someone Who'll Watch Over Me*; Marina Carr, *Low in the Dark, Portia Coughlan*; The Gallery Press, for Brian Friel, *The Enemy Within*; Thomas Kilroy, *The O'Neill, The Death and Resurrection of Mr. Roche, Tea and Sex and Shakespeare, Double Cross, The Secret Fall of Constance Wilde, The Shape of Metal*; Marina Carr, *The Mai, By the Bog of Cats, Ariel, Woman and Scarecrow*; Methuen Drama, an imprint of A&C Black Publishers Ltd., for Brendan Behan, *The Quare Fellow*; Tom Murphy, *On the Outside, A Whistle in the Dark, The Gigli Concert, Bailegangaire, The Wake*; Stewart Parker, *Pentecost*; Christina Reid, *Tea in a China Cup*;

Sebastian Barry, *The Steward of Christendom*; Martin McDonagh, *The Beauty Queen of Leenane, A Skull in Connemara*; Nick Hern Books, for Conor McPherson, *The Weir*; Gary Mitchell, *In a Little World of Our Own, As the Beast Sleeps*. I would like to thank the author for permission to quote from Tom Murphy, *Alice Trilogy*.

Part of Chapter 6 first appeared in Hiroko Mikami, Minako Okamuro, Naoko Yagi (eds.), *Ireland On Stage: Beckett and After* (Carysfort Press, 2007). I am grateful to the publisher for permission to revise and re-use that material.

Introduction

Ireland has always had a particular affinity with drama. The Irish Literary Revival channelled much of its energies into the creation of a National Theatre and produced such challenging, enduring works as Synge's *The Playboy of the Western World* and O'Casey's *The Plough and the Stars.* The riotous reactions gave compelling evidence of those early audiences' close involvement with what they witnessed on the stage and the extent to which the plays bore on issues of national and personal identity. In earlier centuries, a vivid succession of Anglo-Irish dramatists – Farquhar, Goldsmith, Sheridan, Wilde and Shaw – had played with the comic conventions and norms of English society to subversive effect. So it should come as no surprise that contemporary Ireland should have experienced renewed theatrical activity in a series of remarkable works from living playwrights who – particularly in the case of Brian Friel – are also winning through to a world audience. In Ireland, the framework, the life narratives offered by Church and State, have become increasingly threadbare and inadequate. Irish drama in the second half of the twentieth century regained its urgency, as it did in the approach to Irish independence, as the site in which old models could be broken up and reshaped, re-imagined through the medium of play. In the larger worldwide crisis, of politics and of spirit, contemporary Irish drama has much to offer.[1]

The question any critic considering such a subject must ask is twofold: when to begin and what to include. Richard Pine has settled on the date 28 September 1964 as the inaugurating moment; the occasion was the premiere of *Philadelphia, Here I Come!* and with it the emergence of Brian Friel as 'the father of contemporary Irish drama'.[2] While I agree with Pine on the lasting significance of Friel's play, 1964 seems rather belated in terms of the emergence of a contemporary drama in Ireland. Friel had already written several plays prior to *Philadelphia* and if we go back five years to the late 1950s, when he was publishing short stories and having plays performed on the radio, Friel no longer seems such an isolated figure. Rather, he emerges as one of a succession of leading Irish playwrights who first came to prominence at that time. Brian Friel, Tom Murphy, Hugh Leonard and John B. Keane, who between them account for a great deal of what is covered by the term 'contemporary Irish drama' in terms of performances, publication and audience appeal, all began writing and having plays staged within a few years of each other; and Thomas Kilroy was already writing theatrical criticism in that same year (1959) before becoming a practising playwright in the 1960s. It cannot be said that the reason for this simultaneous emergence was due to the encouragement of the National Theatre. Indeed, if anything, the playwrights shared a rejection by the Abbey. Tom Murphy's first full-length play, *A Whistle in the Dark* (1961), was turned down by the Abbey, and when it was later performed at the Olympia Theatre in Dublin, it provoked the remark 'I never saw such rubbish in my life' from Ernest Blythe, the Abbey's Managing Director.[3] John B. Keane submitted his early plays to the Abbey and just as readily had them returned, often, he felt, unread.[4] In the case of Murphy, his play received a staging in London and found its way back to Dublin via that route. Keane's emergence coincided with the beginnings of the influential and popular amateur drama festivals in Ireland; and it was the presentation of his first play, *Sive*, by the players of his native Listowel at the first All-Ireland Festival in 1959 that carried off first prize – and ironically won it a brief run in the Abbey as a result.

Those playwrights who had a play accepted at the Abbey ran into a different set of possible problems. There was always the possibility that a second might be rejected, as was the case with Hugh Leonard. When that happened, the man who was born John Keyes Byrne adopted the *nom-de-théâtre* of Hugh Leonard from the character of a psychopath in the rejected play. Subsequently, Leonard was to work in close tandem with the annual Dublin Theatre Festival, which had also begun in the late 1950s, and premiered a play there every year for several decades, either an original work like *The Au Pair Man* and *The Patrick Pearse Motel*, or an adaptation like *Stephen D*, based on Joyce's *A Portrait of the Artist as a Young Man*. The fate of the early Friel is the most complex and interesting, and receives extensive treatment in Chapter 2. But for now it is enough to say that the one play of his – *The Enemy Within* – staged at the Abbey in 1962 contains many of the themes but little of the dramatic inventiveness of his later work. It is impossible to imagine *Philadelphia, Here I Come!* achieving its full effect without the intricate and subtle lighting provided by the play's original director, Hilton Edwards, at Dublin's Gaiety

Theatre on 28 September 1964. From shortly after the death of Yeats until well into the 1960s, Ernest Blythe ruled the Abbey with a tight fist and an equal lack of imagination. It was only with the opening of the new Abbey and Peacock Theatres in 1966 and 1967 respectively, and the enlightened artistic direction of such as Tomás Mac Anna, Hugh Hunt and Alan Simpson, that the situation for aspiring playwrights improved considerably. I would single out the production of Tom Murphy's *Famine* at the Peacock in 1967 as a key event. Murphy had been in England during the 1960s, writing plays that had nowhere to be staged. He returned from London to live in Dublin in 1970, as did Hugh Leonard; both returns were made possible by the tax-free status afforded artists and writers by the then Minister for Finance, Charles J. Haughey. But even prior to that the Abbey had signalled its intention to produce his works; and he moved to the main stage with *The Morning After Optimism* in 1971 and *The White House* the following year.

But even if that group of late 1950s playwrights is concentrated upon, and their emergence related to the lifting of the economic embargo of decades and the welcoming of foreign investment, it overlooks a key theatrical development in the Dublin of the earlier 1950s: the brave experiment of the small Pike Theatre, founded by Alan Simpson and Carolyn Swift. What the Pike ought chiefly to be remembered for was its discovery of a contemporary Irish playwright, when on 19 November 1954 (almost a full decade before Friel's *Philadelphia, Here I Come!*) they presented the world premiere of Brendan Behan's *The Quare Fellow*. Although subsequently taken up and re-presented by Joan Littlewood at London's Stratford East, it was the production of the play at Dublin's Pike and the attention it received that drew world attention to Behan. But the Pike also gets the credit for presenting the Irish premiere of Samuel Beckett's *Waiting for Godot* only a week after its London English-language début. What followed was a year-long run; *Godot* in Ireland was to prove a popular rather than an avant-garde success. Out of these experiences Alan Simpson wrote a most interesting book, *Beckett and Behan and a Theatre in Dublin*, offering 'a general background which should serve as a basis for future study of the playwrights [Beckett and Behan] as Irishmen and increase his [the reader's] appreciation of their plays'.[5] My first chapter is written in just this spirit, a joint study of Beckett's *Waiting for Godot* and Behan's *The Quare Fellow* as two plays in which the title character never appears and in which a great deal of the on-stage activity consists of waiting. It might be objected that Beckett's play was originally written in French. I would counter-argue that there are two distinct but related dramatic texts, *En Attendant Godot*, written in French, and *Waiting for Godot*, written in Hiberno-English, and further add that the history of literature in Ireland has much to do with a dual linguistic inheritance, the sense of English as somehow a language which is not entirely native. This hybrid quality is registered in Beckett's text by the ghostly presence of French, as it is in Behan's play by its relation to the Irish.[6]

This study argues that the presiding genius of contemporary Irish drama is Samuel Beckett. His *Waiting for Godot* had as profound and far-reaching an influence on an Ireland emerging from the self-imposed isolationism of several decades as it did on contemporary drama elsewhere. But I will go on to stress the crucial significance of Beckett as an exemplary figure for Irish drama. His plays raise such dramatically relevant issues as translation, exile, estrangement and dispossession in their language and stage situations, themes which are equally at the centre of plays which occupy a more recognisably Irish and local setting. I do not wish, especially now that he is dead, to re-inter Beckett in a comforting Irish critical milieu when he spent so much of his life and art working to free himself from such associations. His is, and will remain, an extreme and singular talent. I wish less to posit a direct influence, although in the cases of younger contemporary Irish playwrights like Frank McGuinness and Marina Carr this has been creatively and explicitly acknowledged. More than any influence, the critical act of placing Beckett in the context of contemporary Irish drama reveals preoccupations that he shares in common with more clearly 'rooted' Irish playwrights: a rejection of naturalism and the linear plot of the well-made play as inappropriate to a post-colonial society like Ireland; a favouring dramatically of an imposed situation in which the characters find themselves and which they either disguise or subvert through rituals of language, gesture and play. Contemporary Irish drama does not so much rely on a plot as on a central situation, whose implications are explored and unfolded in a process which is more likely to be circular and repetitious than straightforward. This is as true for Behan's prisoners in Mountjoy and Frank McGuinness's sons of Ulster at the Somme as it is for Beckett's Vladimir and Estragon in their rendezvous with Godot.

If situating Beckett's plays in an Irish context reveals features of his drama that are more culturally grounded than is usually recognised,[7] the context of Beckett in turn enables the emergence of a more international and purely theatrical dimension of 'Irish' plays. His contemporary Irish successors have sought to bring the dramatic and other lessons of 'our necessary interpreter' (Desmond Egan's phrase) back to bear on a more recognisably Irish milieu and context. The comparison with Beckett will establish the major criterion for inclusion in this study – that all the contemporary Irish plays selected for analysis strike a balance between national and international, between local speech and situations and the universal language of theatrical form. The aim has not been to limit contemporary dramatists like Friel, Murphy, Kilroy and McGuinness by 'fitting' them to Beckett, but rather to create between them a larger cultural context in which facets of very different works are mutually illuminated. This has inevitably led to certain exclusions, of which I must particularly note and regret the plays of Hugh Leonard and John B. Keane, works which deserve and would benefit from a different critical orientation to show their dramatic qualities.[8]

The central character in Beckett, as earlier in Yeats and Synge and later in Friel and Murphy, is on the margins of society, driven to extremes by conditions which the play will examine. A liminal figure like Friel's Frank Hardy or Murphy's JPW King uses all the resources of drama to call the norms of that society into question. The outsider's condition is demonstrated in Tom Murphy's first play, *On the Outside*, written in the late 1950s with Noel O'Donoghue, which concentrates on two young men denied access to a dance hall, and gives a more localised version of the themes of exile and estrangement, displacement and dispossession. The early plays of Murphy and Brian Friel register an increasing dissatisfaction with either the well-bred bourgeois domestic setting or the peasant cottage kitchen as equally demeaning and distorting theatrical stereotypes. Part of the struggle of their early plays is an attempt to escape the 'Irish' label without succumbing to anonymity or the competing materialist ideologies of England or the US.

Such an argument prepares the ground for the middle section of the book, a career-length chapter each on the plays of Brian Friel, Tom Murphy and Thomas Kilroy. In only one of the three cases, that of Kilroy, does each of the writer's plays receive individual attention. This was done in the case of Kilroy's work because it is still (undeservedly) the least known and lacking in critical treatment. In the case of Brian Friel's drama there are over a dozen full-length studies in print and several collections of individual essays have appeared.[9] Accordingly, I have chosen to divide the majority of my Friel chapter between the play which stands head and shoulders over all his early work, *Philadelphia, Here I Come!*, and the play which I consider his masterpiece, *Faith Healer* (1979). I do not omit the extremely important *Translations* (1980), but preferred to place that in the context of the Field Day Theatre Company co-founded by Friel and actor Stephen Rea in 1980. *Dancing at Lughnasa* (1990) is considered as his greatest global success; and *Wonderful Tennessee* (1993) as his most Beckettian.

Friel's work falls chronologically into phases which can be aligned with successive decades. His 1960s plays set the local and the international in geographic terms. His setting is the archetypal Donegal village of Ballybeg (from the Irish, *baile beag*, meaning 'small town'), its community, claims and frustrations set in opposition to the appeal of the US as the site of emigration, projected through media images and enhanced by distance and desire. In *Philadelphia, Here I Come!* Friel's dramatic breakthrough is to find an appropriate theatrical metaphor to realise his theme. He does so by casting two male actors in a continuous double-act in order to represent the divided psyche of his 25-year-old lead, Gareth O'Donnell. Friel's work of the 1970s is increasingly more complex as it comes to focus on issues of language and identity, given local provenance by political developments in Ireland North and South. The culmination is *Faith Healer*, where the four monologues provide perspectives on a situation – man/woman, Ireland/England – that are finally divergent and in which the lure of homecoming turns into a death trap. The 1980s saw Friel's energies devoted to the Field Day theatre company. In two of his three plays for Field Day, Friel moves away from the present to examine key turning points in Irish history – the decline of the Irish language in the nineteenth century in *Translations*, and the defeat of Hugh O'Neill and a Gaelic civilisation at the Battle of Kinsale in *Making History* (1988).

Translations is a great deal more: it draws on and develops the father and son situation from the earlier *Philadelphia* and above all meditates on and contextualises the profound relationship between two civilisations and languages. In 1990, Friel broke with Field Day and returned to the Abbey Theatre for his play, which focused on the repressed energies of five Donegal women from the 1930s, *Dancing at Lughnasa*. The play appeared in the same year as the election of Mary Robinson to Ireland's presidency and went on from Dublin to acclaim and awards in London and New York. When he accepted Olivier Award for the play in 1991 in London, Friel did so by quoting the late Graham Greene to the effect that 'success is only the postponement on failure'. He claimed back his necessary freedom and space as an artist by going on to write *Wonderful Tennessee*, which explored the same themes as *Lughnasa* (such as Christianity versus paganism) but in a more abstract and challenging way. His next two plays, *Molly Sweeney* (1994) and *Give Me Your Answer, Do!* (1997) also unflinchingly confronted the uncertainties of the Irish present. Since then, Friel has primarily devoted himself to a creative engagement with the drama of Chekhov, in translations, adaptations and an original work, *Afterplay* (2002), where characters from *Uncle Vanya* and *Three Sisters* meet up in a Moscow cafe decades later. There have been two late original works, 2003's *Performances*, about the composer Janacek, and 2005's *The Home Place*, in many ways a sequel to *Translations*, with head-measuring replacing map-making.

The other leading contemporary Irish playwright is Tom Murphy, less well known than he deserves to be outside the country, but whose best plays have had a seismic impact on Irish theatre, provoking walk-outs and heated discussion. While much media attention has been given to the North, there has been a 'Southern' crisis over the past four decades which has hardly made it into the international press and of which Murphy has been one of the few and certainly the most incisive of commentators. His first full-length play, *A Whistle in the Dark*, manages to write a genuine tragedy out of a contemporary situation as it looks at one Irish family, the House of Carney, displaced into an English suburban setting and taking out their inherited conflicts not only on the traditional enemy but ultimately on themselves. The intervening decades have not diminished the ferocity of the play and, while Murphy has written subtler plays, *A Whistle in the Dark* retains a primacy among his early work. There are several fascinating plays of the 1970s which have had to be reduced to passing comment while deserving much more, such as *The Morning After Optimism* (1971). The situation of *The Sanctuary Lamp* (1975), whose two male characters *in extremis* work out their hostilities overnight in an abandoned church, lies behind so much of Friel and Murphy's greatest work in the following decade. But after a period in the doldrums, Murphy roared back. He did so first with *The Gigli Concert* (1983), the play in which he went furthest in terms of the characters' extremes of elation and despair, and the wildly disparate elements out of which he was attempting to make a drama. This was followed two years later by *Bailegangaire* (from the Irish, 'town without laughter'), in which he gave pride of place not to men (as he usually had), but to women, in a three-hander of a grandmother and her two granddaughters gathered in a cottage in the west of Ireland. The play staged the most sustained and bizarre juxtaposition of the traditional and the contemporary to be found in contemporary Irish drama, and in the central character of the senile, rambling Mommo gave the late Siobhán McKenna her last role, one which brought her fully into the foreground of contemporary Irish theatre. After 1989's *Too Late for Logic*, which dramatised a writer confronting both domestic and existential crises, Murphy was relatively quiet in the 1990s, turning to prose for his 1994 novel, *The Seduction of Morality*. His dramatisation of that work as *The Wake* in 1998 appeared to re-engage him with the theatre and led to two important works, *The House* in 2000, a definitive representation of the plight of the returned Irish émigrés in the 1950s, and 2005's masterpiece *Alice Trilogy*, which historicises its heroine and her comi-tragic dilemmas from the 1980s into the midst of the Celtic Tiger boom years.

The fourth chapter looks at the dramatic achievement of Thomas Kilroy, his figures of radical individuality set against the surrounding society and the mutual interrogation they undergo in the course of the drama. Kilroy's development is traced from his first two plays of the 1960s, where Kelly in *The Death and Resurrection of Mr Roche* and Hugh in *The O'Neill* try and fail to assimilate themselves to their society. *Tea and Sex and Shakespeare* and *Talbot's Box* in the 1970s set their eccentric, marginalised protagonists in theatrical opposition to society in a shattering of formal constraints. In his work of the 1980s, a version of Chekhov's *The Seagull* and *Double Cross*, the assertion of the individual's freedom is rendered suspect, its implications for the society he disavows more subtly worked out. In *Madame MacAdam's Travelling Theatre* (1991), the cross-dressing theatricality of his travelling troupe is more innocent than the brown and blue shirt transvestism of an Ireland denying any involvement in World War II. *Madame McAdam* signals a crucial shift in Kilroy's plays from a man to a woman as the central character, specifically in relation to agency and the figure of the artist. Oscar Wilde is moved to the sidelines by the dramatic emphasis of *The Secret Fall of Constance Wilde* (1997) and in *The Shape of Metal* (2003) the sculptor Nell Jeffrey has to consider what she has made both of her life and her work.

The fifth chapter concentrates on theatre coming from the North of Ireland, where the charged conditions offer a challenge to its writers and audiences.[10] The first career to be considered is that of the late Stewart Parker, who came from within the Protestant community in Belfast. Parker was alert to the dramatic formulas which readily offered themselves in regard to the Northern situation and fought against them with inventiveness, wit, passion and resilience. His premature death (in 1988) is on a par with the loss of Synge and Behan in relation to Irish theatre, all the more so given the society and people for whom he wrote. But he warned against martyrdom and would prefer that his legacy be a living, lovingly admonishing one. This prophetic legacy is explored in an examination of his last and best play, *Pentecost*. The Northern context is also that in which Irish women playwrights have made the greatest contemporary contribution.[11] Christina Reid's *Tea in a China Cup* offers a dramatic reassessment of the terrain covered by Parker's *Pentecost* from the perspective of the women – three generations of them – in a working-class Protestant community; the drama is more communal than personal, invoking a Kristevan sense of women's time and a radical approach to the structuring and emphasis of experience. Anne Devlin's *Ourselves Alone* appropriates the Sinn Féin emblem for its three women characters and demands their insertion into and disruption of a nationalist history which is male and chauvinist in about equal proportions. The Charabanc Theatre Company drew together a nucleus of women from the theatrical community to collectively write and perform (but not direct) plays which went on tour throughout the North.

This chapter also provides a context for the fifteen-year dramatic experiment of Derry's Field Day Theatre Company, not least of whose successes was the number of first-rate plays it contributed to the repertoire: two by Thomas Kilroy, Stewart Parker's *Pentecost,* three originals and two adaptations by Brian Friel, and versions of classic plays by Tom Paulin, Derek Mahon, Seamus Heaney and Frank McGuinness. In addition to individually examining the plays Kilroy and Parker wrote for Field Day, this study goes on to examine two more in detail, Friel's *Translations* and Paulin's version of Sophocles' *Antigone, The Riot Act* (1984), as representing the treatment of history and myth respectively in the Field Day enterprise.[12] The chapter then considers the formidable achievement of Frank McGuinness, whose career began in the early 1980s and has led to equal acclaim in Ireland and worldwide, both for his fierce and moving original plays and a series of acclaimed adaptations, especially of Ibsen.[13] In *Observe the Sons of Ulster Marching Towards the Somme* (1985), McGuinness showed himself the most disruptive and experimental of contemporary Irish dramatists. 1992's *Someone Who'll Watch Over Me* dramatised the fate of the hostages in Beirut through the Beckettian situation of three men waiting in a cell for murder or freedom and gave McGuinness one of his greatest successes to date. The changing political landscape of Northern Ireland in the 1990s was registered in the theatre. Both Field Day and Charabanc Theatre Companies came to an end; but their impact was registered by – and the baton passed to – the younger companies and new playwrights who emerged in the North. Actor-writer Tim Loane co-founded Belfast's Tinderbox Theatre Company with actor Lalor Roddy, with the declared intention of developing and producing new work that would interrogate and explore life in Northern Ireland. In 1993, they premiered the first stage play by the most important Northern Irish playwright to emerge in the 1990s, Gary Mitchell. The chapter will examine the two plays with which Mitchell came to wider attention, *In a Little World of Our Own* (1997) and *As the Beast Sleeps* (1998). Both plays discomfit any cosy notions that the advent of the peace process in Northern Ireland would transform hardened traditional attitudes and patterns of behaviour overnight.

The final chapter looks at four of the younger generation of Irish playwrights who established their careers in the 1990s – three from the Republic of Ireland (Conor McPherson, Sebastian Barry and Marina Carr), one London-born product of the Irish diaspora (Martin McDonagh). If this was the period when a large measure of peace (not least a cessation of daily violence) came to Northern Ireland, it also witnessed the arrival of unprecedented affluence in the South under the sobriquet of the 'Celtic Tiger'. And yet what is so striking about the contemporary plays by McPherson, Carr and McDonagh is how financially impoverished their characters are. The pub dwellers in McPherson's *The Weir* carefully count out their coins when buying their round of drinks; only the property owner Finbar is flush. And the cuisine of the McDonagh plays is far from *haute*: lumpy Complan, chicken vol-au-vents and Kimberly biscuits are much discussed. What all of them suggest is that the economic benefits of the boom were unequal and that there remained a distinct under-class which was in no position to benefit from property speculation. Even when the plays move on to depict characters who are considerably more affluent, as in Friel's *Wonderful Tennessee*, Murphy's *Alice Trilogy* or Marina Carr's *Ariel* (2002), they also demonstrate a degree of spiritual anomie and of sundered personal relations which affluence has aggravated rather than reduced. This was also the period when the seamier side of Irish society was coming to light; the high incidence of child abuse, for example, which is considered by McPherson in several of his plays and by Carr in *On Raftery's Hill* (2000). The political scandals and corruption which a succession of tribunals brought to light are scrutinised by Carr in *Ariel* and by Sebastian Barry in *Hinterland* (2002). In Irish cities, Dublin in particular, familiar landmarks disappeared and generic business centres and shopping malls took their place. Direct representation of such a landscape would show little to distinguish contemporary Ireland from any European equivalent, although this was more of an issue for film-makers. Perhaps as a consequence, plays of this period often eschewed conventional dramaturgy and opted instead for the one-on-one intimacy of the monologue play, where a single character stood on stage and addressed the audience directly. Brian Friel's *Faith Healer* and *Molly Sweeney* were influential here, but behind them in turn were the monologues of Beckett and Pinter. Young urban playwrights like McPherson and Mark O'Rowe made a particular feature of the form; and even a McPherson play where the characters meet and interact (as in *The Weir*) still foregrounds the monologues of the individual storytellers.

Equally non-representational, in a way, were the distinctive features that came to mark out an 'Irish' play: a rural setting, impoverished country folk, extremes of drinking and violent behaviour. If these had ever been true of Irish society, they scarcely were so with the largely urbanised and educated population of the 'Celtic Tiger'. But some or all of these features became emblematic of Irish plays in the 1990s, especially in the case of Martin McDonagh. This was no longer a case of representing a society undergoing a rapid change from tradition to modernity but of making play with a set of markers, of dramatic ingredients, which were available to any enterprising writer, urban or rural, indigenously Irish or diasporic. Another way of putting this is to say that there was no longer a single Ireland (predominantly Catholic, almost exclusively nationalist) to represent. This was vividly dramatised by Sebastian Barry's history plays, which imagined back into existence the lives of family members who had been erased from the record because their life narratives did not conform to that of Catholic Nationalist Ireland. What the playwrights now have to contend with is a series of micro-Irelands, all widely differing from each other in terms of the social reality they represent. As Chris Morash puts it, writing of the year 2000: 'there is no such thing as *the* Irish theatre; there are Irish *theatres*, whose forms continue to multiply as they leave behind the fantasy of a single unifying image, origin or destiny'.[14]

But in a way this has always been the case. From the self-conscious start of Ireland's national theatre a century ago, the staging of a new play with its implicit or explicit claim to represent Irish experience was immediately met with vociferous counter-claims that it signally failed to do so. A potentially endless series of plays has always been generated by one playwright reacting to another and challenging his or her version of Irish experience. Then, the issue was likely to be one of Protestant versus Catholic ideology, or of arguments between differing strands of nationalism. Now, the areas of exclusion are likelier to relate to plays dealing with the experiences of and written by women, gay people or, in an Ireland that now has a significant multi-ethnic population, immigrants.

What has always marked and guaranteed the ongoing vitality of Irish drama is a strong element of creative contradiction, not just when one play provokes another by way of response, but within the individual plays themselves. There, each version of experience that one character seeks to put forward as definitive is immediately called into question by another, as Beckett's Vladimir finds virtually every time he tries to impart something to Estragon; apart from their agreement to wait for Godot, they disagree on almost everything else. Even when there seems to be a leading man in Irish drama, as when Friel's faith healer is given the stage to himself and allowed to offer his apparently definitive version of events in monologue, his place is soon taken by a woman whose account seriously calls into question much of what he has told us. This principle of creative contradiction extends to the audience, who are drawn into the process of questioning what is being represented. This interrogative process was both registered and resisted by at least some of the early audiences of Synge's *Playboy of the Western World* in 1907. But it also ensures the survival and ongoing vitality of Irish drama, whose life is not bounded spatially by the four walls of the theatre or the two-to-three hours' temporal duration of the play. There is no reason why this principle of creative contradiction, of speaking against received narratives, should not continue well into the new century.

1
Beckett and Behan: Waiting for Your Man

Any study of contemporary drama usually takes as its starting date the close of World War II. Once hostilities had ceased, the renewed availability of human and material resources fed the expansion of the media of cinema, theatre, radio and the new baby, television. And the resources of expression no longer had to be channelled exclusively into propaganda. The sophisticated innocence of the 1930s had become the calculated innocence of wartime propaganda. What emerged in the immediate post-war culture was a darker, more disillusioned awareness and an urge to break up the old moulds of expression. The United States, less affected materially since it was spared bombing and extensive rationing, marked the shift most rapidly, the profound change occasioned by the loss of those who did not return and the altered sensibilities of those who did. *Film noir* essays in romantic despair unreeled from Hollywood. In American theatre, the social dramas of Arthur Miller sought to locate the enemy within, while Tennessee Williams's fragile heroines and their dreams of aristocracy could not sustain the shock of a rapacious capitalism fattened by the war. In England, where the spell of nostalgia was greater, returning wartime entertainers like Spike Milligan found only a modified outlet in radio satire for the personal devastations they had suffered. It was not until 1956 and John Osborne's *Look Back In Anger* that the angry young men broke through.

In Ireland, the situation was different yet again, since the country had never officially entered or recognised the war. Instead, it maintained a studied neutrality behind the impassive gaze and steel-rimmed glasses of its public face, Éamon de Valera. With the outbreak of war, a state of emergency was declared. The Irish State's neutrality meant that the country was not bombed. Rationing was selective, with meat readily available. Beckett, returning from France to Ireland in 1945, was struck by the contrast in conditions. He abstained from all but tea and bread, remarking: 'My friends eat sawdust and turnips while all Ireland safely gorges.'[1] Officially, there was no loss of life while unofficially thousands of Irish men and women served in the Allied Forces and many were killed. Under the mask of neutrality, private sympathies were as likely to go in the German direction, especially when stimulated by anti-English feeling. Since it had suffered least and enjoyed instead a 'cosy insularity', Ireland was least likely to mark the end of World War II with a profound change. Instead, it sought to maintain the conditions and climate which had prevailed during and before the hostilities. The North of Ireland served and suffered in both world wars; and the differing stance in this regard is still one of the greatest dividing factors in contemporary relations between North and South.

Samuel Beckett and Brendan Behan, unlike the majority of their fellow countrymen, suffered during World War II.[2] Beckett did so by returning to Paris on 3 September 1939, and Behan through the Free State's determination to crack down on Republican diehards. Prior to playwriting, both men's personal and politicised experience were vastly different from the bourgeois humdrum of Ireland's six years of cosy insularity. While Beckett was assisting the French Resistance, Behan was serving out his time in prison. And both were marking the degree of their alienation from the Irish Free State by schooling themselves in a language and culture to which that mainstream was largely indifferent – Beckett in French, Behan in Irish. It was during the war and in the writing of his novel *Watt* that Beckett made his decision to write in the French prose of the trilogy. With Behan, there was an equal commitment to write in Irish. His four years in Mountjoy Prison from 1942 to 1946 (when he was released under a general amnesty) were Behan's equivalent of Synge's journeys to the Aran Islands. They crystallised his decision to 'adopt a literary rather than a revolutionary career'[3] and brought him into contact with fellow prisoner Sean O'Briain, formerly a Kerry schoolmaster, who coached him in Gaelic language and literature.

During the period of World War II, there was no Irish republican equivalent to the mythic confrontation between good and evil staged on the European battlefield. The lines of conflict were muddied, and had been since the Civil War. Local bombings and random shoot-outs, like that which saw the young Behan sentenced to 14 years' penal servitude in April 1942, were the order of the day. And when in 1949 the Taoiseach John A. Costello declared Ireland a Republic, he did so in the hope 'that the declaration of the Republic might take the gun out of Irish politics'. Instead of inaugurating a new model society, as Pearse's

1916 document was intended to do, the 1949 declaration was simply a stopgap strategy designed to ameliorate an existing state of affairs, a formula of words whose declaration altered very little and passed almost unnoticed. One of its few material benefits, as we shall see, was to allow Ireland an independent production of Beckett's English-language *Waiting for Godot* beyond the legal control of the United Kingdom and the theatrical impresarios of London's West End.

A marked difference between an England at war and an Ireland in a state of emergency was that the Dublin theatres remained open. Far from suffering a decline, the demand for theatrical resources was greater than ever.[4] The Gaiety and Olympia Theatres, now deprived of touring productions originating from London and including Dublin as one of their 'provincial' stops, were forced to look for shows closer to home. As a result, Mícheál Mac Liammóir and Hilton Edwards moved from their smaller Gate Theatre into the Gaiety. But if the larger stage allowed more scope for Edwards's directorial flair and Mac Liammóir's sweeping style of acting, especially in their Shakespearean productions, the need to draw a much larger audience led to a decline in artistic standards in the Gate repertoire. Instead of presenting the best of contemporary European drama and an occasional new Irish play, the partners were forced into repeating popular productions and adapting novels like *Wuthering Heights* in 1944, all the more circumscribed by having to follow the Hollywood contours of the 1939 William Wyler film.[5]

The Abbey Theatre was in even worse condition, artistically, during the 1940s. By the outbreak of World War II, both Yeats and Lady Gregory were dead. The conservative takeover of the board signalled by Frank O'Connor's departure in 1940 was sealed by the appointment of Ernest Blythe as Managing Director in January 1941. Originally described as a 'temporary appointment', Blythe's directorship lasted for another 26 years. While Minister for Finance, he had secured a government subsidy for the Abbey in 1924, making it the first state-subsidised theatre in the English-speaking world. But if the Board had made Blythe managing director out of a sense of gratitude, the sentiment was misplaced and instituted a long reign of mediocrity at the Abbey Theatre.

Rejection by the Abbey Theatre in general and by Ernest Blythe in particular came to feature as a shared experience in the career of almost every contemporary Irish playwright. As Hugh Leonard puts it in his prose memoir, *Out After Dark*: 'I realised ... where his strength lay: he did not give a damn for the opinion of any man on earth. He [Blythe] ran the Abbey as if it were an almshouse and he were the Master.'[6] John B. Keane, who had a total of six plays rejected before he stopped sending them, holds that the Abbey Theatre by the 1950s was very much a conservative cultural institution, hostile to anyone like himself with a 'facility in their makeup to be outspoken'.[7] Under Blythe's rule, the emphasis at the Abbey was transferred from the artistic quality of the plays to the degree of their Irishness, in the most literal sense: 'It was by his decree that only fluent Irish speakers could become members of the company...To Mr Blythe one Irish accent was as good as the next, and, in any case, he regarded the performing arts as so much bunkum. The main function of the Abbey, he believed, was to revive the Irish language.'[8] With such criteria, it is not surprising that the Abbey should show little initiative in its choice of plays, relying on increasingly shop-worn productions of the few acknowledged classics and for new work on formulaic kitchen comedies with the requisite Peasant Quality, or that the insistence on bilingualism should scare off many promising actors. In 1947, the poet Valentin Iremonger and University College Dublin lecturer Roger McHugh walked out of a production of O'Casey's *The Plough and the Stars* to protest the extent to which production standards had declined; but although their protest stirred a short-term controversy, nothing was done to rectify the state of affairs against which they had reacted. On 18 July 1951, a fire gutted the original building and the company moved to the Queen's Theatre, losing the small, experimental Peacock in the process. It remained there until the new Abbey Theatre opened in 1966, and a new Peacock the year after. Thus, during the period of the 1950s, while any aspiring young playwright would first send his or her manuscripts to Ireland's National Theatre, a combination of artistic and economic factors decreed that they were just as automatically being rejected.

The need for a new theatre in Dublin in the early 1950s gained momentum with post-war developments in world drama. A playwright like Tennessee Williams had little chance of being produced at the Abbey when the suggestion of Shakespeare roused Blythe to retort: 'The Abbey doesn't do foreign plays.'[9] Nor would Williams have been staged at the Gate, where his plays would probably have been viewed as too coarse and would have failed to offer Mac Liammóir an obvious leading role. And so Alan Simpson, by day a captain in the Irish Army and a former apprentice to Hilton Edwards at the Gaiety, decided in 1953 to open a small theatre in Dublin with the intention of presenting 'plays of all countries on all subjects...which, for various reasons, would not be seen on either the larger or smaller commercial stages'. The aim of the Pike Theatre, according to the manifesto of founders Simpson and his wife Carolyn Swift, was a resolutely contemporary one: 'We hope to give theatregoers [in Dublin] opportunities to see more of the struggle going on at present in world theatre, to introduce new techniques and new subjects in play writing.'[10] But this struggle to introduce new techniques was one that equally engaged Irish playwrights, frustrated by the lack of an outlet other than the Abbey and by that theatre's stereotyped conservatism. While the Pike Theatre during its nine-year existence made a particular cause of producing Tennessee Williams's plays, it won its greatest acclaim and made a profound contribution to the development of contemporary Irish drama with two productions of first plays by Dublin-born playwrights. The Pike presented the world premiere of Brendan Behan's *The Quare Fellow* on 19 November 1954 and the Irish premiere of Samuel Beckett's *Waiting for Godot* on 28 October 1955. The first drew the attention of Joan Littlewood at Stratford East and subsequently enabled an Irish playwright like Behan to go international; the second enabled an international play like *Godot* to come home, to find a large local audience in a remarkably short time. But their production at the Pike is only the first of many characteristics the two plays share, as this chapter will demonstrate, and through their dual effect a new kind of Irish drama can be seen emerging.

Though not essential to my argument, the question of direct influence needs briefly to be considered. At first I thought there could be no question of it, but the biographies of the two writers reveal coincidences which made me less sure. Samuel Beckett and Brendan Behan met on only three occasions but each was significant. The first was in August 1948 when Behan, recently released from Mountjoy Prison, went to live in Paris, conscious of the long line of Irish writers in whose track he was following. As Ulick O'Connor records: 'Beckett...met him at this time and was pleasant to him in a slightly sad, sardonic Dublin accent.'[11] But the time of their meeting in late 1948 was close to that of Beckett's composition of *En Attendant Godot*, which was written between 9 October 1948 and 28 January 1949. Behan's resolve to stay in Paris did not last, and he soon returned to Dublin. But he continued to visit Paris and in November 1952 came to see Beckett, knocking on the door of his apartment at 6.30 a.m. and desiring to be admitted. Beckett found a figure 'covered in mud and blood, but otherwise in cheerful enough mood. Brendan insisted on coming in and chatted amiably for the next three hours. Beckett found the first half hour of the monologue fascinating, but grew understandably fatigued as it continued unabated.'[12] And Behan was not to be stopped in his apparently ceaseless verbal flow by any Lucky-like expedient of removing and trampling on his hat, if indeed he wore one. Instead, Beckett led him through the streets of Paris and dumped Behan on an unsuspecting Christopher Logue while he proceeded to rehearsals at the Théâtre de Babylone for *Godot*'s first production. On a subsequent visit by Behan to Paris in 1958, where two of his plays were receiving separate productions, Beckett managed to avoid a meeting. But in 1962, when Beckett was in London, he accompanied his cousin John Beckett, who had been musical director at the Pike and was a close friend of Behan's, to see him in hospital where he was undergoing aversion treatment for his alcoholism. As Beckett put it: 'He told us a long story about a drugged greyhound but his mind would wander away and leave the story and then come back. I admired him for his marvellous warmth, but I never really saw him sober. When he went on and on he was impossible.'[13] This is clearly the source of Deirdre Bair's account of the relationship, but she conspicuously omits the note of strong admiration to focus exclusively on the negative: 'From his bed, Behan told long, rambling, incoherent stories about drugged greyhounds. Beckett had to admit that he had never before seen Behan when he was sober, and the sight of his hulking carcass lumped under the wrinkled bed-clothes signified a terrible waste of spirit and talent.'[14] When writing to the Pike about their planned *Godot* on 7 February 1955, Beckett showed that he had read Gabriel Fallon's notice of Behan's *Quare Fellow* by adding: 'Remember me to the new O'Casey',[15] which adds a fondness to the admiration entirely missing from Bair's account. And there is something ironically apposite in the situation of Beckett's being trapped as involuntary listener by a speaker who insists on going 'on and on' and becoming impossible. This is so close to the narrative core of Beckett's prose and drama that perhaps Behan should be credited as one of the many anonymous, Irish-inflected voices that come to Beckett's listeners out of the dark. On the other side, Behan's two interventions into Beckett's life occurred when the latter was, first, in the throes of writing *Godot* and, second, attending daily rehearsals for the crucial first production. It is therefore not impossible that Behan gained earlier access than many to this most innovative of contemporary plays and, wittingly or no, drew on that prior acquaintance when drafting his own first full-length play.

But any possible theory of direct influence from Beckett to Behan is seriously put in question by the fact that *The Quare Fellow*, though not produced until late 1954, had been on Behan's mind for a long time, ever since his four years in Mountjoy Prison. As his biographer has noted, it was only under prison conditions that Behan's talent flourished, undistracted by drink and the Dublin pubs. The nurturing in a literary direction came from both sides of the cell door: the prison governor persuaded Sean O'Faolain, the most influential literary figure of 1940s Ireland, to visit Behan in prison and as a result to publish an article by him, 'The Experiences of a Borstal Boy', in *The Bell*. His fellow Republican prisoner Sean O'Briain's Gaelic schooling gave the young Behan, otherwise denied much formal education, direct access to such works as Brian Merriman's Rabelaisian, eighteenth-century long poem *Cúirt an Mheán Oíche* (*The Midnight Court*). Behan wrote his first play, *The Landlady*, in Mountjoy, where it was given a staging by the prisoners; some of their cellmates in the audience objected to portions of it as blasphemous and obscene. But Mountjoy also contributed in a more long-term way to Behan's career as a playwright by furnishing much of the material for *The Quare Fellow*. The title phrase was colloquial Dublinese for any condemned prisoner. At that time in Mountjoy there were two: one got off by reason of insanity; but the other, the quare fellow who was hanged, was Bernard Kirwan. A butcher by profession, Kirwan had so mutilated the corpse of his murdered brother that 'not even the State pathologist was able to identify the sex of the torso found in the bog'.[16] The trial, in other words, was unable to produce the body of the murdered man, just as the play Behan was to fashion from this material never delivers up to the audience the corporeal presence of the title character. While in prison, Behan subjected Ernest Blythe to a series of letters. One of these, dated 18 May 1946, reveals that the writing of *The Quare Fellow*, then called *The Twisting of Another Rope*, was already underway:

> I enclose the first Act of my play – *The Landlady*... I have written one Act of another play. *The Twisting of Another Rope* I call it, because everything is shown in the *black cell* in some prison. Two men are condemned to death and waiting for the Rope – I would send it with this but better not scare the Department of Justice before we have anything done. There is nothing political in it, of course. I'll send it to you, if you like.
>
> Every thanks to yourself, Ernest Blythe
> for your kindness to,
> Brendan Behan
>
> P.S. You know I have no chance of typing the M.S.[17]

Behan chose his original title to refer explicitly to Douglas Hyde's one-act play, *Casadh an tSúgáin/Twisting of the Rope* of 1901. In this he consciously alludes and contributes to a tradition of native Irish drama. But the mordant twist of his retitling equally signals Behan's intention of releasing some of the subversive linguistic, comic and (despite the poker-faced disclaimer) political energies which the National Theatre Movement's latter development had suppressed.

One sign of this was the failure of the Abbey to take the play. In the eight years between this begging letter to Blythe and its staging as *The Quare Fellow* at the Pike (where a new, shorter title was required to save on advertising costs), the play was translated into English and submitted in its one-act form to Radio Éireann and the Abbey Theatre, both of whom rejected it. It was then rewritten as a three-acter and resubmitted to Blythe; again, it was rejected. Since Behan's play pre-existed *Godot* in some form, therefore, the case is more one of simultaneous development; and, as I hinted earlier, the direct influence may well have been two-way, since we know that Behan gave Beckett an earful. The three-act version of the play was offered to the other companies in Dublin during the early 1950s. They too rejected it because, as Carolyn Swift surmises, its anti-hierarchical ensemble offered no leading-man role to actor-managers Mícheál Mac Liammóir and Cyril Cusack.[18] But it was noticed by Sally Travers, MacLiammoir's niece, who was friendly with Simpson and Swift and passed it on to them.[19] The play required an all-male cast of thirty, over half the number the Pike theatre could seat; this was reduced to 22 by doubling. Alan Simpson got Behan himself to record 'The Old Triangle', the song sung throughout by an offstage prisoner whose verses punctuate or rather orchestrate and help to compose the action of the play. The smallness of the Pike Theatre worked to bring the audience closer to the cramped physical conditions shared by the prisoners, evoking a feeling of claustrophobia[20] and all but abolishing the distance usually separating audience and performers by forcing them into close proximity.

Despite critical acclaim and full houses, the Pike had to take off *The Quare Fellow* after a mere four weeks. The size of the cast meant that, even though the largely amateur group was on meagre wages, running costs still exceeded box office takings. The production of Beckett's play, with a cast of only four adults and a boy who didn't have to be paid, presented no such problems. When Beckett had attempted to secure his first production as a playwright, he offered director Roger Blin not just *En Attendant Godot* but *Éleuthéria*, a full-length play with 27 characters which has never been produced and which was only published after Beckett's death. Whatever the respective merits of Beckett's two plays, Blin opted for *Godot* mainly on the practical grounds of the respective sizes of the cast. With the success of the Behan play, the Pike naturally tried to have its production transferred to a larger Dublin theatre, as it was later to do when its *Godot* was moved to the Gate. The Gaiety was not available. The management of the Olympia refused to take the production, apparently on the grounds that a relative of Behan's had worked in the theatre's front-of-house staff and had also sold copies of *The Daily Worker* to customers, thereby breaching the unspoken but entrenched class warfare by which mainstream Dublin theatre remains a middle class institution. In the meantime, a frustrated (and impoverished) Behan had sold the rights to Joan Littlewood. And when *The Quare Fellow* had received the imprimatur not only of a London production but a transfer to the West End, the Olympia was happy to offer its stage to that same imported production. The non-commercial Abbey also underwent a change of heart and, now that the play had been rendered respectable, was happy to mount its own production of *The Quare Fellow*.

In the meantime, the Pike had staged *Waiting for Godot* on 28 October 1955. The London producers tried to stop the Dublin production by pointing out that they held the UK and British Commonwealth rights. Alan Simpson was able to point out in return that Ireland had in 1948 left the Commonwealth and had since 1949 been an independent Republic. Therefore, since Ireland was not legally bound by those rules, the gentleman's agreement he had concluded with the author still held. With Beckett's support Simpson and Swift went ahead with a Dublin production; out of deference to his wishes, they waited a week; and the almost simultaneous presentation of *Godot* in both Dublin and London stands as an important event in the decolonisation of Irish theatre. The Pike production of *Godot* ended up running over a year, during which time it transferred to the Gate Theatre, toured the country, played Dun Laoghaire's Gas Company Theatre, and established a run of over a hundred performances. What was regarded as an avant-garde play in Paris and London established itself from the first as a piece of popular theatre in Ireland, where less cultural and physical distance separated the audience from the characters on stage.

When Behan attended a symposium at the Pike on *Godot*, he described it as 'a piece of class propaganda and a product of the Welfare State'.[21] But Behan's *Quare Fellow* reveals profound, far-reaching and illuminating points of comparison with Beckett's *Waiting for Godot*, even though the one seems resolutely specific (and specifically Irish) as to its setting and characters, while the other names its characters from diverse European nationalities, is set in a void, and undermines every particular of time and place.

What I now propose doing is to consider them intertextually and by so doing to unfold a series of correspondences in the deep structure of the two plays that will help to mark out the terrain of a contemporary

Irish drama. Such a comparison is part of this study's aim to reveal Beckett as not only an international (or French or English) but as an Irish playwright. But it also hopes to show Behan as much more an accomplished international playwright and much less the inspired local amateur that he has often been taken for. I would argue that this double perspective is essential if we are to arrive at anything like a proper awareness of their dramatic achievement, a sense of Beckett's Irishness as the base from which he has extracted or abstracted his art, and a sense of the degree to which Behan has transformed his materials into international and enduring works of art.

The original title of Behan's play, *The Twisting of Another Rope*, was changed to the shorter *Quare Fellow*, as we noted. But the new title, referring to the condemned prisoner whose imminent hanging generates the activity of the play, brings it more into line with Beckett's *Godot*. For one of the most distinctive features of both plays is that neither of the title characters, the Godot whose imminent arrival is continually anticipated or the quare fellow whose grave we see being dug, ever makes an appearance on the stage. Neither is ever seen or heard by the audience. The plays do not disclose this from the start. To do so would rob them of much of their dramatic tension, since a considerable air of anticipation is created by the expectation of their arrival on stage. This is a classical stage device: to hold off the entrance of a major character, someone who wields a dominant influence on the lives of the play's people.

In Act One of Beckett's *Godot*, encouraged by the promise in and of the title, the audience initially shares Vladimir's confident assumption that the act of waiting for Godot will be dramatically rewarded. As time passes and Estragon's sceptical questioning reveals how little either knows about the man or his intentions, doubts are sown, especially when the elaborate device of taking up lookout positions at either side of the stage turns out to be a false alarm. Just at that point in Act One when we first begin to think that no Godot is going to show, Pozzo marches on exhibiting just the air of authority and command with which Godot has been invested by them. And Estragon initially mistakes him for Godot:

ESTRAGON: [*Undertone*]. Is that him?
VLADIMIR: Who?
ESTRAGON: [*Trying to remember the name.*] Er …
VLADIMIR: Godot?
ESTRAGON: Yes.
POZZO: I present myself: Pozzo.
VLADIMIR: [*To* ESTRAGON]. Not at all![22]

If Pozzo turns out to be an extended red herring, then we might be tempted into assuming that Godot does not exist, that he is merely the wishful fabrication of these two desperate characters, a projected father figure who will come along and relieve them of the burden and responsibility of their existence. While this may be subjectively true, the objective reality of Godot within the world of the play is reasserted at the end of each Act. The Boy appears to bring Mr Godot's message that, though he has not come today, he surely will tomorrow. As the audience goes through the virtually parallel sequence of the play's Act Two, it does so at one level to confront the increasing recognition that, despite all assurances to the contrary and the traditional satisfactions of theatrical conventions and plot, Godot will *not* appear and that, denied this source of minimal solace, audience and characters alike will just have to get by on their own. This realisation has a great deal to do with the more sombre tone, the air of greater desperation, palpable in *Godot*'s second Act.

Behan's *The Quare Fellow* devotes the first of its three Acts to what seems at first conventional exposition and dramatic build-up. Set in the cells among the men *not* condemned to die, they are not just speculating on the event of the hanging, wondering whether a last-minute reprieve will be granted, but intensely imagining what the quare fellow must be going through:

> To be lying there sweating and watching...and you lie down and get up, and get up and lie down, and the two screws not letting on to be minding you and not taking their eyes off you for one half-minute, and you walk up and down a little bit more.[23]

Their number is increased by the unexpected arrival of one of the two original condemned men, who has received a reprieve. His presence serves a number of dramatic functions. As with the deliberate Pozzo/Godot overlap and blurring of identity, he occupies the role of the title character without managing to fill it. He may rather be said to travesty it. As a surrogate for the man about to be hanged, Lifer (so called because his death sentence has been commuted to life imprisonment) brings the hanging closer and manifests a greater sense of revulsion at hearing it talked about. But Lifer's reprieve for the same crime serves to point up the arbitrariness of the sentence and to expose the official act in all its civilised barbarity. So too does the hangman in precisely the opposite way – in the lack of gravity with which he approaches his task, making pleasant remarks while taking clinically precise measurements of the head, the weight of the body and the distance of the drop. In Act Two, as the scene moves from the claustrophobic confines of the cells into the apparently greater openness of the prison yard, it is only to witness the grave being dug of the soon-to-be-dead man. The Chief and Warder Donelly await the emergence on-stage of the quare fellow after his tea, with the same mix of morbid curiosity and anticipation as the audience does:

CHIEF: Is the quare fellow finished his tea?

WARDER DONELLY: He is. He is just ready to come out for exercise, now. The wings are all clear. They're locked up having their tea. He'll be along any minute.

CHIEF: He's coming out here?

WARDER DONELLY: Yes, sir.

CHIEF: [*exasperated*]. Do you want him to see his grave, bloody well half dug? Run in quick and tell those bloody idiots to take him out the side door, and exercise him over the far side of the stokehold, and tell them to keep him well into the wall where he'll be out of sight of the cell windows.

(*CP*, 88)

At this point we begin to realise that we are never going to see the condemned man on the stage, that he will be kept out of sight of the cell windows and of the proscenium frame, and also why. The case for or against capital punishment can equally be made in his absence but, more to the theatrical point, what is central to the play is the effect of the quare fellow's death on the other characters, especially on his fellow prisoners. Again, as in *Godot*, the later stages of the play progressively darken as it becomes clear not only that the title character is not going to appear on stage but that the quare fellow will not be returned to the stage of life, as the slender thread of hope for a last-minute reprieve is definitively snapped.

The prospect of hanging naturally looms large over a work originally titled *The Twisting of Another Rope*. By the mordant twist he gives to Hyde's one-act peasant play, Behan shows a keen awareness of the degree to which hanging spins itself through Irish drama as the fate and figure of contemplated action. In Hyde's comedy, the poetic outsider is extricated from the community values he threatens through the device of having him spin a rope and back himself out through a door which is promptly and decisively slammed. In Act Three of Synge's *The Playboy of the Western World*, itself a development of Hyde's little drama in which Synge discerned 'the germ of a new dramatic form',[24] the villagers once more slip a rope over the disruptive intruder and drag him outside – but now with the explicit intention of hanging him. Christy Mahon has already been verbally threatened with this fate by a piqued Pegeen Mike, whose description mingles physical punishment with suppressed sexual desire: 'the hanging of a man. Ah, that should be a fearful end, young fellow... to think of you swaying and swiggling at the butt of a rope, and you with a fine, stout neck, God bless you! the way you'd be a half an hour, in great anguish, getting your death.'[25] At the end of Friel's *Translations*, schoolmaster Hugh recalls how he and his friend Jimmy Jack set out as young men in 1798 'with pikes across their shoulders' to fight in the rebellion, but then got homesick for Athens and returned to the talk with which the play concludes. 'My friend,' Hugh says to Jimmy Jack, 'confusion is not an ignoble condition.'[26] The hanging which would inevitably and historically have been their fate had they proceeded is not necessarily noble, despite its romanticisation in patriotic ballads and such plays as *Cathleen ni Houlihan*: 'I will go cry with the woman, / For yellow-haired Donough is dead, / With a hempen rope for a neckcloth, / And a white cloth on his head.'[27] The prescribed fate for direct political or independent action in a colonial society is hanging, the legacy a condition of confusion.

The widespread hangings after the 1798 Rebellion were designed to demonstrate publicly the fate of all would-be rebels. The hanging was a public spectacle, which it had been in England; but in Ireland there was the added political complication that all those who viewed it were potential terrorists. The racialist and class elements which fed into this public spectacle were in Ireland intended to apply to the entire native population by the colonialist administration. With the advent of the Irish Free State after the Civil War, it might be assumed that hanging would no longer serve that purpose and had outlived its usefulness. But as Dunlavin trenchantly puts it in *The Quare Fellow*, giving the lie to Behan's claim to Blythe that there were 'no politics' in the play: 'When the Free State came in we were afraid of our life they were going to change the mattresses for feather beds. ... But sure, thanks to God, the Free State didn't change anything more than the badge on the warders' caps.' (*CP*, 59)

The continuation of the practice of hanging as a colonialist legacy is stressed by the dramatic point that the hangman, the necessary instrument of the practice, is an Englishman. Much is made in the play of the hangman's coming over from England to perform the deed, along with his sober hymn-singing assistant, like a pair of music-hall comedians on a tour of the provinces. The hangman is unable to use Irish assistants because they incapacitate themselves for the task through drink. As the governor is forced to admit:

> We advertised for a native hangman during the Economic War. Must be fluent Irish speaker. Cáilíochtaí de réir Meamrám V. a seacht. [Qualifications in accordance with Memorandum Seven.] There were no suitable applicants.
>
> (*CP*, 112)

In 1954, the same year as the Pike premiere of *The Quare Fellow*, the British hangman Albert Pierrepoint executed the last man to be hanged in Ireland on 20 April at Mountjoy Prison. Capital punishment was not removed from the books until 1990. It had remained on, like so much pre-independence legislation, rarely if ever acted upon, but continuing to cast a shadow over people's minds.

In *The Quare Fellow*, one of Behan's aims is to expose the hypocrisy by which the physical act of hanging is neutralised by a brief newspaper announcement. The colonial legacy is obscured by the blank facade of bureaucracy. Warder Regan is consistently the author's voice for demystifying such practices: 'I think the whole show should be put on in Croke Park; after all, it's at the public expense and they let it go on. They should have something more for their money than a bit of paper stuck up on the gate.' (*CP*, 114) Regan's proposal here is reminiscent of the scene in Joyce's *Ulysses* where a hanging is put on as a public spectacle. But in Joyce the techniques of Boucicaultian nineteenth-century melodrama are used parodically and satirically to explode the subject. In Behan's later treatment, the dramatic approach is more deliberately Brechtian: to acknowledge the spectacle as a theatrical event and penetrate through it to the motives which put it on, the social practices which authorise its continuance.

Part of *The Quare Fellow*'s socialist design is to bring its bourgeois audience beyond the public facade inside the prison, to chart in excruciating detail the processes involved in the hanging of a man. At one level, this process is procedural, exposing the outer mechanisms and elaborate ritual by which the law marks out the last days of one of its elected victims. The drama's setting in Act One contains a multiplicity of signs designed to coerce, efface, or cover up, but which the playwright in turn is objectifying in order to expose. '*Silence*' is not only a key and frequent stage direction, in Behan as in Beckett, but a visible sign '*printed in large block shaded Victorian lettering*' (*CP*, 39) on the wall confronting the audience when the curtain goes up. Outside each man's cell are cards giving the name, age and religion of the occupant which, from the point of view of the audience, might as well be blank since they are indecipherable. The inhabitants' private history is now a blank, their only identity a public one constructed from the record of their crimes. The obscene neutrality of the newspaper account – 'Condemned man entered the hang-house at 7.59. At 8.03 the doctor pronounced life extinct' – is countered by a graphic account of what is physically involved in the act of hanging: how an officer has to catch the rope and stop it wriggling; and how, after an hour, 'they cut your man down and the doctor slits the back of his neck to see if the bones are broken. Who's to know what happens in the hour your man is swinging there, maybe wriggling to himself in the pit' (*CP*, 45–6). That act of imagining is carried out by the prisoners as they take it in turn to speak out what the condemned man must be experiencing simultaneously in his cell, in the passage quoted earlier. When the quare fellow is finally hung in Act Three, his execution takes place off-stage to a dual, comic/tragic response. First, one of the prisoners offers a parodic radio commentary describing the hanging in elaborate social terms as a public celebration and spectacle, a day at the races. But as the hour strikes, the collective response from the prisoners is a '*ferocious howling*' (*CP*, 121), the point in this loquacious play at which language breaks down. The inmates can no longer articulate their reaction at the moment of their fellow's death and vent instead a mixture of rage, grief, despair and protest.

What most disrupts the 'official' hanging at the very end of *The Quare Fellow*, complicating both the course and the nature of the play's dramatic action, is the unofficial hanging with which Act One concludes. There, the condemned man who has been reprieved, the other quare fellow, is discovered unexpectedly to have tried to hang himself in his cell:

> [WARDER DONELLY *bangs on* LIFER'S *closed door, then looks in.*]
> WARDER DONELLY: Jesus Christ, sir. He's put the sheet up! Quick.
> [REGAN AND DONELLY *go into* LIFER'S *cell. He is hanging. They cut him down.*]
>
> (*CP*, 69)[28]

Where the following day's hanging is foreknown and planned, providing a strict chronological sequence and progression all the way through the play, Lifer's action is unexpected, unannounced, random and quintessentially private. His attempted hanging is entirely lacking in external authorisation. In terms of plotting, it offers no more than an isolated incident. But in its resemblance to the official event it is threatening by virtue of its identity-in-difference. For at the formal level it is a replication of the act of hanging, the very fate from which Lifer has been officially reprieved and which he now appropriates to himself. Within the play as a whole, it introduces the principle of repetition since two of the three Acts now end with a hanging, one private and one public, but dramatically indistinguishable from each other. If the official hanging represents the persistence of colonial acts of legislation after the announcement of independence, the unofficial shows even more the incorporation within the individual colonised subject's psyche of that legacy of hanging as a mode of escape from an intolerable situation.

Behan's play contains both acts of hanging, one questioning and cross-examining the other. But there is also much talk of hanging in Beckett's *Waiting for Godot*. Indeed, it emerges as the only form of self-generated 'activity' considered by Vladimir and Estragon other than the legendary waiting of the title. Here, in Beckett, the gap between an independent action and the hanging to which it would inevitably lead is foreclosed into a situation where the colonised subject hangs himself. As with Behan's Lifer, the external authorities enforcing such an outcome are internalised, and the only form of escape and possible freedom is perversely the very act which validates that authority. 'What about hanging ourselves?' (*CDW*, 18). This first reference to hanging occurs early on in the play when it has already been established that the two tramps are to wait for Godot but not what they are to do while waiting; that is, what the nature of the dramatic action is to be. The prospect of hanging is raised significantly at the close of *Godot*'s two Acts:

ESTRAGON: [*Looking at the tree.*] Pity we haven't got a bit of rope.
VLADIMIR: Come on. It's cold.
[*He draws* ESTRAGON *after him. As before.*]
ESTRAGON: Remind me to bring a bit of rope tomorrow.

(*CDW*, 51)

and

[ESTRAGON *draws* VLADIMIR *towards the tree. They stand motionless before it. Silence.*]
ESTRAGON: Why don't we hang ourselves?
VLADIMIR: With what?

(*CDW*, 87)

There then occurs the comic business with the twisting of another rope and the falling of the trousers before they conclude:

> VLADIMIR: We'll hang ourselves tomorrow. [*Pause.*]
> Unless Godot comes.
> ESTRAGON: And if he comes?
> VLADIMIR: We'll be saved.
>
> (*CDW*, 88)

It is as necessary that the hanging be postponed as that the arrival of Godot not occur, to open up a space in which the two tramps can continue to function, even in the most circumscribed of ways. Their attempts at hanging themselves are represented, not realistically, but as comic knockabout. The idea of climax is mocked and exaggerated through their efforts not just to achieve some sexual gratification but to give a frustrated audience the satisfactions of plot denouement, a spectacular climax, through the act of hanging. In Beckett's play it is twice contemplated as not only the outcome of the action but as constituting the action itself – contemplated but evaded, elided as a fate by means of the comically subversive, self-defeating pantomime with which they fail to hang themselves.[29]

With this talk of hanging, it is worth going on to consider the related question of the extent to which the dramatic situation of Beckett's tramps resembles that of Behan's prisoners. Though the setting of *Godot* is apparently open, Vladimir and Estragon turn out to be no less prisoners of their stage space than the Mountjoy inmates, unable to leave despite repeated efforts to do so. Their condition is highlighted by the last moments of each Act where, released for the time being from the admonition to wait for Godot which has held them in place, Vladimir and Estragon are no more able to move or remove themselves from the stage, despite their joint verbal agreement to do so, than prisoners in a cell:

> ESTRAGON: Well? Shall we go?
> VLADIMIR: Yes, let's go.
> [*They do not move.*]
> *CURTAIN*
>
> (*CDW*, 88)

The stage itself is a kind of prison in Beckett. Often, those who hold the prisoners subject are unseen, off-stage presences. In *Play* (1964), for example, we have to assume an authority figure who works the roving spotlight that compels each of the three victims to spit out more of their life narrative; the set-up has reminded more than one viewer of the World War II prison camps. It is a condition which the audience in Beckett is brought to experience and share, joint entrapment with the people on-stage for the allotted sentence of the evening's performance, one which a great many of the Miami audience for *Godot*'s American premiere sought to flee by managing an early escape. As Martin Esslin recounts, *Godot* only really achieved success in the US when Herbert Blau's San Francisco Actors' Workshop staged the play for 1400 convicts in San Quentin Penitentiary. A teacher at the prison was quoted as saying: 'They know what is meant by waiting...and they knew if Godot finally came, he would only be a disappointment.'[30] An inmate of San Quentin who viewed Blau's production and as a result founded a theatre workshop there, Rick Cluchey, became, after his release, one of Beckett's chosen handful of interpreters. The play has subsequently been put on by the prisoners themselves, as Behan's received its earliest performance from the prisoners in Mountjoy. And Beckett himself lived for many years in Paris overlooking a prison, which he could view from his study as he wrote.

Behan's *The Quare Fellow* is very precisely situated in an Irish prison and crammed with local Dublin idiom and reference. Mixed with this is cockney slang. They have been in gaol in England, too, these inmates. But while bringing its audience into contact with the sights, sounds and smells of the prison, the play chooses for its form the routines and rituals that make up a prisoner's life. This applies not just to the rigid, regimented techniques of the authorities, the series of dehumanising line-outs and inspections to which the prisoners are subjected, but even more to the ingenious ways in which the prisoners manage to subvert these drills with their own improvised set of counter-rituals. So, in the medical inspection, Dunlavin directs the doctor's gaze to his right leg while he drinks as much as he can from the meths bottle in full view and with the shared complicity of prisoners and audience. In a place of such deprivation and hardship, freedom is relative, measured by the space in which to smoke a cigarette. The sign of another man's humanity can be his willingness to share his 'snout'.

The prisoners also have their own internal means of communicating. One of these private languages or counter-codes is their tapping on the pipes:

WARDER 2: Yap, yap, yap. It's a wonder the bloody old hot-water pipes aren't worn through.

[*Tapping.*]
WARDER 1: Damn it all, they've been yapping in association since seven o'clock.

(*CP*, 105)

The Irish language also serves the same function in the play. If private communication between the prisoners shuts out the authorities, Gaelic goes even further by breaking down the absolute separation between warder and prisoner and the system of order maintained by such segregation and hierarchy. In Act Two, the play switches to Irish for a 'necessarily' brief dialogue between Warder Crimmin and Prisoner C, a young man from the Kerry Gaeltacht. The point is that it *is* dialogue, a fluent and equal exchange between two individuals on a first-name basis ('a Thomáis'). It contrasts sharply with the usual mode of exchange between warder and prisoner, a series of imperatives to which the only possible response is a prescribed 'yes' whose submission is ratified by the even more obligatory 'sir'. When Prisoner C is addressed by Crimmin, his 'Seadh' is questioning and wondering, an invitation to discourse which the use of the first name further encourages:

CRIMMIN: [*calls* Prisoner C] Hey!
PRISONER C: [*comes to him.*] Seadh a Thomáis?
CRIMMIN: [*gives him cigarettes and matches*]. Seo, cúpla toitín. Táim fein is an screw eile ag dul isteach chuig an oispeadéal, nóimeat. Roinn amach na toitíní siúd, is glacfaidh sibh gal. Má thagann an Governor nó'n Chief nó an Principal, ná bíodh in bhur moill agaibh iad. A' tuigeann tú?
PRISONER C: Tuigim, a Thomáis, go raibh maith agat.
CRIMMIN: [*officially*]. Right, now get back to your work.
PRISONER C: Yes, sir.

(*CP*, 91–2)

Gaelic here provides this prisoner and this warder with a medium for a more private and authentic exchange than the English language and voice, with its imperialist and authoritarian overtones. If the Irish first name provides a basis for abolishing distance and permitting the illicit distribution of *toitíní*/cigarettes, the English public titles of 'Governor', 'Chief' and 'Principal' stubbornly refuse to be translated and stand out as alien terms in the discourse. Ironically, in the case of an Irish as much as a foreign audience, they also provide the listener's only point of recognition and contact in this otherwise unintelligible brief exchange.

For all the dissimilarities between Behan's *The Quare Fellow* and Beckett's *Waiting for Godot*, my argument has converged on their shared linguistic doubleness. Each has written a play of two languages (English and French, English and Irish) with one text haunted by its other ghostly double. Each helps to define the limits of the other and in so doing to foreground the situation whereby the plays' speakers inhabit what Fredric Jameson has termed the 'prison-house of language'. The prison-house context already invoked for both plays also shows other shared contexts. Like the inmates of a prison, Didi and Gogo subsist on a near-starvation diet and inhabit an all-male environment. Their activity is principally made up of waiting and a series of formal, futile exercises. It is an environment in which the rituals they propose announce their own insufficiency and hollowness, as mere exercises to fill up and deaden the passage of time. If they choose to name the ultimate and unseen force which keeps them prisoner 'Godot', the absence of physical restraining bonds shows the imprisonment to be psychological and internalised. But a more explicit and visible context of imprisonment arrives in the form of Pozzo and Lucky. Pozzo is dressed and addresses others with all the verbal and visual trappings of authority. Lucky is restrained by a length of rope '*passed round his neck* (*CDW*, 23) – endlessly being hanged, as it were – by the crack of the whip, and by verbal commands. As with the warders in Behan's play, the arrival of Pozzo and Lucky brings the idea of social hierarchy into the stage space with profoundly disturbing consequences. It upsets the dramatic equilibrium of Didi and Gogo, as the warders break up the ensemble of the prisoners, replacing the verbal cross-talk with a series of questions and answers and lengthy set-pieces.

What both *Waiting for Godot* and *The Quare Fellow* lack in traditional theatrical terms is a leading man. It is in this feature that the anti-hierarchical nature of a post-colonialist Irish drama is most apparent. On the rare occasion when a central character seems to be foregrounded, his majesty the self is likely to be subjected to all kinds of dramatic cross-questioning and undercutting. As one of many possible examples, I would cite Frank Hardy in Friel's *Faith Healer*, who gets to speak the first and last of the play's four consecutive monologues. Any dramatic authority this would seem to confer on the title character, and Hardy's own meta-theatric role as faith healer, is undercut by the two intervening accounts of the same events by his wife Grace and manager Teddy, which vary significantly in their details. More often, as in Friel's earlier

Philadelphia, Here I Come!, the leading self is split in two and dramatically distributed between two actors, in this case to represent Gar Public and Gar Private. I will have more to say about these psychic doubles in the course of the book. But they are following a theatrical line in Irish drama from the deeply divided Marlow in Goldsmith's *She Stoops to Conquer* (1773), through the Fool and the Blind Man in Yeats's *On Baile's Strand* (1904), to the two tramps in *Godot* who, while physically separate, manifest a psychic interdependence in nothing more than their inability to split up. The bifurcated dramatic structure of the double-lead means that every pronouncement by the one is likely to be countered or questioned by the other. Vladimir only gets a chance at a Shakespearean monologue when Estragon falls asleep. More characteristically, any attempted speech and its assumed centrality of cultural norms such as Western Christianity is interrupted and put in question by a refractory auditor:

> VLADIMIR: Ah yes, the two thieves. Do you remember the story?
> ESTRAGON: No.
> VLADIMIR: Shall I tell it to you?
> ESTRAGON: No.
> VLADIMIR: It'll pass the time. [*Pause.*] Two thieves, crucified at the same time as our Saviour. One –
> ESTRAGON: Our what?
> VLADIMIR: Our Saviour. Two thieves. One is supposed to have been saved and the other… [*He searches for the contrary of saved.*] damned.
> ESTRAGON: Saved from what?
> VLADIMIR: Hell.
> ESTRAGON: I'm going. [*He does not move.*]
>
> (*CDW*, 14)

In the Behan play, Dunlavin and Neighbour perform a similar double-act, each rarely appearing without the other and always keeping the conversational ball in play. In his book, Alan Simpson made a sustained comparison between Dunlavin and Neighbour's reminiscences of the prostitutes with whom they worked – 'Ah, poor May, God help her, she was the heart of the roll' – and the dialogue between the two old men in Beckett's translation of Robert Pinget's *La Manivelle/The Old Tune* (1963) – 'Gertie Crumplin great one for the lads she was you remember, Gertie great one for the lads.'[31] What this indicates more than mutual influence is Beckett and Behan's joint indebtedness to Sean O'Casey and his garrulous male pseudo-couples, like 'Captain' Boyle and Joxer Daly in *Juno and the Paycock*. Dunlavin and Neighbour do not, however, appear continuously throughout *The Quare Fellow*. Rather, the play's form is of a continuous alternating ensemble with various fluid groupings occurring onstage, where there is little dramatic continuity from one prisoner or warder to the next. In both cases, the central character is an absence, an empty presence gestured at in the title, as we have seen. The closest the plays offer to a leading man is Pozzo, who continuously tries to seize mastery of the stage space and does so through a series of old trouper's tricks: using a vaporiser to increase the sonority of his voice; trying for a long soliloquy which he is unable to sustain; making a big fuss about his entrances and exits.

The absence of a leading man in contemporary Irish drama is bound up with the absence of an informing or centralised plot by which the action is propelled. I have already approached the question of why this should be so through the discussion of hanging as the closest Vladimir and Estragon come to initiating a course of action. The central activity of *The Quare Fellow* and *Waiting for Godot* is waiting, or more precisely, as the French title of Beckett's play indicates, what to do while waiting: for Godot to show up or the quare fellow to be hanged. In this respect, the plays show scant regard for, or interest in, the engineered developments of plot as the driving force of drama. Indeed, when Behan's play was premiered, it was criticised for its lack of plot, which was put down to his lack of theatrical experience; and the original script of *The Quare Fellow* had many of its repetitions eliminated by Swift and Simpson in the interests of a more linear, traditional plot. But a lack of interest in plot is characteristic of much of modern drama, going back at least as far as Chekhov, and is often a recognition that God is not in his heaven, having died or absconded, and that all is far from right in the twentieth and twenty-first century worlds.

The absence of a traditional plot is particularly prevalent in Irish and post-colonial drama. In the space of a single month (June 1990), the theatre critic of the *Los Angeles Times*, Sylvie Drake, made the same complaint of Derek Walcott's *Viva Detroit* as of Brian Friel's *Aristocrats*: 'Virtually *nothing* has happened by the end of Act 1 and not much more by the end of Act 2.' This is an echo, unintentional or not, of what has been most remarked about *Waiting for Godot*, following the late Vivian Mercier, that Beckett's is a play in which 'nothing happens *twice*'. It dispenses almost entirely with a conventional plot of mounting expectation which relies on a sequence of narrative cause and effect. The same is true of a great deal of Irish drama from Yeats and Synge, through Behan and Beckett, to Frank McGuinness's *Carthaginians* (1988). In these plays, the notion of a plot is played with, the audience teased with the expectation that something eventful or of great external significance is going to happen, like the rising of the dead in a Derry graveyard in *Carthaginians*. What the audience and the characters learn to share is the experience of disappointment or disillusionment as these conventional hopes are repeatedly frustrated. It is not that plot is entirely absent from Irish drama. It is continually anticipated, watched for, talked about. Rather, it never entirely coincides with the stage space, never sweeps the central characters up in its agency, but always seems to be somewhere else, off-stage. Subordinate characters do arrive with a great deal of bustle, who appear to be agents of a plot – like Pozzo in *Godot*, much given to consulting his watch and talking about business matters; or Bartley in Synge's *Riders to the Sea*, off to sell horses at the Galway fair. And these particular characters, no less appropriately, become the victims of plot: Pozzo by being struck blind; Bartley by falling into the sea and being drowned. These events occur off-stage and are almost incidental to the energies of the central characters, which are given over to survival, both physical and spiritual, to improvising a mode of endurance.

The central activity in most Irish plays raises the dramatic question of what to do while waiting. Since the space is cleared of traditional plot expectations and activities, other forms of theatrical activity are free to enter the vacuum or void thus created. If Estragon is the one who complains that 'nothing happens' in the conventional sense, and so makes the audience all the more aware of the fact, it is also he and Vladimir who direct our attention to the more lowbrow, popular, alternative forms of theatre which the play incorporates:

VLADIMIR: Charming evening we're having.
ESTRAGON: Unforgettable.
VLADIMIR: And it's not over.
ESTRAGON: Apparently not.
VLADIMIR: It's only the beginning.
ESTRAGON: It's awful.
VLADIMIR: Worse than the pantomime.
ESTRAGON: The circus.
VLADIMIR: The music-hall.
ESTRAGON: The circus.

(*CDW*, 34–5)

Didi and Gogo, as has often been observed, resemble nothing so much as music-hall comedians who go on to a virtually bare stage with nothing more elaborate by way of set-up than the ability (and necessity) to generate entertainment out of random chit-chat between each other. Their stage 'action' alternates between rapid-fire verbal exchanges and pantomimic set-pieces of elaborate comic business: the elaborate exchange of hats or the outsize, falling trousers. Beckett is presumably drawn to music-hall comedians by the strong comic element, their ties to popular culture, the inner resources by which they make light of their impoverishment (as with Chaplin and his own tramp persona) and the courage with which in theatrical and existential terms they perform without a net.

The double emphasis on 'the circus' in Didi and Gogo's exchange supplies a context for the whip Pozzo brandishes. He is the ringmaster leading on the performing beast of burden, Lucky, into the ring of the circular structure which is Beckett's play as experienced by the two tramps. What Pozzo offers to them as entertainment – 'What do you prefer? Shall we have him dance, or sing, or recite, or think, or – ?' (*CDW*, 39) – brings up a form of theatrical entertainment peculiar to Ireland, the *feis ceoil* (musical festival), especially when recitation is offered as one of the categories of performance. The 'thinking' as a public performance which Lucky eventually puts on appears to have strayed in from Trinity College, Dublin, and its *viva voce* examinations for postgraduate students. Pozzo himself, with his travelling props and his diminished troupe of one, closely resembles one of the old actor-managers who used to tour around Ireland and England with the fit-ups, like the Shakespearean Anew McMaster with whom Mac Liammóir, Edwards and Harold Pinter all began their acting careers.

In the case of Behan's playwriting career, what most drew the wrath of certain Irish critics in regard to his one other full-length play, *The Hostage* (1958), were the strong music-hall elements added to Behan's Gaelic original, *An Giall* (1956). These were viewed as 'alien' ingredients imported from England and sullying the purity of an Irish dramatic text.[32] All Behan's poetry was composed in Irish; and he followed up *The Quare Fellow* with a play in that language commissioned by Gael-Linn (*An Giall*). There is an even more unstable textual relationship between these two 'versions' of the same play as to priority than there is with Beckett's *Godot*. Again, Behan was his own translator into his 'native' English, but he did so under the encouragement and with the assistance of Joan Littlewood, his director at London's Stratford East. And so those who dislike the camp elements of *The Hostage* can blame them on the

Englishwoman's influence and claim *An Giall* is more authentic. Declan Kiberd cites Behan directly to refute this widely held view:

> I saw the rehearsals of *An Giall* and, while I admire the producer, Frank Dermody, tremendously, his idea of a play is not my idea of a play. He's of the school of Abbey Theatre naturalism, of which I'm not a pupil. Joan Littlewood, I found, suited my requirements exactly. She has the same views on the theatre that I have, which is that the music-hall is the thing to aim at in order to amuse people and any time they get bored divert them with a song or a dance... and that while they were laughing their heads off, you could be up to any bloody thing behind their backs – and it was what you were doing behind their bloody backs that made your play great.[33]

Behan's strategy towards the audience resembles that of his character Dunlavin during the medical inspection when, directing the doctor to the pain in his left leg, he shows him the right instead and, behind this piece of comic business, is able to drink off some of the meths while winking at the audience. The on-stage joke is as contrived as the pains in Dunlavin's legs. But the larger piece of comic business has the more serious aim of subverting the prison authorities and keeping some life and warmth in the old man's famished body.

For *The Quare Fellow* is as indebted to the music-hall as its more notorious successor, *The Hostage*, Beckett's *Godot*, or the plays of Sean O'Casey. All of them stand in marked contrast to the stolid, sincere naturalism which has held the Abbey and Irish stages almost from the beginning. The potent theatrical fusion of music-hall and politics in the O'Casey plays of the 1920s finally did not succeed in taking over from or displacing the kitchen comedies which had become the Abbey's stock-in-trade. The mixture of high and low art which the socialist playwrights O'Casey and Behan favour along with Beckett has always offended equally the caretakers of high culture, the bourgeois middle-class audience, and the cultural nationalists whose view of an Irish theatre insists that it come emblazoned with certain insignia. Even Alan Simpson objected to the presence of old jokes in *The Quare Fellow* and instanced this as a sign both of Behan's laziness and a lack of technical competence, which is to miss the point almost entirely. The joke Simpson specifically refers to is the discussion of the Bible in Act One where the pious objection to 'talking disrespectfully about the Bible' is countered by the claim that the Bible often gives 'consolation to a fellow all alone in the old cell' by offering sheets of paper in which the tobacco can be rolled to furnish cigarettes.[34] In *Godot*, the Bible is upheld by Vladimir but countered as cultural artefact by Estragon, who talks of it as a cross between a child's picturebook and a travel guide for honeymooners:

VLADIMIR: Did you ever read the Bible?
ESTRAGON: The Bible... [*He reflects.*] I must have taken a look at it.
VLADIMIR: Do you remember the Gospels?
ESTRAGON: I remember the maps of the Holy Land. Coloured they were. Very pretty. The Dead Sea was pale blue. The very look of it made me thirsty. That's where we'll go, I used to say, that's where we'll go for our honeymoon. We'll swim. We'll be happy.

(*CDW*, 13)

Both 'jokes' ignore the verbal content of the Bible in favour of its surface materials, but both plays exhibit a constant indebtedness to the same source in their language and argument.

In a situation where most forms of agency are directed or proscribed, the emphasis is thrown on talk, on the orality of speech as a site of potential freedom and self-realisation. The delight in talk for its own sake which is so marked a feature of Irish drama gains particular prominence from the threat of silence in both plays. If the jokes serve the more immediately subversive ends of opening up a space in an oppressive situation, there is another, more sustained level of discourse at which an argument or debate is carried on throughout. This is true of many British left-wing plays also; but the argument in Irish plays tends to have no overt or immediate social application, to eschew the didactic. Behan's *Quare Fellow* could so easily have become a tract in favour of the abolition of capital punishment, Beckett's *Godot* a specific account of life during wartime. Contemporary Irish dramatists subscribe more to Russian film director André Tarkovsky's view that 'the allotted function of art is not, as is often assumed, to put across ideas, to propagate thought, to serve as example. The aim of art is to prepare a person for death.'[35]

Through much of the opening section of *Godot*, what Vladimir insists on discussing, despite Estragon and the play's best efforts to interrupt and deflect him, is the case of the two thieves hanged on either side of Christ. What Behan's prisoners and warders alike return to discussing after all the jokes and non-sequiturs is the case of the two quare fellows in Mountjoy Prison, both condemned to die. What distresses Didi about the two thieves is what generates so much discussion in Behan's stage prison, not alone that man has to die but that according to the story in the Bible, at least as told by one of the four Evangelists, one of the two thieves 'is supposed to have been saved and the other... damned' (*CDW*, 14). Similarly, although both quare fellows have been condemned to hang, one of them is suddenly and unexpectedly reprieved from that fate in Act One. Dunlavin, who reports the reprieve, is unable to give any rational explanation for the decision: 'though why him any more than the other fellow is more nor I can tell' (*CP*, 42). In the ensuing debate about one prisoner being 'saved' while the other is 'damned', the prisoners are as bothered about the logical gap or flaw opened up by this as Vladimir is and attempt to fill it with a variety of explanations, mostly having to do with the higher social class of the reprieved prisoner and the especial ferocity with which the 'damned' quare fellow treated his victim. They conclude, in other words, that he only got off because he was a better class of murderer. What the debate engages in both cases is the prospect that the elaborate hegemony of Western Christianity, and the legal system by which criminal offenders are judged and punished, are merely empty formalisms from which any discoverable or coherent first principle has long since disappeared, that it is chance and randomness which decide a man's fate. At the comic level of music-hall, it becomes a gamble, a throw of the dice on the odds for salvation or damnation. Dunlavin and Neighbour wager a non-existent piece of bacon on the outcome, as to whether the quare fellow will get a reprieve or be hanged. In his play, Beckett does not come down on either side. For him it is the equilibrium struck between the poles of affirmation and denial, presumption and despair, that gives the 'shape' to his drama. Behan's drama is likewise shaped out of such contradictory impulses and profound ambivalence. The vocal impulse in such a double perspective is equivocal; its dramatic shape requires at least two characters to share the stage equally. For, in the words of Beckett's favourite book of the Old Testament: 'Two are better than one... for if they fall, one will lift up his fellow' (Ecclesiastes, Book 4, Chapter 9).

The single prospect in each case is the grave. And the image of the grave looms ever larger as each play proceeds. When Behan's play moves out of the prison in Act Two, there is no escape or sense of release, not only since the grounds of the prison extend there, but even more because in the centre of the stage the condemned man's grave is being dug. And the dramatic *memento mori* continues after the hanging, which is offstage, with the chiselling of an anonymous 'E777' on his tombstone and the dropping of his personal effects into the grave;

there, they are gambled over by the four prisoners detailed to bury the dead man in a scene with Christological overtones. The final moments of *Godot* also disclose the prospect of the grave: 'Astride of a grave and a difficult birth. Down in the hole, lingeringly, the grave-digger puts on the forceps' (*CDW*, 84). But neither play ends there for, as Vladimir goes on to reflect, 'Habit is a great deadener.' *Godot* concludes, as we have seen, by bringing down Estragon's trousers just before it does the curtain and by having the two tramps verbally resolve to go on. The Behan play concludes with another burst of the song which has recurred throughout, extolling the 'female' prison where there are seventy women. The grave has opened and closed, without claiming either the plays' characters or the audience. The vertiginously dizzying prospect of the grave induces the sense of tragedy. But the survival of gaining the other side is comic and adds bite to the laughter and song with which the drama concludes. It is grim; it is black; but it is finally the comedy of survival.

At an Irish wake, it is traditional to tell a story about the dead person as an act of remembrancing or to sing a song as a way of uplifting spirits. In the music-hall a not dissimilar mix of jokes, storytelling and songs is met with. And the loose rather than tight way in which contemporary Irish plays are structured not only finds room for a song but is itself often structured along musical lines, with recurrent verbal motifs and the orchestration of related themes. But orchestra is the wrong term. For the development of unaccompanied singing in Ireland has much to do with the communal nature of most forms of entertainment and the lack of wealth and status which made heavy or big instruments a liability. With singing, performer and instrument are one. Each Act of *The Quare Fellow* begins and ends with the voice of a prisoner singing the ballad 'The Old Triangle'. In the original Pike production, it was Brendan Behan's own, a 'rich North City voice resonantly recorded'.[36] Each verse takes us through a later stage of the day and adds the refrain: 'And that old triangle / Went jingle jangle, / Along the banks of the Royal Canal' (*CP*, 39). On each occasion in the play that a verse is sung, the singer is shouted down and silenced by the warders, the song interrupted, his voice cut off. In the short (dramatic) term, the effect is of incompletion, fragmentation. And as each verse comes to an end, the question of what comes next immediately raises its head. But in the long run, the cumulative course of the play's three Acts, he proves irrepressible, like the human spirit in both plays, and returns to sing again at the break in each Act. It is worth recalling that there is also a crucial song in *Godot*, the German round-song which Beckett himself translated and which Vladimir sings *'loudly'* at the beginning of Act Two:

> A dog came in the kitchen
> And stole a crust of bread.
> Then cook up with a ladle
> And beat him till he was dead.
>
> Then all the dogs came running
> And dug the dog a tomb –
> ...
>
> And wrote upon the tombstone
> For the eyes of dogs to come:
>
> A dog came in the kitchen.
>
> (*CDW*, 53)

And off it goes again. In so doing, it announces circularity and repetition as the dramatic principles by which Beckett's drama is going to proceed into the second Act, as Behan's song does in *The Quare Fellow*. Beckett, when asked how a country as small as Ireland could have produced so many great writers in such a short time, is said to have replied: 'When you are in the last bloody ditch, there is nothing left but to sing.'[37]

While Beckett against all the odds was to go on and on, neither Behan nor the Pike survived far into the 1960s. Behan was done in by excess and success, the lethal toll of a flow of drink on his diabetes; the Pike Theatre, by the arrest and imprisonment of Alan Simpson for his 1957 production of Tennessee Williams's *The Rose Tattoo*, where the alleged dropping of a condom on-stage brought a summons, an arrest, a trial, and the ostracism of Simpson, Swift and the Pike by most of the theatrical community and public (membership dropped from 3000 to 300). The person who brought it to the attention of the authorities that the play contained 'objectionable passages' was never named or identified, nor were the particular passages specified. He or she, unlike Simpson, never appeared in court. The charge remained impersonal, its source off-stage and unknown.[38]

2
Friel's Drama: Leaving and Coming Home

The playwright who has most established the prominence of contemporary Irish drama on the world stage is Brian Friel. The three plays which have most secured this position are *Philadelphia, Here I Come!* (1964), whose theme of emigration awakened a strong response in Irish and American audiences; *Translations* (1980), whose theme of cultural erosion and linguistic displacement has found equivalents the world over; and *Dancing at Lughnasa* (1990), where five women trapped in an Ireland of the 1930s and silenced by official history are liberated into expression through the medium of Friel's theatre. *Philadelphia* was first staged at the 1964 Dublin Theatre Festival and later ran on Broadway for almost a year. *Translations* in 1980 launched the Field Day Company in Derry, played through Ireland to immediate acclaim, secured Friel's place on the London stage, first at the Hampstead, then at the National Theatre, the first Irish play to enter their repertoire since O'Casey, and progressed in the subsequent ten years to productions around the world. Friel's third coming has been with *Dancing at Lughnasa*, from its 1990 Abbey premiere, to London's National Theatre and a long run in the West End, and then on to Broadway, where it won a Tony award for Best New Play. Nor has that international success been achieved at the expense of his status in his own country. As Seamus Heaney puts it, Friel's constant renewal of his dramatic art is a profound record of 'what it has been like to live through the second half of the twentieth century in Ireland'.[1]

By 2008, 12 full-length studies of the playwright had been published.[2] All of them give detailed accounts of the entire career and *oeuvre* (with an excess of attention to the short stories, in my view). All but Richard Pine apportion more or less equal space to each of the fifteen plays, and even he short-changes *Philadelphia* in his haste to get on to the later works. And yet *Philadelphia* stands out from the first half dozen or so plays.

It is even more piercing and illuminating after all these years, and the only one from this period to be retained by Friel in his 1996 *Plays: One*. There is much to be said in support of Pine's claim that 'Contemporary Irish drama begins in 1964 in *Philadelphia, Here I Come!*'[3] and in accounting for the excitement that Thomas Kilroy recorded at the play's premiere: 'I remember when I first saw *Philadelphia, Here I Come!* what excited me most, as someone who wished to write plays, was the delicate use of the stage as a place of illusion, play-acting, make-believe.'[4] The first half of this chapter, therefore, will concentrate on *Philadelphia* while *Translations* will be discussed in the fifth chapter.

But a most difficult task for me was choosing another, later Friel play to accompany the analysis of *Philadelphia*, to show the span of achievement while suggesting family resemblances. *Aristocrats* and *Faith Healer* (both 1979) vied for the place, the former because of its complex reworking of Chekhov in an Irish context and its belated international success, the latter as the most theatrically experimental of Friel's plays and the one which engages most profoundly with the theme of risk and failure. Finally, it had to be *Faith Healer*, answering to my conviction that it may represent Friel's best work and that its impact has been greater at 'home' in Ireland than anywhere else. For Friel's recent success must be seen in the context of failure, as the playwright himself insisted when accepting the *Olivier* award for *Lughnasa* as the Best Play of 1990. His characteristically brief address on that occasion quoted the late Graham Greene to the effect that 'success is only the postponement of failure'. This context is necessary, not only to recall other occasions on which Friel plays have met with incomprehension or indifference, as so often in the 1970s; or even the extent to which the playwright's own failure resulted in a five-year silence in the 1980s; but above all to stress the extent to which the plays take failure as their theme, constituting their strongest link with Beckett.

The first complicating factor to consider in the emergence of Brian Friel as a contemporary Irish playwright is that he began his career by writing in the diverse media of radio, drama and prose. And during the 1950s and early 1960s his greatest success was as a short story writer. Friel's individual stories were published by the *New Yorker*, eventually forming the substance of two collections, *The Saucer of Larks* (1962) and *The Gold in the Sea* (1966); and it was this success, financial and otherwise, in the short story field which encouraged him to resign his job as a teacher and become a full-time writer.

Friel's move into full-time writing marked also his greater and absolute move away from prose fiction into the medium of drama. Richard Pine concludes that Friel grew frustrated by the limitations of the *New Yorker* format: 'Triteness, encouraged by the *New Yorker* formula, is also a reason for Friel's eventual dissatisfaction with the limiting conventions of the short story.'[5] But there was a short story formula closer to home which Friel has cited as the shadow from which he wished to emerge, which was cast by the formidable cultural presences in the 1950s of Sean O'Faolain and Frank O'Connor: 'I was very much under the influence, as everyone at the time was, of O'Faolain and O'Connor, particularly. O'Connor dominated our lives. I suppose they [Friel's short stories] were really some kind of imitation of O'Connor's work.'[6] One potent myth which O'Connor and O'Faolain had worked hard to disseminate as a means of culturally bolstering their own practice was the idea that Irish writers had a particular affinity with the short story, and that this in some sense mirrored the shape and nature of the society, fragmented and destabilised. The theory was used most of all to oppose the notion that the Irish could or should write novels, since they lacked the stable, hierarchical society that most of the classic English novels of the nineteenth century had been built around. Such a radical theory, however, served a conventional notion of the short story. O'Connor's influence as cited by Friel may well account in part for what George O'Brien remarks about the short stories, the singular lack of development displayed across the two volumes: 'There is little sense of development in Friel's story-writing. He came to his world and its themes early and, rather uncritically, remained with them until committing himself completely to the theatre.'[7] The stories, in this respect especially, stand in opposition to the plays. The latter are marked throughout Friel's career by a continuing pressure of formal experimentation and development. This development is not necessarily linear; rather, it is the case with each play that Friel is responding to and writing himself out of the limitations he has inadvertently come up against in the previous one. Put in broader cultural terms, Ireland at the close of the 1950s did not need one more short story writer, even one as accomplished as Brian Friel. It stood in much greater need of a playwright to take the measure of and to find forms adequate to a society undergoing a profound transition.

When it came to considering himself as a potential playwright, what perspective and options presented themselves? This is the point at which to take notice of Friel's position as a Northerner. Born near Omagh, Co. Tyrone, in 1929, Brian Friel was the son of a Catholic schoolmaster who moved the family to Derry ten years later. The Troubles which ignited in 1968 have thrown that particular area into lurid relief ever since; but the primary impact of Northern Ireland on Friel would have been over the previous 30 years and could be described in a key term for his theatre and this book, a sense of doubleness. Bifurcations are an all too inevitable component of identity in the North, where every label has its binary opposite: Catholic/Protestant, Unionist/Nationalist. A Catholic's position in Belfast may fairly be described as beleaguered; but in Derry the city not only has a large Nationalist and Catholic population but opens on to the hinterland of Donegal, part of the province of Ulster and since 1922 part of the Southern state. This dual allegiance inculcated a stronger sense of enabling identity in Derry Catholics but did not do away with a legacy of unemployment and the status of second-class citizens. Donegal always had a special claim on Brian Friel because his mother came from there, and the family's summers were spent with his aunts in Glenties. Most of Friel's plays are set in Donegal, in the fictional locale of Ballybeg (from the Gaelic *baile beag*, or 'small town'). Brian Friel now chooses to live just over the border in County Donegal, on the peninsula of Inishowen, the island's northernmost point, on the margins of both Irelands.

As an emerging playwright in the 1950s, Friel as a Derry man could look both to Dublin in the Republic and Belfast in Northern Ireland as possible sites for staging his plays. The first to have an impact, *Philadelphia, Here I Come!*, was not the first play of Friel's written and staged, but the fourth. For a long time Friel refused to allow the first three to be published. He finally relented in 1979 regarding *The Enemy Within* (1962), but has remained adamant so far about *A Doubtful Paradise* (*The Francophile*) (1960), and *The Blind Mice* (1963). The former was first staged as *The Francophile* at the Group Theatre, Belfast, and was reworked under its new title as a radio play broadcast by the BBC Northern Ireland Home Service in 1962. Although Belfast served as a valuable outlet for his radio plays, production and reception of Friel's stage work there were far from satisfactory, as more than one Northern playwright has found since. The Abbey Theatre and Dublin offered at some level a more promising venue, with historically a commitment to staging new plays by Irish writers. But as I examined at the beginning of Chapter 1, plays which sought to reflect a changing Ireland through formal innovation and an enlarged subject matter were not likely to be accepted by the National Theatre during the increasingly *ancien régime* of Ernest Blythe. The dilemma facing an emerging Irish dramatist in the late 1950s may be said in some sense to crystallise itself around rejection by the Abbey. Like most dilemmas in Ireland this one cut both ways. For if being turned down by the Abbey posed one kind of obstacle to a potential playwright, so too did acceptance. You could, for instance, as happened to Hugh Leonard, have a second play accepted, but not a third. Or you could, as was the case with Brian Friel and *The Enemy Within*, have a play produced by the Abbey at the Queen's Theatre, Dublin, on 6 August 1962, by Ria Mooney, and still chafe at the constraints imposed by such a 'success'.

This play of Friel's produced by the Abbey would have merited scant attention were it not for the fate of his subsequent plays. *The Enemy Within* can be compared thematically to such later works as *Faith Healer* and *Translations*. St Columba, the central character, is a deeply divided man, as the title suggests. He is torn between his commitment to the monks of Iona in Scotland and to the aboriginal loyalties of Ireland and home. But the form in which Friel dramatises this clash between the public and private personae of Columba is crude and externalised. In a way that resembles medieval allegory, Friel has to find external, discrete characters to embody the different forces influencing Columba:

> [COLUMBA *is in the centre,* GRILLAAN *on one side of him,* BRIAN *on the other. He is torn between the two.*]
> GRILLAAN: The last tie, Columba. Cut it now. Cut it. Cut it.
> BRIAN: They are your people. It is your land.
> GRILLAAN: A priest or a politician – which?[8]

And his younger, more idealistic self is represented by the novice, Oswald, who disappears for much of the play but is retrieved redemptively at the end. The play is fascinating in its thematic material but remains static in its form, awkwardly groping for a theatrical syntax in which to articulate Columba's inner conflict and forced back on mere externals. In *Philadelphia*, Friel was to find a flexible theatrical solution to this problem, to break with the narrow tradition of Abbey realism through the wonderful stage device of choosing two characters (and actors) to represent the divided psyche of Gar O'Donnell.

What intervened between the writing of *The Enemy Within* and *Philadelphia, Here I Come!* was a crucial sojourn. Friel spent three months at the invitation of Sir Tyrone Guthrie at the theatre Guthrie had just founded in Minneapolis. Guthrie functioned as something of a mentor in the career of the young Friel. Though born in Tunbridge Wells, he regarded himself as Irish, since his mother came from County Monaghan and he himself used the family home of Annaghmakerrig as a base from which to plan and make his worldwide theatrical sorties: to Belfast, to work on radio drama, as Friel was to do; to Scotland, to help develop a national theatre; to ten years at London's Old Vic with all the knights (Olivier, Richardson and Gielgud); to Canada, to found the Shakespeare

Festival at Stratford, Ontario; and in the early 1960s to Minneapolis to found his own theatre. Guthrie's autobiography is divided up into chapters on places, and none is more revealingly rueful than the chapter on his 'home', Ireland, from which almost all his working life was spent in exile. Late in life he sought a belated involvement in the theatrical present and future of his country through sponsoring and encouraging two young Irish playwrights: Eugene McCabe, also from Monaghan, whose play *Swift* Guthrie directed at the new Abbey Theatre in 1969; and Brian Friel, another Northerner who lived along a partitioned border. Guthrie's role in relation to Friel as a dramatist is akin to that of Yeats when he advised Synge to go to the Aran Islands. Here, the situation worked in reverse. What Friel got from Guthrie's advice and initiative was what he then most needed:

> Those months in America gave me a sense of liberation – remember this was my first parole from inbred claustrophobic Ireland – and that sense of liberation conferred on me a valuable self-confidence and a necessary perspective so that the first play I wrote immediately after I came home, *Philadelphia, Here I Come!*, was a lot more assured than anything I had attempted before.[9]

During the period he spent in Minneapolis, Friel saw productions by Guthrie which were to have a direct, discernible influence on his own later work: Shakespeare's *Hamlet*, whose presence is felt in Friel's 1975 play, *Volunteers*, through Keeney's reiterated posing of the question: 'Was Hamlet really mad?';[10] and Chekhov's *Three Sisters*, a version of which Friel was to write for Field Day in 1981. Guthrie worked most of all on Chekhov's plays, writing his own translations and directing them. And Chekhov exerts a deepening influence throughout Friel's career, in the way the Irish dramatist will structure plays like *Aristocrats* around departures and arrivals at a big country house and in his increasing use of the extended family as an ensemble. Friel even went so far in 2002 as to write a 'new' work called *Afterplay*, imagining a meeting in a Moscow café between Andrei Prozorov from *Three Sisters* and Sonya Serebriakova from *Uncle Vanya* many years after the Chekhov originals in which they featured. But apart from these specific debts, the Minneapolis apprenticeship with Guthrie made available to Friel the techniques of world drama, a lexicon of alternatives to the traditional manner of staging plays which had taken over in Ireland in the 1930s.

The genius of *Philadelphia, Here I Come!* is that it deploys a number of innovative theatrical effects, not for their own stylistic sake, but as a means of breaking open the hidebound Abbey stage and exposing what the latter claimed to represent – the inner lives of Irish people. The first of these is the device whereby the same character is played by two different actors. In *Philadelphia* Friel came up with the idea of expressing the deeply divided feelings of a 25-year-old Donegal man the night before he emigrates to the USA through having two different, interrelated aspects to his stage protagonist. On the one hand, there is Public Gar, 'the Gar that people see, talk to, talk about. Private Gar is the unseen man, the man within, the conscience, the *alter ego*, the secret thoughts, the id.'[11] The double relationship is crucial to the success of the whole play, the dynamism on which it runs, ventilating and giving light, colour, humour and a variety of perspectives to the human dilemma at its core, freeing the representation of that dilemma from an archaic, no longer adequate form.

I would stress what this study has sought to establish through the plays of Beckett and Behan – that the tendency in Irish theatre has been away from the idea of a single leading man and towards the sharing of the stage space between two male protagonists, neither of whom predominates. With the exception of the opening exchange between Gar and the housekeeper Madge, this is the case in Friel's play. Once Gar has entered his bedroom, he is joined on stage by his Private counterpart for the rest of the play. One never leaves the other alone, whether engaging him in dialogue or, when Public Gar is directly engaged, serving as an ironic choral figure or voicing alternative lines to the realistic speech. Throughout all the double acts in Irish drama, the close physical interdependence has always implied an equivalent psychological interrelation; and Friel is here making explicit something that has been latent throughout the tradition. He is in a sense forced to do so by the setting of *Philadelphia* within the realistic milieu of a Donegal shopkeeper's residence rather than the nowhere land of Beckett's plays and by the concomitant demands for a strong measure of bourgeois characterisation. If Gar Public is drawn into realistic interchanges with father, friends, housekeeper, priest and teacher, Gar Private is there to draw him into a more purely theatrical realm, a drama of the mind.

The interplay of the subjective and the objective, of private and public, that Friel requires for his drama is not entirely met by the device of splitting his protagonist in two. The setting or stage space also needs to undergo a transformation. The main stage space is divided between the kitchen and Gar's bedroom, both features of a realistic Irish play. But he has moved them upstage and in so doing, as Richard Allen Cave has observed, has distanced and isolated the traditional setting from the spectator.[12] The downstage area, the apron, is left in the dark. As Friel's stage directions put it, the 'remaining portion is fluid' (27). This fluidity is specified in terms of place, since at one point it represents 'a room in Senator Doogan's home'. What the play itself will reveal is that this dark, open space also permits chronological fluidity as the site in which *Philadelphia*'s key flashbacks occur. Friel by this means frees his stage from one particular time and place and can range with a greater degree of freedom than the traditional scenario permitted.

In *Philadelphia*, as I have argued, the opening scene functions as a traditional play with and between the housekeeper and Gareth O'Donnell. The transition from one order of theatrical reality to another occurs when Gar moves into the bedroom and begins to alternate highly stylised criss-cross lines with an off-stage presence who then enters. The bedroom has been 'in darkness'; clearly, a more complex cue than just raising the lights is required to signify that Gar is crossing a threshold and encountering his *alter ego*:

> [GAR *sings alternate lines of 'Philadelphia' – first half – with* PRIVATE *(off)*].
> PUBLIC: It's all over.
> PRIVATE: [*Off, in echo-chamber voice.*] And it's all about to begin. It's all over.
> PUBLIC: And it's all about to begin.
> PRIVATE: [*Now on.*] Just think, Gar.
> PUBLIC: Think ...
> PRIVATE: Think ...
>
> (31)

The formal doubling of the lines indicates the close psychological bonding, as it does in Beckett, and also the move away from conventional characterisation towards a more theatrical improvisation (and problematisation) of identity. The double act is initiated by the sharing of a song and rapid-fire cross-talk, and shares many of the features of the music-hall that have already been investigated in relation to Behan and Beckett.

In the private space of the bedroom, Gar Public and Gar Private perform the first of at least three dramatic functions crucial to the overall effect of the play. They encourage and inspire each other to a greater degree of self-expression than is possible in the public or family domain. Some of this freedom is verbal. It is not long before the 'big bugger' of a jet has trained its machine guns on 'a bloody bugger of an Irish boat out fishing for bloody pollock' (31) and we are a good way beyond *Pygmalion*'s first night and the fuss that attended Eliza's 'Not bloody likely'. Irish theatre has always enjoyed a greater verbal freedom than its English counterpart, at least until the demise of the Lord Chamberlain in 1968; but this has often seemed to operate at the expense of and as compensation for the restrictions and inhibitions on physical expressiveness. Here, the greater measure of theatricality introduces the possibility of mime. With Gar Private making all the verbal running, Gar Public is free to act out and choreograph his fantasies: as a footballer kicking the winning goal; as (impossibly) a conductor and performer in the same orchestra. The climax of these fantasies occurs with a wild outburst of singing and dancing from Gar Public, the brief moment of breakthrough when speech and action are synthesised, reconciled, but also the expressive point at which mere description breaks down into a range of almost pure celebratory sound: 'Ya-hoooo! [*Sings and gyrates at the same time.*] Philah-delph-yah, heah Ah come, rightah backah weah Ah stahted from, boom-boom-boom-boom-' (31–2).

The line, of course, comes from and is an adaptation of the popular song, 'California, Here I Come!', something which most audiences recognise even as they note the displacement from California to Philadelphia. But the second line is much more disturbing in the issues it raises: 'Right back where I started from'. For Gar is *not* a native of Philadelphia and his emigrating is not a return to origins but a departure from them. The assertion that the opposite is the case, emanating from an American song, serves to indicate, along with all the other fantasy elements acted out between the two Gars, the extent to which the USA has (and had already by the start of the 1960s) come to Ireland through the proliferating media of music, cinema and television. The cultural ghetto insisted on by the likes of S. B. O'Donnell and the Canon, a house in which there seem to be no books and in which the playing of records is tolerated as a benign idiocy or 'noise' (90), created in Ireland a vacuum which coincided with and was rapidly filled by the Hollywood sound era and by jazz, popular music and the ubiquitous dancing which accompanied it – all forces which the clergy tried to stamp out. Gar's bedroom is a domesticated Ballroom of Romance where he and his *alter ego* improvise a number of romantic scenarios as so many turns on an imaginary dance-floor.

If 'Philadelphia' and all it signifies in terms of fulfilment and desire are what he wants to go to, 'where I started from' is what Gar is anxious to escape. The early intimation of what form an exchange with his father might take is borne out by the brief intrusion of S. B. O'Donnell into the fantasy stretch of Episode One. What proves so shocking is nothing in the normative appearance of the old man himself but the negative transformation it wreaks in Public Gar. He is turned back into the tongue-tied, stuttering lout, as Christy Mahon is by the reappearance of his father in Synge's *Playboy of the Western World*, in an exchange that consists entirely of monosyllables as Gar struggles to remember how many coils of barbed wire were delivered that day. The exchange between father and son is entirely materialistic, confining itself to the deliberately restricted range of utilitarian matters, reminding us that the relation is more that of employer to employee than of father to son. If the first brief intrusion registers a reduction in Gar's expressiveness the next indicates the second of Gar Private's dramatic functions: to fill in the gap opened up by the silences between S. B. and Gar Public. In the play's public domain as in the private, silence is the enemy, a persistent threat Gar Private is set on defeating. So he accompanies the silent and long-drawn-out entry of the father with a comic commentary, treating the event as if it were a fashion show with S. B. O'Donnell as the model Marie-Celeste. In one sense, this follows on from the bedroom scenes, with physical mime once more accompanied by parodic commentary. But here the verbal and the physical are deliberately at odds. S. B. remains resolutely untransformed in gender, nationality or attitude.

The simple opposition between 'Philadelphia' and 'where I started from' is not continued beyond the first episode. The prospect of what awaits Gar in the USA is complicated in Episode Two by the flashback to the visit of Aunt Lizzy; and the question of Gar's origins is complicated by the silence not only of the father but of the mother, here the absolute silence of the grave. In the face of his father's obduracy, Gar's recoil is not just to a future life in Philadelphia but to reconstructing a past. These efforts are centred on the figure of his mother Maire, the missing other half of his personal history.

The key text in the play for the idealistic aura with which Gar invests his absent mother is the ten-times-reiterated quotation from Edmund Burke's *Reflections on the Revolution in France*: 'It is now sixteen or seventeen years since I saw the Queen of France, then the Dauphiness, at Versailles.' On the sixth iteration, this is further extended: 'And surely never lighted on this orb, which she hardly seemed to touch, a more delightful vision. I saw her just above the horizon, decorating and cheering the elevated sphere she just began to move in' (78). The first time Gar cites the Burke quotation, it is entirely without context. The second, it is explicitly aligned with another textual foregrounding, '*a sheet of faded newspaper* [...] The *Clarion* – 1st January 1937 [...] the day they [his parents] were married' (37). On the third, this association with the mother is extended to Kate Doogan, as Gar makes the Freudian transition from the death of his mother to the loss of a girlfriend and prospective wife. In a way most of the audience can only intuit rather than work out rationally, the quotation from Burke about Marie Antoinette is intimately associated by Gar with his dead mother, primarily through the Catholic imagery relating mother and Virgin Mary with which the familiar passage is saturated. The most succinct and probing comment on its relevance to the play is from Seamus Deane: 'the opening lines of Edmund Burke's famous apostrophe to the *ancien régime* of France, written in 1790 by an Irishman who had made the preservation of ancestral feeling the basis for a counterrevolutionary politics and for a hostility to the shallow cosmopolitanism of the modern world'.[13]

The 'preservation of ancestral feeling' is something which, as I have been suggesting, Gar attempts alternatively to found on successive failed attempts at communication with his father or on the evocation of his absent mother's presence as the source of feeling in the family line. Since the mother is never openly discussed and there is no visible evidence of her in the house, all efforts at summoning her up are perforce arbitrary, artificial and willed; like the first appearance of the Burke quotation, they come from nowhere. For much of the first two episodes, the task on which Gar Private concentrates is on reconstructing his mother's history. Like Oedipus, he is both the detective determined to expose what has been covered up and the guilty target of his own inquiry, his birth having been 'responsible' for killing his mother. In this pursuit, Gar is particularly reliant on the oral testimony of Madge: 'She was small, Madge says, and wild, and young, Madge says, from a place called Bailtefree beyond the mountains' (37).[14] His reverie on her is provoked when he opens the cardboard suitcase and lifts out the sheet of newspaper dated for her wedding day. Here as elsewhere, every reference to the mother is heavily textualised rather than naturalised, attention drawn to its implausibility. The honeymoon reminiscence concludes with the near simultaneity of Gar's birth with his mother's death: 'and maybe it was good of God to take her away three days after you were born. [...] It is now sixteen or seventeen years since I saw the Queen of France, then the Dauphiness, at Versailles' (37–8). This passage establishes the crucial emotional resonance of the Burke passage and relates it to the 'Philadelphia' refrain examined earlier. Here, the apparent gap between what Gar is going to and what he is leaving is filled by the evocation of the mother and filled out by the two flashbacks which follow.

The introduction of the mother disrupts the chronological progression of the play. It is Gar Private's third and most important dramatic function, after encouraging self-expression and filling in the silence, to goad

Public into remembering: 'Remember – that was Katie's tune' (38). He cues him to re-experience by re-enacting the scenes with two women which lie behind his impending departure, both associated intimately with his mother: his fiancée Kate Doogan and his aunt Lizzie (Elise) Gallagher. If Gar Private has first encouraged Public to move away from family feeling into the 'shallow cosmopolitanism of the modern world', he now encourages the reverse procedure by forcing his *alter ego* to confront feelings he would rather suppress and deny. Where Private had earlier encouraged Public in fantasies constructed out of the images and idioms of popular culture, these are replaced by flashbacks which reflect on his own lived experience, in which the sense of painful self-division is more apparent as its sources are approached, and which are dominated by three women and the concept of (a) 'home'.

The crisis point in Gar's relationship with Kate Doogan occurs on the night he wished to enter the household of her parents, Senator and Mrs Doogan – 'Mammy and Daddy. They're at home tonight' (41). The objection she raises is economic and social: 'How will we *live*?' Gar's potential to become a father, to inaugurate his own line, does not survive the interview with Kate's father, Senator Doogan, who intimidates him with the socially disabling contrast between what Gar can offer and what his daughter has been bred up to expect, a 'home' defined in terms of class and bourgeois respectability: 'Mrs Doctor Francis King' (44). Gar may relapse gratefully from the smugness of Doogan 'and thank God for aul Screwballs' (45). But the encounter represents a social as well as a personal defeat and furthers the drive to a country where Gar can entertain the possibility of becoming a member of the Senate: 'The US Senate! Senator Gareth O'Donnell, Chairman of the Foreign Aid Committee!' (57). The fantasy now refracts a situation where, as Doogan outlines the meritocracy through which the father of a potential son-in-law of his would have had to graduate – 'school ... university ... law' (43), the forces driving Gar into exile emerge as economic as well as personal.

The groundwork is thereby laid for the play's second flashback, and arguably the most important scene: the return of Gar's Aunt Lizzy and Uncle Con from the US with the offer which he has decided to accept. The scene is one of emotional and economic blackmail. At its climax, Lizzy recalls what has motivated them to do so: '"We'll go home to Ireland," I says, "and Maire's boy, we'll offer him everything we have"' (65). Gar's initial aim in the flashback is to draw upon the privileged witness that Lizzy can offer to his dead mother as the sole survivor of the five sisters: 'And poor Maire – we were so alike in every way' (62). Specifically, he is urging her to provide a first-hand account of his parents' wedding day in the Bailtefree chapel. Lizzie is launched into this narrative as the scene commences, a narrative she never gets to complete. It is cut across by the present in a variety of ways, all serving to fragment and deny the pristine unity that Gar wants to afford this privileged narrative. Friel deliberately makes it the same day as Kate Doogan's wedding, not only to account for S. B.'s absence, but to make clear the emotional connections that are operating within Gar. What primarily gets in the way of a forced memorial reconstruction of a wedding day 26 years in the past is not just a competing wedding day in the play's present but the intervening occasion when Lizzy Gallagher and her husband 'sailed for the United States of America' (61).

Of the three possible relationships embodied for Gar by each of the contending narratives in this scene, this is the most possible, or rather the least impossible. There is the Freudian wish-fulfilment of his mother's marriage, his marriage to her which can be reconstructed in the imaginary, with Gar displacing his father; there is the now purely fantastic marriage with Kate Doogan, all the more compelling as a figure of desire because she is no longer socially real; and there is Aunt Lizzy, married to Con and a good Catholic, but willing to flirt continuously with Ben Burton and to admit Gar to her affections. The space for Gar is conspicuously designed as the gap, the absence in their perfect American home which only he could fill:

> LIZZY: ... We have this ground-floor apartment, see, and a car that's air-conditioned, and colour TV, and this big collection of all the Irish records you ever heard, and fifteen thousand bucks in Federal Bonds –
> CON: Honey. [...]
> LIZZY: And it's all so Gawd-awful because we have no one to share it with us ... (*She begins to sob.*)
> CON: (*Softly.*): It's okay, honey, okay ...
> LIZZY: He's my sister's boy – the only child of five girls of us –
>
> (65)

Up to this moment, Lizzy's vulgarity has kept Gar at bay. But as she broadens her identity into the litany of the five Gallagher girls, Maire and Una and Rose and Lizzy and Agnes, 'either laughing or crying' (65), and makes the severe contrast with the O'Donnell line, 'kinda cold', Gar is stung to acceptance by the fear of losing his capacity to feel. The scene reveals what he has inherited from both sides of the family in ways we can only understand with the appearance of Lizzy, taking over from

Madge as the true mother-surrogate. Gar Private stands by helplessly, understanding how emotional blackmail was used, as Public succumbs '*with happy anguish*' (66) to a mother's embrace and the term 'my son'. Aunt Lizzy has sought a new life and identity in the United States, refashioning herself as Elise, speaking a wonderfully impure mix of Hiberno- and American English. But she has returned with an implicit declaration of failure, in search of her earlier buried identity of Lizzy Gallagher which her nephew will help restore. Gar is no longer caught between the ideal mother and the too-real father, between the seductions of a Hollywood America and the daily stink of Donegal fish, but between two surrogate mothers: the Madge of Ballybeg and the Lizzy/Elise of Philadelphia.

In striving to remake himself, Gar has sought to reconstitute his origins through a series of substitute relations, through furnishing himself with alternative parents. Episode One concludes with the arrival of the teacher Boyle, who 'knew all the Gallagher girls' (53) and first introduces their names into the acoustic space of the play. Boyle provides a more intimate connection than Madge, as the question Private poses makes clear: 'And Maire, my mother, did you love her?' (53). The teacher is a preferred image for the aspiring Gar than that of his own father. The possibility that Boyle 'might have been my father' preys on Gar throughout the play and is finally answered by Madge: 'She married the better man by far' (87). Gar shrinks from the teacher's parting embrace, showing the extent to which he is indeed his father's son.

The issue of surrogate parents and of displaced relations affects almost every area of *Philadelphia, Here I Come!*, indicating the extent to which the question of 'home' is complex and ultimately political. In Ireland, the promotion of the family by the 1937 Constitution and by the Catholic Church, the intimate small-scale way in which political affairs are conducted, the absence of a settled body politic, have all thrown a disproportionate emphasis on the family as constituting the central principle of political identity and as mediating between private life and the affairs of the state. To go one step beyond the symbolic identification of Ireland as a mother is to valorise it as 'home', as emigrants like Aunt Lizzy traditionally do, and to bolster a powerful patriarchy which wields actual power while the woman's role is restricted to mother and converted into spiritual, visionary terms. This is the potent image which is installed at the centre of the affections: 'And surely never lighted on this orb, which she hardly seemed to touch, a more delightful vision. I saw her above the horizon, decorating and cheering the elevated sphere she just began to move in.' The vision is not one Gar can share with his father. But it is the closest text for a language of the affections that Gar has, as its desperate reiterations in the silence of the early morning with his father reveal, as he tries to use the quotation to break through their exchange of mutual, distancing banalities:

> PRIVATE: Say, 'It is now sixteen or seventeen –' Say – oh, my God – say – say something.
>
> (92)

Where Gar has sought to find his way 'home' in the first two episodes by reconstructing the memory of his mother, in Episode Three he turns on and to his father, questioning him, trying to get inside his head to find some mutually re-enforcing memory from the past on which a liveable future relationship could be built: 'Screwballs, do you dream? [...] God – maybe – Screwballs behind those dead eyes and that flat face are there memories of precious moments in the past?' (82). Looking for a shared memory, Gar seizes upon an 'afternoon in May' (95) 15 years earlier when he and his father had gone fishing. What they share is the loss of the mother; all they have is each other; and the protecting gestures now acknowledge the loss rather than attempt Oedipally to compensate for it: 'and you had given me your hat and had put your jacket round my shoulders because there had been a shower of rain' (83). In place of the words which have held communication at bay – 'although nothing was being said' (83) – a way was being found of saying, and hence denying, nothing. Central to this epiphany, and its acoustic climax, is the song, 'All Round My Hat', which Private imagines and re-creates his father singing. The details of the song are true at both the familial and nationalist level, with the protagonist wearing a green-coloured ribbon 'because my true love is far, far away'.

The play's crisis occurs when Gar tries to share this private recollection, to extend and translate it from the dialogue between his two selves into dialogue with his father. It is the moment when Gar Private is threatened with extinction, since Gar Public has so fully taken over the former's role. It is notable on-stage in Episode Three how quiescent Private is, how close Public is to operating on both levels of reality and in both registers of speech simultaneously, to what extent full psychic integration may be possible. The two Gars are poised equally between hope and fear: 'He might remember – he might. But if he does, my God, laddo – what if he does?' (94). In this final act of prompting, Gar Private combines all three of his functions, cueing the enactment of Gar's desire, the initiation of dialogue with the father, and the exposition of possibly painful memories from the lived past.

What follows is a tragicomedy of miscomprehension, with all the roles displaced. Gar Public is now attempting to speak his private thoughts in the public domain, conjuring his other half – now, S. B. O'Donnell in the role of Public – to remember and confirm his recollection. S. B. struggles to recall but every detail he retrieves fails to resonate and confirm Gar's narrative. The blue boat is a brown boat and crucially 'All Round My Hat', which he says he never knew, becomes 'The Flower of Sweet Strabane'. We are left with conflicting versions of the truth, a private narrative which fails to achieve authoritative ratification. When the private is made public in this fashion, the result is embarrassment, Gar's most painful so far, and all the defences are rapidly put back in place. It has taken so long for the two cagey adversaries to get to the negotiating table and attempt to engage in meaningful dialogue, but they don't sing the same songs and the symbolic emblems don't match up.

And yet, a few minutes later, the returned Madge hears the memory of S. B.'s which has finally surfaced, of a day he brought young Gar to school: 'the two of us, hand in hand, as happy as larks – we were that happy, Madge – and him dancing and chatting beside me' (97). Only the audience in its privileged position can comprehend and set S. B.'s epiphany against Gar's and estimate the degree to which, however much the surface details might vary, the two memories are complementary in the feelings they honour.

There remains to the end no direct communication in words between father and son. Only the form of the play and our reception of it can manage that, by conveying to us the shape of their individual experiences and the intimate interrelations we can perceive where the characters see primarily the formal ones. S. B.'s late epiphany, in a play where he has remained the observed while we have attended to Gar, resembles that moment in *Hamlet* where Claudius is discovered at his prayers, minutes before Hamlet enters, and we are given direct and privileged access to his innermost thoughts at a moment of pained recognition. In both cases, the play is tilted towards a more objective, broader perspective. In *Philadelphia*, S. B.'s reminiscence provides an answer at the private level to Gar's own epiphany, but does nothing to resolve matters at the formal or public level. No answer is directly forthcoming in words or dialogue. The play ends in apparent indecision, with Gar unable to formulate why he is going:

PRIVATE: [...] God, Boy, why do you have to leave? Why? Why?
PUBLIC: I don't know. I – I – I don't know.

(99)

As Yeats put it in the last year of his life, man can embody truth, but he cannot know it. And the embodiment of the truth of Gar's predicament is not to be found in any formula of words but in the shape of the play itself, in all the dramatic elements which go to form it. There is, as I suggested, a greater degree of objectification towards the close at the moment of S. B.'s epiphany; but there has been a core of theatrical objectivity all the way through, especially in the device of the two Gars. The final distancing element in defining the shape of the play is offered in technical, almost Brechtian terms by Gar Private's last cue to both Public and the audience: 'Watch her [Madge] carefully, every movement, every gesture, every little peculiarity: keep the camera whirring; for this is a film you'll run over and over again – Madge Going to Bed On My Last Night At Home' (99). The film metaphor here is explicitly to *silent* movies as the final textualised element of *Philadelphia, Here I Come!* indicates: the card which interrupts the action in silent film and Brechtian drama to draw attention in the latter case to the *process* of what is being enacted. The psyche is being technologised here as in *Krapp's Last Tape*, though now the adaptation is to film projector rather than tape-recorder. The rotation of the spool required to project the recorded experience is a reminder that the experience is always, already belated, never primary; that it has already been recorded and so can be disrupted in a variety of non-naturalistic ways: by being stopped, slowed down, speeded up and, most of all, played or 'run over and over again'.

We have already had two long examples of the replay of psychic experience in the flashbacks to Kate Doogan and Aunt Lizzy, and the concomitant disruption of linear chronological time. The fluidity of Friel's staging has made that possible. But fluidity, like so much else in the play, works both ways. And even the so-called 'real' time of the present begins to come unstuck. For one thing, when watching the play, this viewer's experience is that the night seems to go on an unnaturally long period of time, even by Donegal standards, with people still coming and going at a very late hour. For another, Friel plays with that clock called for in the stage directions and at times we have something more than real time, sometimes less. But the time is more reel than real, filmic rather than actual. The appearance of Kate Doogan on Gar's Last Night At Home has always puzzled me. Isn't she confined to flashbacks? Her appearance and the melodramatic scene which ensues amplify the sensation that what we are watching is indeed a film of 'My [Gar's] Last Night At Home', being played over and over again. The dramatic implications of such a perception are manifold: that the 'present' in the play is already the past, with the flashbacks just so many further rewindings of a Krapp-like tape; that the actual events of the night are already 'being distilled of all its coarseness; and what's left is going to be precious, precious gold' (77); that Gar is already gone, Ballybeg already a memory enclosing a question, the only answer to which is another repetition of the experience, its compound of the actual and the imagined. In Beckett's *Footfalls*, one character asks the other: 'Will you never have done ... revolving it all? [...] In your poor mind.'[15] One Gar might well ask the other the same question and the only possible answer would be: 'No.'

In the four-plus decades since *Philadelphia, Here I Come!*, Friel has written (to date) 19 original plays in addition to versions/translations of Anton Chekhov, Ivan Turgenev and the eighteenth-century Donegal playwright, Charles Macklin. There are, as I stated at the outset, three undisputed highlights: 1964's *Philadelphia, Here I Come!*; 1980's *Translations* and 1990's *Dancing at Lughnasa*. One belated success was *Aristocrats* (1979), given a premiere at the Abbey like most of Friel's plays in the 1970s but (also like them) disappearing after its initial run. Then, in 1988, a London production by Robin Lefevre walked off with the *Evening Standard* award for Best Play, and went on to productions in New York, Los Angeles and back in Dublin. *Aristocrats* forms a triptych with *Faith Healer* and *Translations* in a number of ways. It was written in the same intensely creative period of the late 1970s as the other two – a fact alone which would serve to secure Friel his place and reputation, and which makes *Lughnasa* all the more of a bonus and a gift. All three 1979/80 plays, from very different perspectives, converge on the same concern: a homecoming to Ballybeg, Friel's mythical 'small town' in Donegal. In *Aristocrats*, the O'Donnells return to the family estate in Ballybeg for what they expect to be the marriage of the youngest daughter but instead are faced with the death of the family patriarch. In *Faith Healer*, all the narrative twists and turns of the four monologues, and their progress through the winding backroads of Wales and Scotland, lead back finally to *bás in Éireann*, the night of Francis Hardy's final performance in Ballybeg. And in *Translations* the hedge-school master's son comes home to Ballybeg and brings with him the forces that will accelerate its destruction.

In *Brian Friel and Ireland's Drama*, Richard Pine discusses both *Faith Healer* and *Translations* as plays of homecoming, one private, the other public.[16] I would find his omission of *Aristocrats* puzzling in this thematic context, were it not that the two other plays are richer. I too am going to exclude it, partially for reasons of space, but also because I want to focus on *Faith Healer* of the three as the 'sequel' to *Philadelphia*, the return of the exiled prodigal son who turns out to be his own fatted calf.

Translations I will consider, in the chapter on Field Day, as concerned with what it means to write an Irish history play. This chapter will conclude with some remarks on *Dancing at Lughnasa* and Friel's dramatic activities since.

Aristocrats is Friel's most Irish version of Chekhov, with its own three sisters in Alice, Claire and Judith O'Donnell, on the decaying estate. There is an echo of *The Cherry Orchard* in Eamon, the worker from the village who has married into the Big House with deeply ambivalent feelings. Friel has reworked this Chekhovian paradigm in an Irish context through two crucial displacements: of time, into the present, thus filling the play with deliberate archaisms, and from Protestant to Catholic, so that the family undergoes a double dispossession. Perhaps because he put so much into this first version, Friel's reworking of *Three Sisters* two years later seemed tired, pedestrian. And it is as an Irish reworking of Chekhov that *Aristocrats* has travelled. A world audience has been sufficiently secured by that culturally recognisable scenario, the wistful cadences and dying falls of those exquisitely overbred people protesting their annihilation by historical process, for it to sniff at the exotic Irish references. These are teasingly intruded in Casimir's recurrent association of objects scattered around the living-room with great names in Irish history (Yeats, O'Casey, Daniel O'Connell, etc.), and find their most contemporary and disturbing locus in Eamon, with his talk of the Derry Bogside and his later discourse on colonialism.

If *Aristocrats* found its resonance through Chekhov, Friel's greatest successes have done so by offering audiences an image, a composite of cultural icons, which they can identify as 'Irish'. I needn't repeat these in relation to *Philadelphia, Here I Come!* since a good deal of my analysis was founded upon their treatment. In *Translations*, the audience is presented with the Irish-man as peasant, ensconced in a pastoral locale (a stable), dressed in the colourful costume of the early nineteenth century (from the period when visual representations of the Stage Irishman were most prominent), addicted to endless talk (in several languages) and frequently drinking whiskey; they are opposed by English redcoats who initially charm but turn predictably nasty by Act Three; at the close, the two older Irishmen stagger in footless and only avoid saying 'The world's in a state of chassis' by resorting to talk about Greek gods and by quoting Virgil.[17] Both *Philadelphia* and *Translations* play with these images of Irishness. And no one was more aware than Friel himself of the potential for the first play to be merely a tract about emigration and the second to be a lament for the destruction of Gaelic civilisation (he warns against the latter in a diary kept while writing the play).[18] No one was more horrified when the success of *Translations* was attributed to its supposed validation of an ideal and idyllic Gaelic order.[19] In both cases Friel responded with a savage farce – *The Mundy Scheme* (1969), where Irish-Americans are to be sold a piece of the old sod in which to be buried; and *The Communication Cord* (1982), where a venal senator restores a peasant cottage and blathers on about local pieties to a logorrheic linguistics lecturer. These two farces are less funny than they might be precisely because they are serving a personal need for exorcism on the playwright's part. And to its creator's chagrin what has primarily appealed about *Dancing at Lughnasa* to audiences worldwide is an Irish cottage kitchen in the 1930s with not one but five Irish colleens baking, sewing and generally generating *craic*, against the visual backdrop of a huge golden wheatfield and the verbal aura of the narrator's childish wonderment and nostalgia. While pursuing a similar excavation of paganism, 1993's *Wonderful Tennessee* did so in a bleak present in which none of the three married couples' anguish was resolved by the fact that they came on stage singing and dancing. All three plays – *Philadelphia, Translations* and *Lughnasa* – are set in the historic past and offer consoling, nostalgic, unified images of Irishness. Their fate and reception raises – through their very success – the problems associated with writing an 'Irish' play, with what delineates an on-stage definition of 'Irishness'.

The play in which Friel has most daringly dealt with the question of presenting a drama of national identity is *Faith Healer*. The play was not a worldwide success, rather a failure when produced on Broadway in 1979 with James Mason in the title role and direction by José Quintero, and with only a limited run in London. It is in Ireland that *Faith Healer* has had its greatest appeal and found its meaning and *raison d'être*.[20] Produced at the Abbey in 1980, it provided a necessary counterpart and complement to *Translations* in the same year. *Faith Healer*'s impact transformed Irish theatre in a number of ways over the coming decade: by helping to create an audience for spare, demanding plays of spiritual and emotional crisis where, indeed, a great deal of endurance was demanded from that audience (here, the other great exemplar is Tom Murphy, subject of the next chapter); by presenting Donal McCann, that most intellectually passionate of actors, with his greatest, most demanding role (he returned to it in the Abbey in 1991); by forming his creative acting partnership with John Kavanagh; and by problematising the role of woman in relation to this male dual partnership. Its direction marked a high point in the career of Joe Dowling as Artistic Director of the Abbey. It also managed to produce an original playwright since it was the experience of seeing *Faith Healer* in 1980 that persuaded Frank McGuinness, discussed in Chapter 5, to start writing for the stage. For these and other reasons *Faith Healer* became the play with which this central Friel chapter would have to deal.

It is the most theatrical of Friel's plays. As Frank McGuinness put it in a lecture: 'Friel bows to the theatre's demands in *Faith Healer.*'[21] This claim would appear to be contradicted by the fact that in many ways the play seems undramatic since it never manages to bring any of its characters into direct dialogue, the setting is virtually non-existent and the form is monologue, which might as well be read in private or on the radio. But the experience of viewing a production of the play reveals its profound dramatic qualities. Like Beckett, Friel has increased, not lessened, the dramatic potential by reducing the setting and props to a minimum. Those few signifiers he admits, therefore, are all the richer in possible signification.

The first thing that strikes someone viewing the play in performance is that the stage is not entirely bare but rather contains a large poster and several rows of chairs. The poster reads: 'The Fantastic Francis Hardy / Faith Healer / One Night Only'.[22] Within minutes Frank explicitly identifies himself with it by way of introduction. His gesture establishes the self-consciously theatrical nature of the proceedings we are about to witness, particularly in relation to Francis Hardy's sense of his own identity. The poster has been rewritten. It, and he, were originally titled '*Seventh Son of a Seventh Son*'; but that was not only too expensive but 'a lie' (332–3). He then starts to tell us who he was 'in fact' but shrugs it off with 'that's another story'. Frank Hardy, as the play unfolds, is revealed as a storyteller, one whose lies are to be believed and whose facts are suspect. What he aims for is accuracy in and with words themselves. And the poster is the fiction that has stuck, the sign of his best self, marking a desire to transcend the multiple imperfections of the flesh, his own but particularly those who bring their grotesque bodies before him, in a pure, absolute statement. The banner, however, is 'tatty' (332), frayed, and although it survives through the narratives of Grace and Teddy to suggest the persistence of Frank's presence in their memories, its disappearance at the start of Part Four is shocking, the clearest indication of the fate that Frank is about to undergo.

And this is where the chairs come in. Taken together with the poster, they indicate the extent to which the faith-healing performance described by all three characters is being re-enacted before us. This in turn makes the audience itself a crucial participant in the faith-healing, extending the drama from the confines of the stage across the footlights to embrace the entire auditorium. As Frank says when describing their nightly ritual: 'We'd arrive in the van usually in the early evening. Pin up the poster. Arrange the chairs and benches' (335). But what he is narrating in the fictional past is also simultaneously there before us in the visible present-ness of the poster, the chairs and the faith healer himself. Other props necessary to the night's performance are missing – for example, Teddy and 'his amplifying system'. Those missing elements will be distributed throughout the course of the play. When Teddy finally appears in Part Three, he is accompanied by his record player and '*a recording of Fred Astaire singing "The Way You Look Tonight"*' (353–4); and in Part Two, Grace is there at her table. Each will offer in turn their own partial perspective on the night's performance and the interrelated events of a shared lifetime.

What Frank in the opening monologue does concentrate on is the audience, especially the huge expectations with which they come. He defines a gap, between what they hope for and what they expect, stressing the elimination of that hope in order to bring about some kind of resolution. *Faith Healer* clearly contains fragments of both pagan and Christian beliefs, in its reference to harvest festivals and Christmas decorations; and the longing Frank divines is the quasi-religious desire to be made whole, on the part of a maimed people. The primary dramatic experience of his audiences, therefore, is that of waiting and waiting – for a miracle, which may or may not occur. And as Frank himself tells us: 'nine times out of ten nothing at all happened' (334). This is Vladimir's description of the action of *Godot*; but against it needs to be set the line from *Endgame*: 'Something is taking its course.'[23] Occasionally, just occasionally, according to Frank, 'the miracle would happen' (337) and the subsequent flooding of hope would evoke more terror than pity. But what creates the conditions in which the miracle is possible is the presence of the audience and the raising of their hopes. The lengthy description by Frank in Part One has the double function of raising those hopes and showing the audience all the reasons why they should not do so, why they are foolish to be so wooed, an act of calculated theatrical defiance. Frank's pitch leaves the audience free to regulate the degree of its credulity at will but also sets its cap at transcendence. With that much freedom and instability insisted on from the audience, it's no wonder there is no unanimity of response to the play, that some remain unpersuaded and unmoved by *Faith Healer* while for others it has the conviction of a spiritual experience.

The audience in the theatre, therefore, stands in for the congregation in a Welsh chapel or Scottish kirk; and the play is a rehearsal of the very process it describes. But in the larger overarching play that is Brian

Friel's *Faith Healer*, the audience also has a crucial role to play in relation to the interwoven, damaged lives of its three protagonists. They are telling their life histories to us, and each of their monologues is shaped rhetorically and emotionally by this directed appeal. Frank opens his eyes at the start and declares: 'I'd get so tense before a performance, d'you know what I used to do?' (332). At the heart of its jokey apparent contradiction (how can we know what we haven't been told yet?) is not only the promise of a lengthy narrative but the sense that it will evoke a suppressed, half-known truth from the listener's own experience, requiring not only their assent but active participation.

Grace's monologue is by far the most private of the three. She is the silent partner in public, though central to the private dynamics of the faith-healing act. She is the only one who doesn't speak (in) a public voice, either in warming up or addressing the audience. So hers is the quietest address, the one the audience must strain most to hear. It also exposes a different story from her husband's and so reveals its contradictions. But Grace's monologue too is rhetorically shaped by and directed towards an audience, since the immediate dramatic context is her recovery from a nervous breakdown. The interaction is that between psychiatrist and patient, with Grace doing the talking while we the audience are the silent, judging witness to her entreaties: 'But I *am* getting stronger, I *am* becoming more controlled – I'm sure I am' (341). The implicit 'Amn't I?' leads us to want to give a verbal reassurance while we inwardly register a depressed negative.

Of the three, Teddy's address to the audience is the most direct in the sense that it is the most purely theatrical, the most rhetorically explicit: 'What about that, then, eh? [...] Tell me – go ahead – you tell me – you tell me – I genuinely want to know – what sort of act is that to work with, to spend your life with? [...] You tell *me*. I never knew!' (354–7). The main tone throughout his non-stop speech is one of jokey confidentiality, cheerful buttonholing, deepening into moments of personal reverie concerning Frank and, especially, Grace, which break the 'professional' tone and relationship, for all his protests to the contrary. At one point, and it is the key point when Teddy asserts that 'you must handle them [artists] on the basis of a relationship that is strictly business only' (357), he urges 'believe me'. And this is the appeal which underwrites all four of the monologues and their three speakers: believe me, believe in me, in who I say I am and what I am, based on my version of events.

As anyone who has read or seen the play will know, two of the many placenames yield three individual accounts: Kinlochbervie and Ballybeg. Both places provide the points of greatest verbal identity in the separate monologues, since both are initially described with the very same words and order in all the accounts:

> Kinlochbervie, in Sutherland, about as far north as you can go in Scotland.
>
> (337, 344, 362)

and

> So on the last day of August we crossed from Stranraer to Larne and drove through the night to County Donegal. And there we got lodgings in a pub, a lounge bar, really, outside a village called Ballybeg, not far from Donegal Town.
>
> (338, 351, 367)

The accounts of what happened that fatal night in Ballybeg do not fundamentally contradict each other but supply three differing but complementary perspectives. But there is a radical disjunction between the first and second accounts of the Kinlochbervie incident, Frank's and Grace's. Frank has used it as the point of departure for a return to Ireland and the death of his mother, merely a place where 'we were enjoying a few days rest' (337). But when Grace first says, 'Kinlochbervie's where the baby's buried' (344), any audience I've been part of responds with a palpable start, a gasp. They realise that Frank has by no means told us everything and that another story, that of Grace and her stillborn baby, is about to emerge. By the time we get to Teddy and his version, we're ready for anything when he and the dramatist teasingly play on our knowledge about Kinlochbervie: 'there we were away up in Sutherland – what *was* the name of that village?' (362). What Teddy then says appears to confirm Grace's narrative about the fact of her pregnancy, the agonised delivery and the 'tiny little thing' (363) they buried there. But it contradicts her story, which gives pride of place to Frank, by insisting that Frank disappeared and left Teddy to be midwife and chief mourner. When Frank comes on for the last time, the story is not retold a fourth time, as the pattern might have led us to expect. But as he enters, once more intoning place-names, he fixates on the name of the place, tolling it three times like a death knell, just as Grace has in her narrative: 'Aberarder, Kinlochbervie, / Aberayron, Kinlochbervie, / Invergordon, Kinlochbervie ... in Sutherland, in the north of Scotland' (370). In place of the narrative that this oral formula has always cued, and where we might now be expecting a definitive version of what took place, Frank breaks off with: 'But I've told you all that, haven't I?' This remark can be read psychologically as a denial. He is not capable, even if he wished, of delivering the whole truth. It has taken the drama in its multiple narrative unfoldings to tell us all, the summoning up of more than one ghostly presence, with each shade's claim to authenticity thrown in doubt by his or her successor.

The drama of *Faith Healer*, then, can be read in two ways, or in two directions: the dialogue that is set up between the characters and the stories they tell; and the dialogue that is set up between the storyteller and the audience. In this process of interaction between the characters and the audience, Friel's *Faith Healer* reveals a great affinity with Beckett's *Play*. There we also confront three characters engaged, while alive, in a triangular adulterous relationship of mutual need from which they cannot free themselves, even in the death-like state or limbo in which they currently reside. And we learn in the course of *Faith Healer* that all three characters are dead. First, the likely fatal outcome of Frank's meeting with the incurable McGarvey and the four wedding guests is confirmed by Grace's reporting of the doctor's question: '"And what was your late husband's occupation, Mrs Hardy?"' (346). If Grace confirms that the speaker of the first monologue is dead, Teddy in turn describes how he was asked to visit the morgue and identify Grace's body after her suicide: 'And there she was. Gracie all right. Looking very beautiful' (369). And Teddy? Frank is too concerned narrating his own end to pay much heed to Teddy's; on that last night, he has cured his manager of his heart-sickness and dismissed him from service. But Teddy has no life other than as a theatrical creation. As Frank says when he first introduces him to us: 'I never knew much about his background except that he had been born into show business' (334). And when Teddy in his own monologue gestures at the poster and says, 'A lifetime in the business and that's the only memento I've kept' (365), we realise that he has no life outside the 'business' of the play and so in his 'retirement' must keep playing it over and over again, like his worn-out recording of Fred Astaire singing 'The Way You Look Tonight'.

For all three, Frank, Grace and Teddy, are kept animated in a perpetual present by the story they have to tell. Each is radically incomplete, both in themselves and the story they narrate. This incompletion is made more formally explicit in Beckett's *Play*, where the monologues occur simultaneously rather than sequentially. Hence they are fragmented in the audience's reception of them, and so we have literally and painstakingly to piece them together as well as interpret them. To aid us in this process, the entire act is repeated word for word. By his title, *Play*,

Beckett has posited an essential theatricality in some of the ways I have tried to suggest are relevant to Friel's play. But Beckett's own drama, as the first chapter of this study has argued, bears an Irish cultural and theatrical dimension that has for too long been overlooked. And it is the role demanded of the audience that I would most emphasise, the rupturing of a rigid separation between those onstage and those in the audience, and the subsequent act of mere empathy. Rather, in the breakdown which is encouraged, the audience is required to participate in the construction of the play's activity and meaning.

What I would like to concentrate on, in terms of cultural analysis, is the play as a drama of national identity, an issue that is consistently and consciously engaged with by and between the lines. Each of the characters constructs his or her own version of reality in what they say; each tries to tell a story, denies layers of complicity, compromise and betrayal, in order to present a version in which they look good and occupy the key role. And I have said that the play is quintessentially theatrical. It is so in part because each of the characters derives his or her strongest measure of personal identity from the role they play. When Frank is on stage, especially on the rare nights he knows he can make a change, he is assured, at his ease, with all the maddening questions that plague him in private life temporarily silenced. Grace plays a role in which she both supports and is supported, as Frank's wife, muse, lover, aide; it is a role which offers her freedom from the sterile silence enforced upon her by her father but which offers only a surrogate fulfilment nonetheless. And Teddy is always the manager, never the star, always playing second fiddle to Frank as performer and as husband/lover to Grace, only kissed by each to confirm his exclusion.

But this triangle is also a play of national and cultural identities, since *Faith Healer* is so insistent about specifying backgrounds for its creations in these terms. I wish to examine the different stories told by the characters on this score, not only in the Kinlochbervie but also in the Ballybeg narratives, going beyond the personal (the thwarted love, the buried child) to the equally crucial and more distinct narratives accompanying the birth trauma and the fatal homecoming. I wish to consider their stories as contending, conflicting dramas of national identity canvassed one against the other and disturbing most of the received ideological notions of what constitutes an Irish theatre.

The life that is described in the opening invocation is one that addresses the kind of theatre that was available in the Irish countryside earlier in the century, the life of the fit-ups, companies that travelled around from village to village, working in barely adequate conditions, living off cheap whiskey and bags of chips. The life of the itinerant theatrical is epitomised by Teddy thus: 'the smell of the primus stove and the bills and the booze and the dirty halls and that hassle that we never seemed to be able to rise above' (367).[24] In an earlier play, *Crystal and Fox* (1968), Friel more directly dramatised such a travelling show, with its impure and downmarket mix of music-hall, circus, pantomime and conventional melodrama. Indeed, the Melarkeys' troupe featured an earlier incarnation of Teddy's performing dog.

What is unusual about Frank's description of life on the road is that their beat is exclusively restricted to Welsh and Scottish villages, that it is figured primarily in geographic terms and related to the cultural make-up and temperament of the trio: 'Seldom England because Teddy and Gracie were English and they believed, God help them, that the Celtic temperament was more receptive to us. And never Ireland because of me' (332). All the later monologues reveal that Grace comes from an upper-class (probably Protestant) background in Northern Ireland. Frank's distorting simplification removes her from the island of Ireland in a single stroke and simultaneously confirms his sense of his own Irishness.[25] Teddy is all too audibly an Englishman, the only one who can speak the language properly, as he protests. He does so to such an ostentatiously theatrical degree that he constitutes Friel's revenge on centuries of stage misrepresentation of the Irish by concocting this Stage Englishman, the loquacious lovable cockney comedian. But nothing in the Teddy we subsequently meet squares with Frank's efforts in his opening monologue to put Teddy across as a down-at-heel Arnoldian: 'I think he had a vague sense of being associated with something ... spiritual and that gave him satisfaction' (334). Grace and Teddy are never more transparently Frank's fictions, serving his psychic needs, than when he culturally constructs them here as 'English' and thereby activates the binary oppositions which will confirm in him a certain opposite and apposite Irishness.

If Grace is English, as Frank insists she is, then she can embody the traits of rationality and orderliness with which Matthew Arnold endowed them to set against the wayward irrationality of the Celt: 'And there was Grace, my mistress. A Yorkshire woman. Controlled, correct, methodical, orderly. Who fed me, washed and ironed for me, nursed me, humoured me. Saved me, I'm sure, from drinking myself to death' (335). The 'Yorkshire' touch in Frank's fiction is surely provided by the popular entertainer Gracie Fields, later invoked by Teddy as one of the 'great artists' (355) – a purely theatrical invention. But this 'Yorkshire woman' is still a transparent veil for the stereotype by which the irresponsible, drunken Irish-man is to be 'saved' by the redeeming qualities not just of womanly domesticity (they are secondary and come second) but of 'civilisation', all the practical qualities in which the Celt is lacking and which render him ungovernable. The later cultural development out of Arnold and modified by Yeats was to further 'spiritualise' the qualities of the Celtic temperament – vague, dreamy, idealistic – and to castigate the English qualities as urban and materialistic. Arguably, this was a step towards decolonisation; or it could simply confirm the stereotypes at a deeper level. The earliest Celtic Twilight phase of Yeats's career, and the first plays of the Revival, do this. In the 1890s, Yeats could play the Celtic poet in London and exploit an exoticism of the periphery, particularly through the word magic of the Irish place names. In the 1950s, Joan Littlewood could and did act as a London promoter of Brendan Behan; and television was available as the new forum for the Stage Irishman. In Frank Hardy's construction, his English wife and cockney manager manage and make presentable the gifted, feckless Irish scapegrace.[26] They confer on him a certain kind of identity that in its very artificiality gives him a crucial measure of freedom from the biological and cultural facts. But all along Frank knows that his identity is only provisional and has to be proved over and over again. His ultimate audience is his own people and it is inevitable that he will return in the end to Ballybeg; it is equally inevitable that the return will spell his end.

In Frank's drama of exile, the only movement outside of the oscillating rhythms of off-stage and on-stage ('I can't go on, I'll go on') are the returns to Ireland. And each is associated with death. First, there is the death of a parent, transposed by him from the actual death of his father to the symbolic death of his mother. The Oedipal scene he confronts there raises, as it did for Gar in *Philadelphia*, the question of action: 'Jesus, I thought, O my Jesus, what am I going to do?' (338). But the mother is dead, not alive, when he arrives, her beauty and hence her enthralment of him preserved intact. And so the question of action is suspended, deferred. In the second homecoming, to Ballybeg, Frank is no longer responding to a purely personal scenario, one of his own devising. (There are of course allegorical dimensions to the hold of the dead Irish mother on him but these are never made explicit.) What Frank says of the four Donegal wedding guests in relation to their friend McGarvey is no less true of himself in relation to them. 'They created him' and, as they did so, 'I saw McGarvey in my mind' (340). Hardy is here caught up as an agent in a larger drama, involving a fall and a paralysis – the cultural diagnosis is acute. And as he tells us of this 'first Irish tour! The great home-coming!', Frank, in one of the very few stage directions specified by Friel, draws *'as close as he can be to the audience'*. As he does so, we may notice for the first time the *'vivid green socks'* (331) in which he nightly treads the boards. His story has become increasingly present to us as it homes in on Ireland, implicating the audience in his own fate and what is going to happen, the 'nothing' and the form it will take. What Friel represents is, to draw a term from Joyce's *Finnegans Wake*, the 'abnihilisation' of the faith healer. He is not only the sacrificial scapegoat for a community's inherited ills, as Christy Mahon was before him; but the play's closing act, which is both an act of destruction (annihilation) and re-creation from nothing, is one rife with possibilities for a new post-colonial identity and drama.

The question of whether Frank Hardy has the last word can be considered in relation to Synge's *Playboy*. The latter closes with Pegeen Mike's cry of desolation: 'Oh my grief, I've lost him surely. I've lost the only playboy of the western world.'[27] And Pegeen is on-stage at the beginning and end of the play. This should prevent us from privileging Christy Mahon's resolution over Pegeen Mike's; but too often the critical attention has focused primarily on Christy and has seen him walk away with the play at the end. So, too, it could be argued, does Frank Hardy, who is given the last word in *Faith Healer*. But Pegeen's brief closing self-lament is amplified into Grace's 12-page monologue.[28] It's as if a sequel were written to the Synge play, a 12-months-after to show how Pegeen had got along in the meantime. I may be responding here to Declan Kiberd's suggestion that *Faith Healer* is a version of Deirdre – specifically, Synge's *Deirdre of the Sorrows* which (unlike Yeats's) dramatises the period of exile in Scotland as well as the tragic homecoming.[29] And we have seen Christy work his own brand of faith-healing, based on the power of a lie, amidst the physically and psychically maimed community of County Mayo.

What I want to concentrate on in Grace's monologue in terms of the possibilities for an Irish drama is not what was foregrounded in Synge but what he consciously suppressed: his own Anglo-Irish background and identity. For what Grace returns to when she goes home to Northern Ireland is a prototypical Protestant Big House, symbolically represented by 'the gates' (347) as it is in the Jennifer Johnston novel of the same name.[30] These lead into formal grounds, the avenue 'flanked with tall straight poplars', lawn and Japanese gardens. But there is one area of chaos in the vegetable patch, the 'feminine' area to which her mother's disturbed life is restricted and with which it is symbolically identified. The cast is introduced: a mentally unstable mother, an Irish Catholic housekeeper (Bridie) who 'reared' Grace, and a threatening father, a judge high up in the professional classes.

The members of Synge's family rented on the property they had previously owned, and in one of his essays he writes: 'If a playwright chose to go through the Irish country houses he would find material, it is likely, for many gloomy plays that would turn on the dying away of these old families, and on the lives of the one or two delicate girls that are left.'[31] Yet when Synge tried in his first play, *When the Moon Has Set*, to give a dramatic representation of this Big House background, Yeats and Lady Gregory rejected his play outright and suggested that he concentrate in future on representing 'the people'.

In treating a 'delicate girl' from the 'Irish country houses' as a possible source of Irish drama, Friel downplays the usual religious association – i.e. that they are Protestant – by omitting it entirely from Grace's account. Her father's accusation is not that Frank is a Catholic but that she has married a 'mountebank' (348) and so betrayed her class. In *Aristocrats*, Friel writes his own Big House drama and pointedly makes the O'Donnell family Catholic. When Eamon in that play talks of the ambivalent appeal that the place exerted on his childhood imagination, he does so in terms of cultural prestige, which he has sought to acquire by marriage. Frank has moved in the opposite direction and denies that he is married to Grace or what she represents. The aristocracy he seeks to assert is to be defined by a set of private standards, not those bequeathed by history and oppression.

Grace has come back home, she thinks, to ask her father's forgiveness. But she ends up being reduced to a child in his presence and condemned, judged. Frank has remade people by 'some private standard of excellence of his own' (346); he has offered her this freedom. For what the Big House and her father represent is a taboo on speech as self-expression. Words come tumbling out of her father's mouth, 'not angry words but the tired formula words of the judge sentencing me to nine months' (348). The urge is to curse him, 'assault and defile him with obscenities', to adopt the language of the outlaw. Grace claims she hasn't done so and is glad; we know she has and is ashamed. She stands accused of a betrayal of class, of sexuality (an act of *mésalliance*, as in Yeats's *Purgatory*), a betrayal finally of language. Frank's envy of her father is of the word as law, of formal cadences rendering judgment. As so often in Beckett, the words of the father are those of the Bible, especially the prophetic books of the Old Testament. They are now appropriated in and by the drama, parodied, blasphemed, denied authority. Grace talks back, verbally retaliates, with 'words he never used; a language he didn't speak; a language never heard in that house' (348), the opposite of official standard English. The tendency of an Irish writing enforced to work within English is to free words from their traditional unitary allegiance, to refashion them according to some private standard of excellence. When Synge came out of an operation, he is reported to have said: 'May God damn the bloody English language in which a man can't swear without being vulgar.'[32] Grace's mother is 'a strange woman who went in and out of the mental hospital'. Her daughter, in her own bid for existence and survival, has jumped the walls of the estate and embraced a life of 'squalor' (349).

Teddy offers the most complete picture of their life on the road. For him, it *was* his life, as I argued earlier. When he diverts into other narratives, it is to tell shaggy dog allegories that reflect back on his relationship with Frank and Grace. He is the least pure element in this Irish play, both because he's low class and because he's foreign, i.e. English; he also tells jokes and is entertaining. Collectively, these were the elements which Irish cultural guardians for decades sought to quarantine and ban, and to represent as the opposite of what was Irish, as I argued in relation to *Philadelphia*. There, the referents were to the US, the cinema and dancing; here, the context is primarily English, the music-hall comedian and the tabloid newspaper. In both cases, this cultural propaganda was undercut by the considerable popularity of the very same elements with Irish people. It is the lingua franca of these that Teddy speaks and represents. Also with his repeated references to Sir Laurence Olivier (and surely Olivier's come-back performance of Archie Rice in John Osborne's *The Entertainer* is one source of this showbiz persona) and to the world of the theatre, Teddy is also a crucial reminder that an Irish drama in the English language has always existed in some relation to the London stage. The extended run of *Dancing at Lughnasa* in the West End represents a crucial stage in the level of the play's acceptance.

What Teddy therefore displays in his stories of Frank and Grace is the most theatrical as well as the most romantic perception of them. He is outside the circle created by their aura, just outside it, as Grace feels herself outside Frank's rapt self-attention. These are Dantean circles of a showbiz hell. But the gaze that he directs at their relationship is one informed by theatrical energies. His monologue presents accordingly two plays-within-the play *Faith Healer*: the 'Scottish' play and the 'Irish' play. My first term is deliberately intended to recall Shakespeare's *Macbeth*, as does Teddy's description of what happened as curtain-raiser to that memorable night in Llanbethian when Frank miraculously cured all ten people present. At first the house is empty and the only audience that storms in is Grace, calling on her 'Irish genius' to appear (358). What we get in the full-blown exchange between Frank and Grace, from the insistent cue of 'Out! Out!', is their reworking of *Macbeth*, now displaced from Scotland to Wales, from one of their stomping grounds to another. Frank, 'looking like he's about to die' ('brief candle'), is beset by Grace, the wife transformed into a three-in-one witch who taunts and torments him to do the impossible deed he dreads. She has run 'mad' and is 'mocking' him, taunting his manhood and nationhood: '"I came to see the great Irish genius. Where is he?"' (358). What she is both challenging and calling for is an investiture even more self-appointed than Macbeth's. The description emphasises the play of 'voices' echoing around the rafters of the church, inspiring Frank Hardy to act and, by acting, to silence those same tormenting voices protracted to eternity ('it goes on and on and on'). And as we know from the retrospective prophecies we have heard, Frank and Grace have no children ('Kinlochbervie'), no successors, no inheritance to pass on. All that remains is the onwardness. Their private drama is here played out before an audience of one, Teddy, and then recedes before and is replaced by the public. When Frank goes to meet his public, a miracle occurs; Birnam Wood does not come to Dunsinane, but all those present have their physical imperfections cured. In this drama, 'Hardly a word was spoken' (359) and the entire sequence climaxes in a dance between Frank and Grace as they sing 'The Way You Look Tonight'. Teddy is excluded from their circle, as Grace (except at such rare moments of epiphany) usually is from Frank. The play is a series of widening concentric circles, in which they – and we, the audience – come close but cannot touch. They are like the ghostly lovers in Yeats's *The Dreaming of the Bones*, caught in the circle of their unsatisfied desire.

The 'Scottish' play has as its sequel the 'Irish' play, represented by how Teddy sees Frank and Grace on the fatal night in Ballybeg. His gaze represents them as 'Mr and Mrs Frank Hardy. Side by side. Together in Ireland. At home in Ireland' (367). Here, we are being shown the couple, not as they are/were, but as they might have been. That kind of ideal Irish relationship, of personal and cultural harmony, is a view possible only to outsiders. It is, as Friel will develop through the figure of Yolland in *Translations*, an ideal relationship only visible briefly, and on stage, only possible through an Englishman's gaze, and only by ignoring a wealth of tawdry circumstance. The Englishman, whether Teddy or Yolland, can only see the beauty, not the deadliness, because given his position he cannot concede the history, the impermanence, the rifts of division.

But who heals the healer? It is a physical shock when Frank returns after Teddy's monologue. We may have forgotten his promise that he will give a full account of his homecoming later. His return both closes a circle and suggests that another may be initiated. The dimension of *Faith Healer* that comes over least on the printed page is the physical presence of the actors. This truism is sharpened to an existential knife-edge by the play, however, since each actor is required to fill the void with his or her presence, to give the play life while they are on. They do so not only in dialogue with the audience, as I have suggested, but in dialogue with their off-stage partners whom we have just seen. Grace's monologue strives to overcome the fact of Frank's absence, to make him present in a space which (as we can testify) still bears the visceral trace of the actor playing Hardy. Teddy tells us that Grace has just died; but we have just seen her and thought she was alive. Well, try calling her back, then. In this final monologue, Frank describes walking around the Ballybeg pub, taking his leave of Teddy whom he has cured. He does not directly take his leave of Grace but of everything that relates and is related to her: of the children they never had and of her parents; the father he envied; and the mother who was beyond the reach of his powers. The first monologue relied on Frank's acting ability to create Grace and Teddy for us, in subtle and not-so-subtle acts of mimicry: 'Or as Teddy would have put it: Why don't we leave that until later, dear 'eart?' (341). When Teddy appears he almost exactly lives up and conforms to the portrait presented by Frank. On the other hand, what Grace says forces the audience almost completely to revise and reverse Frank's account of her. But in each case there is a model set up for fulfilment by the first monologue.

In the fourth, Frank switches his histrionic attention to focus on the four wedding guests and McGarvey. They have already been described for us by Teddy, in words and phrases that decades of tabloid journalistic coverage of Northern Irish events have made familiar: 'those bloody Irish Apaches' (366). And lest this be taken too unequivocally as the typical British stereotype of the Irish, Friel has the local barman describe them in almost identical terms when he warns Frank: 'I know them fellas – savage bloody men' (374). A clear relation is set up between the group with whom we have shared the intimate experience of the play and the group by whom Frank is going to be killed: 'We had ceased to be physical and existed only in spirit, only in the need we had for each other' (376). In one sense, what follows is a replication of what has gone before, if we regard Frank's monologue in Part Four more as the first term in a new series than as the last term in the previous one. Once more, he comes on-stage and evokes the presence of the other characters in his drama. But since this is the end of the play and no more is to follow, the denouement places a greater strain on the medium. It requires of Frank Hardy and the audience that we imagine those people into some kind of presence by making them real. And within the 'fiction' of the play, Frank does not 'know' this group as he 'knew' his wife and manager: 'Ned was on the left of the line, Donal on the right, and the other two, whose names I never knew, between them' (375). The act he is required to perform calls for him to draw, not parasitically on the lives and energies of the two people who served his every artistic whim, who sacrificed their lives and needs to him, but – in the ultimate self-immolation – on himself.

Frank Hardy ends by performing the ultimate act of theatre, by realising imaginary beings with his own flesh and blood. The final homecoming is to himself. The act annihilates the distance, closes the gap necessary to artistic expression, and results in the death of Frank Hardy as a dramatic creation. In doing so, he feels he is 'renouncing chance' (376). The play itself takes many chances. It is the riskiest and most courageous in Brian Friel's career. It finally expresses the need to look in the face and to acknowledge what Seamus Heaney has called 'the exact / and tribal, intimate revenge',[33] by imagining and dramatising those who commit such acts. The form Friel has chosen is not documentary. It insists on working through drama and does so by working through the various dramas that have developed in and around Ireland. (Stewart Parker did something similar in a more schematic way in *Northern Star* (1984), where the styles range from Farquhar to Behan.)[34] *Faith Healer* makes starkly apparent that there are no easy options when it comes to the vexed matter of theatrical representation where Ireland is concerned, and that any one dramatic solution is at best provisional. As it says on Frank Hardy's poster: 'One Night Only' (331).

Friel followed *Faith Healer* in 1980 with *Translations*. Its opening in Derry's Guild Hall inaugurated his ten-year involvement with the Field Day Theatre Company, which he co-founded with the actor Stephen Rea. The Field Day plays, by Friel and other playwrights, are discussed at length in Chapter 5. With *Dancing at Lughnasa* in 1990, Friel chose not to offer the play to Field Day and instead sought the greater resources of producer Noel Pearson. It was first staged at the Abbey in May 1990 in a production by Patrick Mason and the memorable 'field of wheat' design by Joe Vanek, before going on to even greater acclaim in London and New York, where it was awarded a Tony for Best Play in 1991. *Lughnasa* opened to mixed notices and reasonable houses at the Abbey, but then began to build and build as audiences responded to the plight of these five Donegal women from Ireland in the 1930s. The dedication in the published text, 'In memory of those five brave Glenties women', suggested a basis in autobiography.[35] (Friel's mother and her four sisters were from Glenties in County Donegal.) *Dancing at Lughnasa* stands in the vanguard of a powerful development in recent Irish drama in that it moved the energies of women from the margins to centre stage and shared the historical moment with the election of Mary Robinson as the first woman President of Ireland in that same summer of 1990. By distributing the dramatic weight equally among his five female characters, Friel is writing against the symbolisation of woman so central to the nationalist position.

His awareness that he is doing so is clearly signalled by the debate in his previous play, *Making History* (1988), between Hugh O'Neill, Earl of Tyrone, and Archbishop Peter Lombard. Lombard is writing a biography of the O'Neill which is designed to fashion a single Catholic nation out of the warring Irish tribes by giving them a figure around whom they can rally. O'Neill points up and queries the exclusion of his wife Mabel, a woman from a different religion and background, from Lombard's biography, and argues that it distorts the picture. The play is, among other things, a critique of the all-male bias of the Field Day board, the plays they produced and the notorious three-volume *Field Day Anthology of Irish Writing*, which appeared in 1991. The strongest scene in *Making History*, in general a male-dominated play, is that between Mabel and her sister, who urges her to stop living with these barbarians and return to her English family; the scene is certainly one of the seeds of *Lughnasa*.

The replacement of the two Gars and a succession of fathers and sons in the plays of Brian Friel by the five Mundy sisters marked a major change and development in his career. But no less than with the male double-act, this development of a more woman-centred focus in contemporary Irish drama was anticipated by Samuel Beckett in some of his most important later plays. After the male double leads of *Waiting for Godot* and *Endgame*, Beckett's drama undergoes a profound transformation in the sudden and unexpected emergence of strong female characters and voices: Maddy Rooney in the radio play *All That Fall* (1957) and Winnie in the stage play *Happy Days* (1961). The women's way of speaking (especially in the radio play) bears a very different emphasis from the earlier plays, not least because Beckett at the same time (re)turned to writing his plays first in English rather than in French – a return to the mother tongue.

The all-male domain of Beckett's first two plays is only interrupted briefly when Nell half-emerges from her dustbin in *Endgame*. Although called by husband Nagg, Nell pays scant attention to everything he says, and even less to the peremptory Hamm. Her one word to the outside world is addressed to Clov when she advises him to desert. Instead, Nell speaks a reverie triggered by certain words she abstracts from the male discourse: 'Ah yesterday!' (*CDW* 99). Her share of the dialogue with Nagg is an act of memory concentrated on recapturing her past. Beckett's women characters enact a process of remembering – and forgetting. It is a past that they imagine and remake, rather than directly recall, a cultural as well as a personal past. The virtual monologue by Winnie in *Happy Days* is made up of verbal fragments, half-remembered quotations ('What is that wonderful line?') mingled with details from her past. And in Beckett's *Not I* (1972), the character's search is explicitly for an alternative mode of expression, of speech and storytelling, to that imposed on women by a patriarchal society, in the supermarkets, in the law courts, in the streets. Beckett's Mouth resists the first person singular and insists on the third person 'she' in telling her story. This can be read as the play's resistance to treating a woman's experience as merely personal by insisting that it has a more representative status in terms of gender.

If in *Philadelphia, Here I Come!* Friel struck a first blow against the traditional dramatic hierarchy by splitting the male lead in two, he subverts that hierarchy more radically in *Dancing at Lughnasa* by moving five women into the dramatic centre. There *are* some men in the play. The women's older brother Jack is a missionary priest who has returned from Africa converted from the Catholicism he went to preach to a faith in the goddess-centred rituals of the Ryangan people. Gerry Evans, the father of Chrissie's son Michael, pays two brief visits to mother and son during the play before leaving for the International Brigade and the Spanish Civil War. Michael narrates *Lughnasa* from his adult perspective and is like Gar O'Donnell in occupying a double role as both witness to and participant in the events he narrates. But the other self is so much younger here that the play ceases to be about him and becomes instead about the shared lives of the women among and by whom he was jointly reared. The male characters in *Dancing at Lughnasa* are now moved to the margins and rendered eccentric, unreal; there are no star parts here for Stephen Rea. Michael is a possible exception to this realignment. As the narrator, he occupies the sidelines of the play and yet is central to its construction, since he is the one actively remembering all the other characters into existence. Michael's ambivalent position could be read as Friel's acknowledgement that, for all of the play's emphasis on women, it is being authored by a man. What is so striking in any viewing of *Lughnasa* is the extent to which these memories elude their narrator, possessing a range and meaning beyond his conscious control.

The divided stage space of the male double-act is replaced in *Dancing at Lughnasa* by a protean, five-woman dramatic ensemble in which they depend on each other and enact their own version of the extended family. The Mundy sisters and Chrissie in particular have defied social convention by electing to keep the 'illegitimate' child (as he would then have been termed) rather than seeking a backstreets abortion in England or (the more likely option at the time) putting him up for adoption. It is not as if Chrissie has not had an offer of marriage from Michael's father. Gerry repeats his proposal when he is dancing with her; she gently but firmly declines, showing a clear-eyed appreciation of the character of the man she loves: 'you'd walk out on me again. You wouldn't intend to but that's what would happen because that's your nature and you can't help yourself' (54). Had he gone through with his proposal, such an act would have made Gerry a bigamist, since Michael later learns that his father is supporting another family in Wales. Chrissie might seem to have a special claim on Michael as his biological mother, but the play rejects the bourgeois notion of 'having' a child by giving each of the sisters a maternal role with the boy. And while Agnes's special concern is for the simple and vulnerable Rose, drawing her into the shared activity of knitting and encouraging her to talk, that care for Rose cannot be limited to one person, as becomes apparent from the pervasive concern when she goes missing. Kate, the eldest and the one who normally gives the orders, is unable to handle the situation and the joker Maggie quickly steps in and takes over: 'That'll do, Kate! Stop that at once! (*calmly*) She may be in the town. She may be on her way home now [...] We're going to find her' (86).

The unity-in-difference of the women is most memorably expressed through the dancing that became *Lughnasa*'s defining moment and most memorable theatrical image. (This was only a few years before the *Riverdance* phenomenon and no doubt contributed to it.) Begun by Maggie, the dance moves like a live current through all of the others until even the strait-laced Kate succumbs. When the dance ceases, Kate urges them to behave with propriety; during the dance itself, she has acted by '*suddenly leap[ing] to her feet, fling[ing] her head back, and emit[ting] a loud "Yaaaah"*'(36). If this passionate outbreak is liberating, it occurs early on in the play and is followed by the depression and anti-climax of its fizzling out. Despite Agnes's brave and heart-breaking assertion that 'I want to dance, Kate. It's the Festival of Lughnasa. I'm only thirty-five. I want to dance' (24), the Mundy sisters do not go to the dance.

The energies briefly liberated fail to achieve realisation in the surrounding society of the 1930s. The play is set in 1936. In the following year the Irish Taoiseach (prime minister), Eamon de Valera, after extensive consultation with the hierarchy of the Catholic church, will enact an Irish constitution in which the role of women is given a single sentence and their sphere of activities confined exclusively to the 'home'. There are two ironies in relation to de Valera and the play. The first occurs when Michael mentions that his aunt Kate served on the Republican side in the Civil War. Now that the former republican leader de Valera has entered constitutional politics, Kate's loyalties lead her to uphold the values of the republic over which he presides, even though it falls far short of the ideals for which they fought, particularly in relation to women. And de Valera is best known for his radio address during the Emergency of World War II in which he extolled a pastoral vision of virile youths playing Gaelic games and comely maidens dancing at the crossroads. Yet the 1930s was the period in Ireland in which both traditional dancing at the crossroads and the latest 'foreign' jazz dances were being proscribed. Dance halls were the preferred locale, a source of more money for the parish and a place where courting couples could be more closely monitored. The influence of the Hollywood musical, the song-and-dance romantic pairing of Fred Astaire and Ginger Rogers (both of whom are explicitly referenced in *Lughnasa*), was more difficult to monitor and resulted in endless newspaper pieces and sermons from the pulpit designed to counter its influence. The play offers the cinematic kind in the modern dancing between Gerry and Chrissie in which the others vicariously participate; it takes the Astaire-Rogers choreography off the silver screen and makes it the characters' own, as they refashion it to their own time and place. Dancing is the play's most potent metaphor because it shows the five women's relation to each other and to the outside world at its most expressive and unified. As Michael's closing monologue intimates, dancing suggests that there is another 'way to speak, to whisper private and sacred things, to be in touch with some otherness. [...] Dancing as if language no longer existed because words were no longer necessary' (108). The ending of *Lughnasa* harmonises with that of *Translations* on the limits of expression for any Irish dramatist writing in the English, and by extension any, language.

There was a three-year interval between *Lughnasa* and its successor, *Wonderful Tennessee*, also premiered at the Abbey Theatre. Expectations were high in the light of the former's spectacular success, unnaturally so. In the event, *Tennessee* proved a disappointment, and only ran for ten performances on Broadway. Whenever one of his plays proves extremely successful, Friel sees its sharp edges blurred, its political and historic ironies flattened or removed, its feelings sentimentalised. The next play he writes is always in some way a reaction against this process, a move to recover his artistic freedom, to experiment rather than to repeat himself. *Translations* was set in the historic past, as was *Lughnasa*, and the reception of the two plays was coloured by nostalgia. Their successors were both set in the present, where the pieties of the previous play were relentlessly demythologised. In *The Communication Cord*, Senator Donovan's attachment to the past is rendered literal when he is chained to the wall of his restored peasant cottage. In *Wonderful Tennessee*, a clear line of connection is drawn between the dance the characters perform when they enter, a '*parodic conga dance, heads rolling, arms flying – a hint of the maenadic*',[36] and the celebrated dance in *Lughnasa* where '*there is a sense of order being consciously subverted, of the women consciously and crudely caricaturing themselves, indeed of near-hysteria being induced*' (37). Where *Communication Cord* responds to its predecessor through the medium of farce, *Wonderful Tennessee* provides a more abstract and philosophic meditation on the themes which had so engrossed audiences in *Dancing at Lughnasa*. Where the political and cultural certainties of the 1930s gave a consistently negative patriarchal force against which to react, the uncertainties of the present, of Ireland in the 1990s, are a different matter. The material circumstances of Friel's characters have certainly improved; where Maggie wonders how she can feed the entire family from a few eggs, Terry complains in the 1993 play that none of the exotic foods supplied in the picnic hampers he has ordered is actually edible. Where Catholicism was a repressive regime which exercised considerable social control, it also encouraged a readiness to believe in the metaphysical whereas now 'people [have] stopped believing' (372). But although it contains some sharp comments on 1990s Ireland, *Tennessee* does not rely on the medium of farce. Rather, the lens through which it chooses to view contemporary Ireland and which informs its dramaturgy is Samuel Beckett's *Waiting for Godot*.

I have been arguing throughout for a Beckettian presence in the plays of Brian Friel. That presence is explicitly acknowledged and confronted in *Wonderful Tennessee*, whose 'action' centres on six characters (three men and three women) waiting for a boatman to bring them to an island off the Donegal coast. He, like Godot, though much anticipated, never arrives and so never makes an appearance in the play. Frank, one of the visitors, returns from the cottage to give a graphic description of the revenant he encountered there: 'Ancient; and filthy; and toothless. And bloody smiling all the time' (385). But can anyone still be living in a landscape which is completely depopulated, one from which the last emigrants have migrated fifty years before? There is only his word that the boatman actually exists. And Frank is the one who gets everyone into a state of readiness and frenzied excitement by announcing that the boatman is 'here! He's bloody here!'(390) before confessing shamefacedly that he is only crying wolf. The boatman bears the overtly mythological and non-naturalistic name of Carlin, which immediately suggests the ferryman over the River Styx. The name has the same supernatural aura as Godot, signifying someone who has a bearing on the fate and fortune of the characters who await him. These are latter-day pilgrims journeying to Oileán Draoichta (or the Island of Mystery)[37] – a destination on whose appearance none of them can agree as they struggle to describe what they see. Since they cannot sail to the island without the boatman, the six characters spend the night waiting for him to appear and occupy the duration of the play, like Beckett's tramps and Behan's prisoners, filling in the time with inconsequential dialogue, aborted jokes and snatches of song. One of the six, George, is a musician, and since the throat cancer of which he is dying makes it extremely difficult for him to speak he tells his story through playing a succession of sacred and profane songs on his accordion.

The acoustic of *Wonderful Tennessee* is markedly Beckettian. The opening stage directions call for '*silence and complete stillness*' (347) as the natural order of things in this theatrical space. No matter how much the characters may talk, silence remains the ground against which they do so. The point is made dramatically by delaying the physical entrance of the three couples, their arrival announced instead by the offstage noise they make and the Babelian chatter of their clichéd phrases. No Friel play is more marked by linguistic insufficiency than *Wonderful Tennessee*, as the opening lines make clear with their insistent repetitions and the fact that most of the lines are interchangeable. Within four pages of dialogue (though one can scarcely call it that), Terry the gambler/impresario leader of the troupe who has brought them there says 'Believe me' at least four times and the refrain 'It is wonderful' not only devalues each of the words but insistently points to their inability to describe or articulate what it is they want. In the course of the play, most of the lines are brief and repetitious. Many of the long speeches are explicitly drawn from (and hence are quoting) written sources: Terry's account of the island; Angela's classical allusions. The other lengthy set pieces are the formal stories told by each of the characters in turn. They articulate irrational, improbable occurrences: a dancing dolphin, a flying house, George's non-appearance on the day of his wedding. The characters in

Dancing at Lughnasa may have experienced a Dionysiac moment in the dance; but none of them would have described it as such. Here, the hyper-educated contemporary characters can supply their own critical commentary: as when Angela, who lectures in classical literature at a university, wreathes George in dried seaweed and proclaims him 'Dionysus!' (359). The only dramatic action anyone performs is when Terry's wife Berna, drinking copiously on top of her anti-depressant tablets and conscious that her husband is having an affair with her sister Angela, attempts suicide by throwing herself off the end of the pier. Frank removes his belt at the close, not to try to hang himself, but to drape it around '*the rotting wooden stand, cruciform in shape*' (344). Time is of a different order in this setting; when somebody asks the time and Berna realises her watch has stopped, she throws it back in the water as no longer relevant.

The six characters in the play are barely that; like Beckett's tramps, they have little differentiating bourgeois detail to give them 'character'; and that is only in part because they have moved from the routine of their everyday lives into a more abstract and abstracted setting. They are all interrelated in a number of complex, almost mathematical ways. Terry and Trish are brother and sister; Angela and Berna are sisters. Terry is married to Berna but having an affair with Angela, who is married to Frank. Frank and Angela bicker but he is always courteous and attentive to Berna. Trish is married to George, who played in a musical group with Terry. The six characters move around each other like figures in a complicated dance, achieving a greater degree of dialogue in the many songs they sing than in the banalities they exchange. It is only in isolated moments that they admit the despair underlying the self-consciously inconsequential banter. Their movements become consciously ritualistic as each in turn drapes an object around the cruciform wooden stand and just before they leave drops a stone in what becomes a mound. Although they say they will return the following year, there is clearly no guarantee that they will. In the final muted exchange between George and Angela, the former finally owns up to the fact that he is dying from cancer when he asks her to come back '*for* me – in memory of me' (445). Christopher Murray has noted the extent to which *Wonderful Tennessee* echoes *Waiting for Godot* but argues that the contrast between the two plays is 'far more significant'. For him, this turns on a reading of the Beckett play as unrelentingly bleak whereas the characters in *Wonderful Tennessee* 'begin to stir towards renovation'.[38] For me, as I have sought to show in my analysis of *Godot* throughout, Beckett's bleakness cannot be all-pervasive because laughter keeps breaking in. The removal of

Estragon's belt to attempt suicide causes his baggy trousers to drop to the ground in the closing moments of *Godot*. And for all of *Tennessee*'s minimal gestures towards renovation, the characters still have to face back into the bleakness of their everyday lives at the close. But I can agree with Murray that it is 'a far better play than its failure on Broadway might suggest'.[39]

More commercially and culturally acceptable than the overtly Beckettian qualities of a 1990s play like *Wonderful Tennessee* are the versions of Chekhov which Friel concentrated on between 1998 and 2002. Chekhov's plays too are marked by an absence of plot and a good deal of waiting for a miracle which never occurs; above all, as Friel himself has remarked, the Russian and the Irish characters are bonded in and by the sense that 'their problems will disappear if they talk about them – endlessly'.[40] In these versions, the Irish context gives way to and is subsumed by the Russian. In his original works it is through the medium of Beckett that Friel has looked most unsparingly at Irish conditions, especially in the present.

3
Murphy's Drama: Tragedy and After

The other leading playwright in contemporary Irish drama is Tom Murphy. Born in Tuam, Co. Galway, in 1935, Murphy has had staged 16 full-length original plays, two one-acters, and a number of adaptations. Much less well known than he deserves to be outside the country, his best plays have had a seismic impact on Irish theatre, extending its boundaries and deepening its concerns, provoking walkouts and heated discussion. Where much media attention has been given to events in Northern Ireland, there has been a 'Southern' crisis since the late 1950s which Murphy's plays have addressed and helped to articulate. During that period, a traditional rural society has experienced accelerating social change through the influx of foreign capital, the advent of television, a move to the cities and a renewal of emigration, the breakdown of a conservative way of life and the undermining of its certitudes. A story Murphy tells (given in full later in the chapter) describes the experience of encountering in the west of Ireland a farmer carrying turf to the accompaniment of a Walkman, while down the road in a multinational hotel a form of strained English etiquette prevails.[1] As Fintan O'Toole put it,[2] Ireland has passed from a traditional to a postmodern society without encountering modernism; and Murphy's plays dramatise this bewildering transition, this sense of endlessly facing two ways, in a funny and fearless manner.

Tom Murphy's is the most restless imagination in Irish drama, as Friel himself has observed.[3] His characters are on the run even when they're standing still, seeking an escape from inherited conditions that threaten to stifle their sense of joy and of possibility, working to find a more unconstrained outlet for their imaginations. But the plays also bring those characters smack up against their own limitations and into some sense of confrontation with what they have sought to elude. This restlessness is matched by an impatience with inherited forms of theatre and language.

Even more so than with the other Irish dramatists this study is considering, there is a strong sense in Murphy of starting over with each new play, of an even more ingrained distrust of established structures as a personal and historical imposition. Part of the lengthy, arduous process of composition for him is to elaborate a theatrical environment in which an extraordinary collision of apparently random elements can coexist: a Japanese factory, a laughing competition and an unwanted Irish pregnancy (*Bailegangaire*); a Catholic church, a circus strong man and a love triangle (*The Sanctuary Lamp*); or a quack psychiatrist's office, a property developer and an obsession to sing like an Italian tenor (*The Gigli Concert*). Murphy's drama is concerned to give to 'airy nothing' not only 'a local habitation and a name' but, even more, a formal justification of cohabitation and a naming which calls on new resources of expression.

Language is rarely adequate to the expressive needs of Murphy's characters. They are drawn to everything from full-blown monologues of funny, quirky parables to a muttered 'yeh know', whose appeal is across the barriers of the unspoken to a kind of shared understanding. The syntax in the speeches is frequently fractured, often incomplete, the energy behind them driven by the need to break down the language in the hopes that some genuine expression might leak through. Murphy's plays are full of what Beckett considered the most important requisite of drama, 'fundamental sounds',[4] and they are texts to be scored for actors fully alert to the music of their distinctive utterance. Hearing Siobhán McKenna as Mommo in *Bailegangaire*, I had the sense of listening and responding to something highly evolved yet irreducibly primal, a language which spoke to many layers of repressed experience in the Irish psyche, but which might well resist translation into the English of global theatre.

Murphy's first play was co-written with his friend Noel O'Donoghue in Tuam in 1959. The two of them were lounging around the town after last Mass when O'Donoghue suggested, 'Why don't we write a play?' Murphy replied: 'What would we write about?' And O'Donoghue responded: 'One thing is fucking sure – it's not going to be set in a kitchen.'[5] And where Friel sought some form of accommodation with the traditional kitchen setting in *Philadelphia, Here I Come!*, Murphy and O'Donoghue defiantly set themselves outside it. When one of his plays does finally feature a kitchen, in *Bailegangaire* (1985), Murphy plants a bed down in the middle of it, and lets the public and the private interact more disruptively than in *Philadelphia*.

Over the span of his career, Murphy's plays have predominantly had two kinds of settings. One group takes place in a number of overtly symbolic locations, at one remove from everyday reality – the psychiatrist's office/home in *The Gigli Concert*, the deserted church of *The Sanctuary Lamp*, the fairy tale forest of *The Morning After Optimism*. These are places of refuge to which a character at the end of his or her tether flees, very much like one of Tennessee Williams's 'fugitive kind'.[6] The other setting of Murphy's plays is in or near Galway, in more recognisably realistic settings – pre-eminently the public house in *Conversations on a Homecoming* (1985) in which the older, disillusioned group of 1960s idealists gather to welcome Michael back from the United States.

The local Galway milieu provides the setting of that first play with Noel O'Donoghue, *On the Outside*, as two young men wait outside a dance hall on a Saturday night. The two women they were to meet have already gone in; Frank and Joe now find they have not enough money to admit themselves, let alone the women. The rest of the one-act play revolves around their increasingly desperate attempts to gain entrance by a number of different stratagems: cadging money from a passing drunk, acquiring a pair of pass tickets from a departing couple, chatting up the woman at the desk. The sense of social exclusion and the attendant frustrations of their lives emerge under the mounting humiliation. The scenario has the Beckettian elements familiar to this study: a drama based on two men waiting; a determined effort by one to keep up the spirits of the other; a belief that, if they get inside, all their troubles will be over. The two fellas excluded from entering the dance hall works as social realism; and the economics of their situation are relentlessly explored: when Frank and Joe finally get some money from the drunk, it isn't enough; the two of them come closest to breaking up when one wants all the money in order to enter on his own and when inside to borrow on the other's behalf. But there is also built up a symbolic sense of a more general exclusion, heightened by Frank's fear of being trapped in a big tank:

> The whole town is like a tank. At home is like a tank. A huge tank with walls running up, straight up. And we're at the bottom, splashing around all week in their Friday night vomit, clawing at the sides all around.[7]

Themes that are going to be more fully developed later in Murphy's career are here touched upon. They want to be someone else – going in on another person's ticket. They want to be somewhere else, speaking of emigrating to England even though they have nothing but contempt when others who have done so return home flashing the money they have earned there. When Frank meets Kathleen emerging from the dance hall with another man, he compensates for his lack of money with a desperately improvised story about a football match, telling a tale to save face. Instead of going home or going in, he and Joe make stories out of the desperate situation of being caught in the space between.

The sense of exclusion faced by the two characters, on the very threshold of the dance hall, was experienced by the putative playwrights themselves in relation to the Irish stage. Murphy and O'Donoghue received a letter dated 8 May 1959 telling them that *On the Outside* had been entered for the All-Ireland Amateur Drama Festival at Athlone. 'There the adjudicator refused to award any prize, saying that no entry was up to standard.'[8] The play had to wait until 1974 to receive its first professional production. O'Donoghue called it quits and became a solicitor. Murphy went on to write *A Whistle in the Dark* (1961), the first of his full-length plays. It was submitted to and rejected by Ernest Blythe at the Abbey, who remarked when it was eventually produced at the Olympia on 13 March 1962: 'I never saw such rubbish in my life.'[9] The play was taken up and produced at Stratford East before going on to London's West End. But after that first flush there was a lack of follow-through for Murphy on both sides of the Irish Sea, fully discussed by Fintan O'Toole in his Introduction to *The Politics of Magic*.[10] I agree with O'Toole that from the English point of view, Murphy, in his two subsequent plays and in his own person, failed to play the wild Irish boy that was expected of him and so kept from emulating Brendan Behan's fate. I do not agree with O'Toole that he was as much overlooked on the Irish side, at least after the doldrums of most of the 1960s. With the opening of the new Abbey in 1966 and the Peacock in 1967, with the moving aside of Ernest Blythe in favour of Tomás MacAnna, a potential home emerged for Murphy's migrant drama; *A Crucial Week in the Life of a Grocer's Assistant* and *Famine* soon received productions. I date my own introduction to Murphy from the memorable staging by Hugh Hunt of *The Morning After Optimism* on the Abbey main stage in 1971. Not every Murphy play since has been 'successful'; the relationships, whether between the playwright and the theatre, or the productions and their audiences, have often been stormy. But through a succession of artistic directors, the Abbey Theatre has shown a sustained commitment to the 'rough and holy'[11] plays of Tom Murphy and his restless odyssey to break new theatrical ground in the staging of the self's confrontation with its demons.

Murphy started his career proper with a full-blown tragedy in *Whistle in the Dark* and a complementary comedy in *Crucial Week*. He moved most fully beyond these into the painful laughter of *Bailegangaire* (Irish for 'town without laughter'), and beyond or 'after tragedy',[12] as he himself puts it, with *The Gigli Concert*. This chapter will look first at *Whistle*, then go on to examine these two crucial plays of his maturity, to suggest the range and scope of Murphy's achievement. Finally, I will consider his late masterpiece, *Alice Trilogy* (2005), which effortlessly extends from the 1980s into the new century. But I am conscious of the many fine plays I am overlooking, and that the drama of Tom Murphy deserves a book, not a chapter.

The theatrical environment of Murphy's *A Whistle in the Dark* has much in common with the Behan and Beckett plays discussed in Chapter 1. Instead of tramps or prisoners, the audience encounters a gang of drunken rowdies which it thought it had left behind on the street. The Carney boys expend much of their energies throughout the play attempting to disrupt the middle class home of their brother Michael and his English wife Betty in Coventry. The all-male environment they encourage and the absence of female companionship in their own lives signal a perversion of creative energies into forms of competition, cruelty and the urge to dominate. When they face others, there is a grouping unified by violence; but when the Carneys are on their own, they vie with each other, jostling not only for control but for a confirmation of identity. The polite, formal exchanges of the well-made play are replaced by shouting, singing, elaborate if not downright violent physical mime, jokes, cross-talk, monologues, and repeated words and motifs. The language of the play is as much at cross-purposes and at odds with itself as the divided and interdependent characters.

During the opening moments of Garry Hynes's 1987 production of *A Whistle in the Dark* at the Abbey, Betty cradled a pillow in silence, a gesture which suggested she was holding and longing for a baby. This opening tableau drew attention to those moments in the play when she attempts to draw her husband away from his father and brothers, up the stairs and into the bedroom; though we know if he went there, Michael would not act but rather talk the night away. This opening image was beautifully balanced by the close, where Michael knelt and cradled his youngest brother Desmond's head. It is Betty who recognises that Michael's thwarted paternity is displacing itself on to Desmond, that the relation is more that of father to son than of brother to brother. One of the most terrible moments, in a play that has more than its fair share, occurs when Michael responds to Betty's gestures of appeal by striking her violently, the only example of a physical exchange between them in the play. Michael's striking his wife aligns itself with the blow he gives Des, the one which kills him, as the death blow not only to their relationship but to the possibilities for creation and growth symbolised by their union.

This displacement of creative energies into acts of violence is nothing new in the history of dramatic literature. It is rendered explicit in Lady Macbeth's injunction: 'Come to my woman's breasts / And take my milk for gall, you murd'ring ministers'[13] and the course taken by the increasingly bloody developments of Shakespeare's play. In Greek tragedy, a sea of blood washes through plays like *Agamemnon* and *Medea* that lifts the protagonists to great heights of passion, language and a kind of frenzied joy. But time and cultural orthodoxy have tamed the wild energies still present in these classic texts, have taken away a sense of their threat and danger. This is something of what Murphy seeks to restore in writing *A Whistle in the Dark* as tragedy.[14] Garry Hynes sought to heighten this immediacy by doing away with the usual Abbey Theatre proscenium and thrusting the stage into the audience. This brought us uncomfortably close to the violent energies of Harry, Hugo, Iggy and Des Carney. For all the play's apparent air of documentary realism, with its setting in Coventry and the England of the late 1950s and early 1960s, with its exaggerated but still recognisable portrait of the drinking, fighting Irish émigrés, Murphy's play is consciously aware of its classical antecedents and tips us off to the parallel in Dada's speech to Betty at the opening of Act III:

> Did you ever read *The History of Ancient Greece*, did you? I'm reading that now, just before I came over. Very interesting on how... Yeh. Did you ever read *True Men As We Need Them?*... No... I bet you never read *Ulysses*? Hah? – Wha'? – Did you? No. A Dublin lad and all wrote *Ulysses*. Great book. Famous book. All about how... how... Yeah... Can't be got at all now. All classic books like them I have.[15]

Dada's description of *Ulysses* falters, not least in deciding whether the story, if you follow Homer, is about how the warrior Ulysses returns to his wife Penelope after the Trojan War and years of wandering, or if you follow Joyce, about the relations between advertising canvasser Bloom, his unfaithful wife Molly and young Stephen Dedalus in Dublin on 16 June 1904. In Homeric fashion, *Whistle*'s House of Carney pits itself in heroic contest against the House of Mulryan, and the warriors talk of their pride in ways that recall pagan virtue more than Christian sin. But the violence directed against the Mulryans in the play is also viewed in Joycean demythologising terms as an undignified set-to in the toilets of a public bar. The deliberately squalid details and setting make a mockery of and seriously qualify the brothers' claims to heroic status.

What is even more disturbing than the external action, and almost directly parallel to the Greek tragedies, is the way Murphy shows the family violence passing itself on cyclically from one Carney brother to another, as from one generation to another. This drama, the heart of the play as a tragedy, centres itself on the young and still relatively innocent Des and on Michael's efforts to save him from capitulating to the barbarities of his brothers. Because Des is still relatively new to this environment, we see possibilities for another kind of development in him: his playfulness with Betty when they first come in together; his sticking up for Michael in the latter's absence when all the others have their minds made up against him; and the respect of the youngest for the oldest brother. In many of the scenes Des is literally or metaphorically the man in the middle, with Michael (and Betty) urging him in one direction while his brothers and his father urge him in the other. But for all Michael's passionate pleading with him, we recognise from early on the inevitability of Des being drawn into the round of violence, of his responding to the claims of blood and kinship in going out to fight. The process has already begun when he enters, as we learn from the discussion of the fight that has occurred in the pub:

> Well, they went for me. Well, I had to defend myself. Well, I had to try to. Two of them. One of them sort of came at me with his head, his forehead – [...] And I got this. (*Bruise.*) And stars for a minute, and then, well, lights out.
>
> (22)

Each of the others then takes it in turn to instruct him in the best method of self-defence: Dada's 'Your back to the wall' (23), Harry's recommendation of a knuckle duster or broken glass, Iggy's preference for the unaided use of his fists. In Act Two, their approach is to turn around Michael's arguments on, say, the unevenness of the odds against them, as positive reasons why Des should join in and make it a matter of family loyalty and self-worth. The blooding which Des has received in the pub acts like a drug. When he returns later from the fight with the Mulryans, his reason has been entirely submerged in blood-instinct; he rounds in turn on each of his brothers and threatens to take them on. No longer unified in the face of a common enemy, the Carneys in Act Three begin turning on and destroying each other. In the final logical twist Michael himself is drawn into the violence he has all along resisted, and is goaded into destroying Des, the very one he has sought to protect. With the destruction of the youngest male heir through Michael's belated decision to act, and his and Betty's lack of children, the legacy of violence has run its course through the family line. The ending leaves all the remaining brothers with the dawning recognition of the ultimate self-destructiveness of their apparently outward and other-directed violence. They begin to realise the extent to which they bear the psychological taint of their father's inadequacies, and how he has compelled them to live out his compensating fantasies of violence.

Murphy's tragedy shows how close the play and Irish society are to the culture and development of the Greeks. In Aeschylus's drama, the characters think and feel in tribal terms; their organisation is based around an extended family with a father or chief at the head. In Murphy, his west of Ireland characters think and feel in tribal terms; their loyalties are organised around the family, with Dada as the hitherto undisputed head or chief from whom they take both direction and orders. This is virtually an all-male society, the men making up the warriors of the tribe, while the women are relegated to the background (as with Betty and the absent Carney mother, back home in Ireland). The society defines itself constantly by measuring itself against other tribes, settling scores and feuds in a way that is violent but which also has a personal integrity since it is delegated to no one else and has the authenticity of action. Harry's greatest shock comes from discovering the dissociation in Dada between word and deed: 'Person'lly, I don't mind a man, no matter what he talks, if he means it, and you can see it, and if he'll stand up for it, and if he's – faithful' (78).

But the faction-fighting tribe of the Carneys has been displaced from its native Mayo, from an Irish rural hinterland to a modern urban setting. In this locale, their activities are no longer appropriate, are rather at odds with the environment. Marriage is at the centre of this new bourgeois ordering, as the guarantor of property. Accordingly, *A Whistle in the Dark* is set in Michael and Betty's house, with its emphasis on manners and the constraints of the new civilisation, against which the Carney men react. They do so by proceeding to wreck the premises and by ostentatiously refusing to conform to these new imposed standards of behaviour. At one level, it is a reaction against an ideology which is doubly alien, not only new to them personally but imposed by the hated English, whose line on the Irish for centuries has been consistent: a wild, unruly race of fighting drunkards who need British rule and order to sort them out, who must be taught above all how to *behave*. In terms of behaviour, for the Carneys to fit into the bourgeois scheme of things is at some level to suppress the distinguishing marks of Irish identity and to become an ersatz Englishman:

HARRY: You're not a Paddy?
MICHAEL: We're all Paddies and the British boys know it.
HARRY: So we can't disappoint them if that's what they think. Person'lly, I wouldn't disappoint them.
MICHAEL: You won't fit into a place that way.
HARRY: Who wants to?
MICHAEL: I do.
HARRY: YOU want to be a British Paddy?
MICHAEL: No. But a lot of it is up to a man himself to fit into a place. Otherwise he might as well stay at home.

(14)

With the Carneys devoting much of their energy to destroying the drawing-room set, I would suggest that this can be taken as Tom Murphy's own rebellion in his playwriting against the constraints of urban bourgeois drama, the type of theatre that has prevailed in England in the 20th century. Refusing to confine passionate speech and action within the polite formalities of middle-class manners and social chit-chat, Murphy is declaring war on the reigning pieties of conventional theatre and attempting to spill some blood in the waxworks museum, to break up what Beckett called the 'complacent solidities'[16] into sharper-edged fragments, a theatre of rough edges which can be put to new uses.

Murphy remembers, and recalls to our awareness, that the essence of theatre is tension, and accordingly structures *A Whistle in the Dark* around a series of conflicts. The most obvious is that between the Carneys and Mulryans, the conflict around which Murphy organises his most conventionally and fully plotted play. There is the suspense of the build-up to the fight, the question throughout Act Two of how the fight is proceeding, and the payoff in Act Three of the returned victors' account of what happened. But the fight takes place off-stage, at several removes from the audience's attention, and is displaced further by the emphasis on the psychological conflicts. Betty spells it out for Michael in these terms at the outset:

> Which comes first, which is more important to you, me or your brothers? [...] To hell with Des and the rest of them! It's us or them. Which is more important to you?

(9)

The approach of the Carneys to the house is positively territorial. They sprawl on sofas, spill beer over the furniture and intimidate their supposed hostess. The hostility is not just directed at the house, but more especially at Betty herself, who is on the receiving end of the Carney brothers' taunts on the double score of her Englishness and her womanhood. The usual social norms are reversed: the guests take over the house while the owners are forced to leave, Michael on the night of Act One, Betty in Act Three. Michael tries throughout most of the proceedings to play a double role: husband to Betty, brother and son to the boys and Dada. Betty loses the battle, partly on the score of numbers (six to one), partly because her sexuality is outweighed by the sheer intimidating physical threat of all those Carneys, and because the ties of a legal marriage do not operate with the same compelling atavistic force on Michael as the blood ties of family kinship. At the end of the play, her husband no longer recognises Betty. As he once more enters into and takes on his familial identity, she becomes a stranger to him.

Even when not fighting another tribe or demonstrating their fear of women and domesticity by jeering and carrying on, the brothers still talk in terms of us versus them:

HARRY: [...] I'm not afraid of nobody! They don't just ignore me! They don't ask me what I had for my dinner!! They don't –
DES: They? Who?
HARRY: Oh, they – they – they – they – THEM! Them shams! You all know who I'm talking about. You know them. You know them. He knows them. [*To* MICHAEL.] You suck up to them, I fight them. Who do they think most of, me or you?
MICHAEL: If it comes to that –
HARRY: Aw, do they now? They think more of you? I can make them afraid. What can you do?

(44)

These and similar speeches in the play reveal the extent to which the true conflict for the Carneys is that between themselves and internalised figures of authority, whom they resent and admire, simultaneously want to do down and impress. In terms of the play's setting, there is the ambivalence for the Irish characters of living in England and earning money there while still trying to maintain a distinct cultural identity, of seeking to impress those around them in ways that only confirm their worst racial stereotypes.

The greatest ambivalence of the colonial legacy, of Irish versus English, comes out in the divided attitude towards the English language. On the one hand, there is a lack of confidence in handling it in, for example, the stuttering of Iggy: 'Are we r-r-ready?' (3). Iggy is a barbarian in speech (*barbar*: someone who stutters) as in action, living up to his name as 'The Iron Man'. There is the breakdown of syntax on those occasions when one would expect it to be most complete: when Dada apologises to Michael or in many of the formal conversational exchanges. Instead, phrases get repeated, like Dada's 'Great woman you got there.' On the other hand, both Dada and Michael in their monologues speak with much greater confidence and eloquence. Throughout, in relation to language, when a character speaks there is a palpable effort of rhetorically overcoming social inhibitions, sometimes successful, sometimes not. The Carneys wreak the same havoc on language as they do on everything else, essentially wanting the language to work for them, to express an individual sense of themselves, rather than their submitting to the language. Combined with this creative disregard is a contradictory and pedantic obsession with formal correctness. This concern for linguistic and syntactic nicety goes much deeper than the scant regard for social norms briefly observed when Dada and Des arrive and greet Betty. That bland exchange scarcely compares with the fierce argument that breaks out between Dada and Mush as to whether 'inst.' or 'ult.' is the correct term; or the repeated verbal mannerism whereby Dada automatically corrects his own grammatically deficient adverbs, from 'proper' to 'properly'. When we understand the cultural forms to which these characters have been subjected, we better understand the urge to break free and the persistent threat to break down.

Even in terms of the Carneys' own family history, there is an us-versus-them legacy. Although they have all physically left Ireland behind them, they still bear the psychological scars inflicted there – by the schoolteacher McQuaide, for example, who never appears on-stage and is many years in their past. But McQuaide is still an active presence, as we realise in Harry's re-enactment of the classroom scene as the site of an original humiliation and the unexpected confession by this pimp-cum-psychopath that he originally wanted to be a priest. Mush early on talks about 'them' in a vague way but, pressed by Harry, becomes more specific:

HARRY: The holy ones is the worst.

MUSH: The ones that say 'Fawther' – like that: 'Fawther' – to the priest. And their sons is always thick. But they get the good job-stakes all the same. County Council and that.

HARRY: Them are the ones that gam on not to know you when they meet you.
MUSH: Even the ones was in your own class.

(13)

In Ireland, the same forces of class and upward mobility are at work as in England, but in a colonised, imposed way and with less of the sense of an absolute impersonal force. Instead, class in Ireland comes across in the more personalised terms of some people getting, and getting away with, more than others. In post-independence Ireland, the structures of the British class system remained in place and could now be appropriated and filled by the native population. The attitude towards those power structures is again double, and is concentrated in the figure of Dada, who has passed on this legacy of division to his sons. Dada claims to pride himself on having been a figure of authority, a police officer; but if as Mush claims the sons of guards always do well, then why haven't the Carney sons prospered in their native place? It emerges that Dada did not remain a guard:

DADA: [...] Did he [Michael] ever tell you I was a guard once? Did he ever tell you I was a guard, a policeman?
BETTY: No.
DADA: No, he wouldn't. No... Well, I was! [...] But they – No, I didn't like it anyway. Packed it in. I resigned! [...] I hate! I hate the world! It all... But I'll get them! I'll get them! By the sweet, living, and holy Virgin Mary, I'll shatter them! They accepted me. They drank with me. I made good conversation. Then, at their whim, a little pip-squeak of an architect can come along and offer me the job as caretaker. To clean up after him! But I'll – I'll – Do you hear me? I hate!

(60–61)

The people at the golf club that Dada alternately courts and despises, whose good opinion he solicits and with whom he seeks to be on equal terms, those who have the power and prestige that he covets, are also those he feels have raised themselves to a superior level which they do not deserve and from which they can look down on him. The colonial victimisation, the inherited sense of inferiority – the national inferiority complex, in fact – is made more complicated by the fact that the positions in society are no longer filled by the foreigner but by one's own, those who know you intimately and can use that knowledge as a psychological weapon. Or so Dada imagines.

England appears to offer the prospect of change, of anonymity and a new beginning. At least it did to Michael Carney, the character in the play most obviously torn between the life into which he was born and the life he is trying to make, between the past and the present. Michael has sought to evade a crippling psychological inheritance by a geographic act of exile, by moving to a new and open space in which he can remake himself. But Michael's past catches up with him in *A Whistle in the Dark* as much as it does with Christy Mahon in Synge's *Playboy*.[17] It is in Act Three that he deliberately re-enters his past in order to complete the exorcism he has only partially performed. This Act most reveals Michael as his father's son, especially when he too can only respond to the call to arms by running off to the pub and taking refuge in drink and talk. Like Dada, Michael is given the dramatic freedom to sit and talk. He does so in one of those great monologues which have increasingly become so much a feature of Murphy's drama as he works free of the naturalistic world of conventional plots towards areas of greater openness and possibility, where the talk becomes the agent of attempted articulation, liberation, and a shared understanding. In what was probably the strongest moment of Garry Hynes's production of *A Whistle in the Dark*, Act Two closed with Michael's confession of the fight with the Muslims. He recognises it as the one time his brothers were for him, and is anguished by his inability to respond in kind. In a speech reminiscent of Dada's, Michael climaxes with the tortured confession, 'I don't want to be what I am' (57), a recognition of the shame and humiliation at the base of his deeply divided self. The decision to act only brings with it a further cycle of violence in his killing of Desmond. But it also serves to expose and break Dada's power, as Hamlet's closing action does with Claudius and the Danish court. In the final tableau of *A Whistle in the Dark*, his older brothers gather around Michael as he holds Des's body in his arms and Dada is left outside their circle, standing alone on a chair. This belated outward action suggests at least a partial acceptance by Michael of his own fragmented identity.

In the fifteen full-length original plays Murphy has written in the half century since, the line between reality and fantasy has become increasingly hard to draw. He has gone from the social realism of a play like *A Whistle in the Dark*, with its family of brawling Irish brothers, to exercises in the imagination like *The Morning After Optimism* (1971), where an aging pimp and a whore seek refuge in an otherworldly forest. Nicholas Grene has expressed the contrast as follows: 'If *Conversations on a Homecoming* may be taken to stand for the representational in Murphy's drama, putting the realities of Ireland upon the stage, then *The Gigli Concert* is in his other idiom of dramatic images dislocated from a social context.'[18] With the first group, the concern is to find the theatricalising in ostensibly naturalistic plays; in the second, to find the formal underpinnings and charged areas of concern in these superficially random texts. The distinction is therefore less clear-cut and absolute than might at first appear. An illustration of this is provided by the two productions of *Conversations on a Homecoming* staged by Garry Hynes in 1985 and 1991. The first drew on the intimacy and in-the-round qualities of Galway's Druid Theatre to stress the representational dimensions of the play; the effect was like being in a pub with the five main characters, eavesdropping on a series of conversations in the course of an evening's drinking. When she restaged the play in Dublin as the Abbey's Artistic Director in 1992, Hynes did not seek to bring the audience closer to the play in this proscenium setting, by taking out seats and thrusting the stage into the auditorium, as she had done with her production of *A Whistle in the Dark*. Rather, she emphasised the distance and stressed the play's symbolic aspects, this bar of lost souls along Eugene O'Neill lines, with Frank Conway's hyper-real bar stretching the length of the stage and the characters isolated in pools of light as they turned outwards to face the audience – and themselves. The Druid production gave more weight to the play's satiric side, its social comedy, as we shared the *craic* and joined in the proceedings. The Abbey production left the audience nowhere to turn from the tragic implications of what was being enacted on-stage, the collective disillusionment and the individual estrangement it bred. This revisioning of the play, while largely successful, was not achieved without considerable effort and some sense of going against its realistic grain.

A greater range of dramatic possibility is offered by *The Gigli Concert* from the very beginning, not least because of its setting. In terms of place, the play is set in contemporary Dublin; but the specific local setting, JPW King's office, is not deployed as a part or microcosm of the larger social context but at one particular remove from it. Part of this impression is communicated because its one inhabitant pays so little heed to the events of the outside world, seems oblivious to its activities, sleeps and eats when and as he chooses; so an alternative order of time and action is suggested. But with the arrival of his one patient, the Irish Man, there is imported into the play the sense of somebody fleeing that outside world and its constraints, seeking a temporary place of refuge, if not a site of healing. The plays in this symbolic category establish a pronounced sense of dialectical opposition between here and there, with 'out there' characterised as the insistently determined world of normative relations and material acquisition, and 'in here' as a relatively more open space in which possibilities beyond that closed world can be considered. *The Morning After Optimism* goes furthest in the direction of out-and-out fantasy with its fairytale forest and its impossibly idealistic young lovers; the contrast with the street-wise talk of the pimp Teddy and the whore Rosie remains stark and oppositional. In its choice of a Catholic church, *The Sanctuary Lamp* manages to increase both its literal and its symbolic properties at the same time; while its 'real' characters who flee there for sanctuary turn out, wonderfully, to work for a circus, Harry as a strong man. With *The Gigli Concert*, the deliberate remove from the social reality of its Dublin setting works as with the previous two plays to assert the unequivocally theatrical nature of what is going to occur. In the space thereby opened up, there is the possibility of more than one single determining dramatic image or frame as a lens through which the night's proceedings can be viewed.

The first cultural frame which the play's opening suggests is that of the private detective. JPW King's desk comes equipped not only with a telephone but with an obligatory and rapidly emptying bottle of spirits. His first move is to roll up the blind, revealing a window with reverse faded lettering, through which he stares at the big city landscape with a mixture of idealism and cynicism. The features of the original Abbey set in 1983 most recalled the *film noir* movies of Hollywood in the 1940s, with Humphrey Bogart definitively limning the tough guy keeping watch over his own romantic tendencies in both John Huston's version of Dashiell Hammett's *The Maltese Falcon* and as Philip Marlowe in Raymond Chandler's *The Big Sleep*. The strong play of light and shadow in these films is deliberately evoked by the arrival of the Irish Man into JPW's outer office, where his silhouette appears through the (anachronistic) frosted glass panels. Since the Irish Man is wearing a hat of *'30s–40s American style – as worn by Gigli'*,[19] the silhouette would be in keeping with the visual evocation of 1940s *film noir*, where whole scenes were filmed in silhouette or with Bogart eavesdropping through frosted glass panels. Much of what follows is keyed by the conventions of the genre, with the stranger cultivating a deliberate air of mystery, refusing to give his name while insisting that he has chosen the right man for the job, and haggling over the *per diem* pay. But when he opens his mouth, we hear that he is indeed an Irish man, developing suburban estates and building houses for the Ireland of the 1980s, and the first of the juxtapositions by which Irishness itself is going to be defamiliarised is put in place.

By suggesting a metaphorical equivalent between Hollywood gang-sterdom of the 1930s and 1940s and the prevailing atmosphere of

Dublin in the 1970s and 1980s, Murphy is following through on his previous play, *The Blue Macushla* (1980). Its story of corrupt dealings in the Dublin nightclubs and back streets, while given a veneer of Irish patriotism, was primarily etched in not only the visual but the verbal idiom of the Hollywood gangster films I have already mentioned. That *The Blue Macushla* did not succeed is part of the genesis of *The Gigli Concert*. Long awaited (it had been five years since the success of *The Sanctuary Lamp*), *Macushla* opened to poor houses and negative critical reaction and was taken off early. A stage adaptation by Murphy of Liam O'Flaherty's *The Informer* in 1981 fared little better. This is the background, of silence and failure, out of which Murphy wrote *The Gigli Concert* during an eight-year gap, and in which Brian Friel came back into critical prominence. But Murphy works as much with the idea of failure as either Friel or Beckett. And part of his process is to rework material that has not succeeded, to take away something from a critical failure like *The Blue Macushla* while searching creatively for a way out of the impasse. The problem with *The Blue Macushla* is that the play is static, held in thrall by its central metaphor of Irishmen masquerading as American gangsters. It never transcends the limitations of pastiche; whereas, in *The Gigli Concert*, the Irishman/gangster metaphor is one of only several means the play has of addressing its central concerns. But even as the 1930s hat is discarded and other frames of reference are engaged, the Irish Man keeps one hand in his pocket, one finger on what JPW imagines to be the trigger of his gun, and so maintains a continuous reference to the gangster imagery. Events in the North of Ireland from 1969 on, and the forces they have mobilised on both sides of the border ever since, have brought the gun firmly into play in contemporary Ireland, particularly in the neighbourhood areas where petty hoodlums determine local business practices.

As the Irish Man demonstrates to JPW in the second last scene, what he has been concealing in his pocket is not a gun but a container of sleeping pills:

> JPW: And that gun you have been terrifying me with.
> MAN: What gun?
> JPW: Your trump card, the final word, that gun in your pocket that you have been threatening to shoot yourself with, or me – I never knew which.
> MAN: These? [*He produces a small cylindrical container of pills.*] Mandrax, sleeping pills.
>
> (22)

The contemporary equivalent of the priest, a figure in whose eyes all people lapsed from a traditional religion are equal and most of all in their spiritual crisis of the self, is the psychiatrist. Murphy continues his responsiveness to Ireland's evolving social practices by the multiplicity of roles offered by JPW King in response to the Irish Man's call.

The private eye's office becomes that of a psychiatrist, and the play is structured around a series of six sessions between doctor and patient: '*The* priority, a good relationship of trust, mutual feed of energy between auditor – that's me – and subject' (169). He is, as the sign on the window proclaims, a 'dynamatologist' (167). The strange term enables Murphy to address the rise of cults in the vacuum left by organised religion. King first makes the point that the organisation he represents is 'not a military oriented movement' (168), not a front for IRA activities, as the 'dynamite' might suggest. As Jimmy tells the Irish Man later, their founder is long gone, the organisation he serves has forgotten all about him and probably no longer exists. Its founder Steve is a 'Revolutionary thinker' but the implications of that revolution must be limited so as not to activate any explicit association with an Irish revolution or its putative army. (The other area of explicit denial, of course, is with the Catholic Church it is meant to supplant.) But no fear of that, since its founder is an American and its local representative an Englishman. And so the Irish prove their modernity and their freedom from an enslaving, superstitious past by importing a body of beliefs, slogans and practices from the USA. The 'ology' in 'dynamatology' refers in general to these self-help movements. Several of the details suggest Scientology, whose founder was also, like Steve, an American (L. Ron Hubbard) and which was likewise subject in Britain to media attacks and a call for banning. Murphy makes particular play with the language of cults, the terminology – 'you may call them particles, but we call them whirlings' (168–9) – with which the process is invested with authority and meaning. But since the aim of the process is 'to project you beyond the boundaries that are presently limiting you', the self-consciousness with regard to the language used has a twofold effect: to free language from its preordained, single referents by professing alternatives (whirlings/particles) but also to suggest language as not only arbitrary but limited in its ability to convey or achieve any ultimate meaning. The way is being prepared for the extremes of silence and singing, paradoxically through the medium of a flow of language.

The Irish Man, when he first enters, alternates violently between asserting his normality – 'I'm not insane!' (171) – by boasting of his success in the material world and by admitting his distress, that a 'cloud has come down on me' (172). When he finally confesses that he wants to sing like Gigli, he expresses the play's central conceit, that he wishes for a miracle, not just to reach a level of pure expression, but to become someone else. As Ivor Browne, the psychiatrist to whom *The Gigli Concert* is dedicated, has said, in comparing the modes of consciousness employed by artist and psychotic: 'There is a failure to clearly differentiate self, to evolve a separate person, capable of achieving adult independence.' But, more positively, Browne continues, 'because there is no clear boundary around the self, no clearly separate person, it is much easier to slip into the fantasy mode of consciousness'.[20] To take the last point first, the Irish Man evidences the extent to which he has slipped into the fantasy mode of consciousness through his fantastic identification with the figure of Gigli.

When JPW as psychiatrist asks for the patient's life story, he and the audience are treated to an improbable narrative, with the speaker transposed from Ireland to Recanati in Italy. JPW asks for and gets the life story, the Italian version. Thus far, the Irish Man has carefully and consistently suppressed any hint of his personal identity or history by refusing to give a first or last name, an address, or any other distinguishing mark. In one sense, this leaves him free while in the removed space of JPW King's office to assume any identity he pleases. But in his narrative he is addressed as Benimillo, a first name, a pet name, after Gigli. As the stratagem makes clear, he has no true identity of his own, and wants to assume one. He is, as he makes it his boast, a self-made man, but one who is profoundly dissatisfied with the self he has made. At one level, the story he proceeds to tell is absurd, since he neither remembers all the details to give the Italian narrative a consistent verisimilitude, nor is he able to prevent 'Irishness' breaking in, as with his father and mother's dialect ('twas their own fault', 'no great harm in it' (177, 178)). But his narrative can be accepted at another level – that of fiction – since it admits that its author is inventing a biography, a life history, through its frequent references to the textual sources out of which it is constructed: 'I read it somewhere.... Another thing I read' (172, 173). JPW at one point refers to the 'Thieving Magpie' and so tips us off to the Irish Man's technique. His reading is not sustained or systematic, but much more akin to that of a magpie. It is dictated, not by an imposed outside concept of social improvement (in which he has failed) but by his psychic needs and by the living myth he is constructing. 'Naked we came into the world', as JPW remarks; and the idea of myth as clothing is picked up and developed when it comes to singing and song, as something 'to clothe our emotion and aspiration' (236) and so make it visible.

When the Irish Man describes his moment of stepping on-stage, he does so as a moment of release, of escape from himself and his embarrassment. He does so in a transvestite role, dressing up as a girl and singing the soprano: 'Then, suddenly, everything was alright and I sauntered back and forth with me parasol' (178).

This moment of on-stage freedom provides the key to the framing narrative, where the Irish Man has already dressed his tatty narrative in the scraps he remembers from his magpie reading; he has already switched nationalities (Irish to Italian) as a prelude and complement to changing gender. And one reminder, of how gender is culturally constructed and assigned, leads us to consider how much the same may be true of nationality, of the 'Irishness' by which the Man is identified in and by the play. The life story he offers is truly post-modern in the way in which its details of Irish home life and speech are bizarrely transferred and surreally blended with those of Italy. Any residual feeling that 'Irishness' constitutes the true or core base of the narrative, the Irish Man's real identity, is undermined by two other aspects of the play: the way in which the Englishman King not only claims a Tipperary grandmother but salts his speech with local Irishisms, down to his grotesque pronunciation of 'Comhlacht Siúicre Éireann' (218). But there is also the second of the three life-stories that the Irish Man tells, the 'true' account of his upbringing that he serves up on the Sunday. For the scenes, like the sessions, are set on particular days; they take their flavour from the day of the week on which they occur. The longest session, with the greatest amount of drinking, takes place on the day before, the Saturday. And JPW registers his protest at being roused at the opening of Scene Five: 'Now it is Sunday morning and you arrived – what? – three hours early and, great lapsed churchgoing people that we are, half of this city is still sensibly in its bed' (214).

Murphy likes the device of the week as a means of structuring or organising his plays. His second full-length play was entitled *A Crucial Week in the Life of a Grocer's Assistant*. Such a mode of organisation is both loose and tight; it gives to each day-long scene the definition and intensity of an entire mini-drama; yet there is a sense of vast scope and creative possibility conveyed by the seven days. In the case of *The Gigli Concert*, the play is significantly one day short of the full week, for in six days God created the world, and JPW King has six days in which to make the Irish Man sing like Gigli. He not only announces it is Sunday, the traditional day of church-going, and thus he is being sought out by one of the country's many lapsed Catholics as a surrogate priest to enact a ritual which will produce some of the guilt-reducing effects of the Mass.

JPW immediately goes on to announce the theme of the session as 'Sex, if you please. [...] Only matters sexual now or I-shall-not-listen!' (214), and kicks off with the first of the Freudian narratives, the rules of which will shape what is said throughout this particular session. Unlike so many other plays which take the encounter between the psychiatrist and disturbed patient as their core (like, say, Peter Shaffer's *Equus* or countless American examples), Murphy's *Gigli Concert* is not going to privilege the Freudian line, so substituting one orthodoxy for another, and allowing for a final, purgative confession which delivers the primal truth in a privileged narrative. Thus, the Freudian confession comes in the middle of the play, not at the end, and is framed by other life-stories which in turn serve to qualify and contextualise it. Besides, Jimmy is no more an orthodox Freudian than an orthodox anything else; as he points out jokingly, the letters 'JPW' are before rather than after his name. His authority is fictive, playful, and most of all in the unorthodoxy with which he switches techniques and modes of inquiry.

Although *The Gigli Concert* all takes place in JPW's office, it manages to spread out and impinge on many aspects of Irish life in the 1980s. One way it manages to do so is in the scenes from real life that the Irish Man brings with him and relates to King and the audience. In seeking the refuge of King's office, the Irish Man is seeking to flee the intolerable situation of the ultimate home he has built, that which houses the nuclear family of himself, his wife and his child. It is a monument to a new-found status and respectability, very like that first enjoyed by Ireland in the 1960s, when people left either the countryside or the city centre and moved into the suburbs. But its strongest feature, like that of the tomb, is its silence. The silence is that of erasure and suppression, of everything that contributed to make up the characters of those who live in that house. In giving his most recent accounts of what occurred to drive him out, the Irish Man reproduces his wife's voice and what she said, and the silence with which she has been met. The effect is like an act of ventriloquism; it stresses the secondariness, the belatedness of what is being said: 'The wife come running. Her concern. Love, love, are you alright. Love. I don't want her concern. She's so – good' (174). For the Irish Man, words of consolation and of love are no longer adequate to break the silence; only obscenities will do it.

The self-loathing with which he is afflicted and which he seeks to shout down is continually seeking a displaced object and threatening to issue in violence. A further displacement from the scapegoating of his wife and child emerges through the encounter with the itinerants, an encounter in which Murphy most clearly identifies the Irish Man's pathology with that of the culture. For itinerants are by definition wanderers, houseless people, and are so construed by the Irish Man as an affront to all he has sought to build up and distance himself from. He goes for them with a slash-hook when they come on 'my territory' (174), and only when he turns to his house after this encounter does he call it his 'home'. The gap has been reinstated; the intimacy with which the itinerants threatened him has once more been fended off.

In seeking to recover himself, the Irish Man looks to do so by constructing an ideal home, not one of the thousands he has built for other families and which are all too 'real', but an imaginary construct by which his own riven household can be saved. And so, in choosing JPW King, he does so at least in part as a 'medium' around and through which the norm can be reasserted; and responses by JPW consciously provide the Irish Man with a verbal counter to the picture of his own home. Crucial to this contrast is the way JPW describes his own home life, the absoluteness of the division into public and private life. His denial that he lives in the office, when we have seen from the opening that he does, leaves JPW free to invent. But since the audience knows Jimmy is inventing, we realise and register the absurdity of his picture of domestic bliss. The question 'Then where do you live?' resounds throughout *The Gigli Concert*. The overwhelming sense it conveys is that life, living, is always somewhere else. When the Irish Man first knocks, JPW is asleep on the sofa and scrambles around frantically, humorously struggling to convert his living quarters into an office. As with so many of the settings in contemporary Irish drama, the characters live and work in the same place, with the boundary line between public and private difficult to locate. The sense is of impermanent, improvident, impoverished, improvised, makeshift lives. The language at a certain level seeks to deny this, by asserting absolute claims about personal and national identity, about 'home'. Jimmy's pretence of living elsewhere is given away by his activities throughout the scenes, particularly his efforts to shave. Anyone else would see through the stratagem sooner than the Irish Man, who registers a need to believe through the blind eye he turns on Jimmy, a readiness to take his declarations at face value. But when JPW confesses that he is shut out from 'the poxy, boring anchor of this everyday world' (203), even the Irish Man sees through him to their essential similarity, that 'This house of yours, sylvan dwelling – this wife of yours' are all invented.

The assertions about national identity emerge at key points in the play, not at these moments of shared recognition, but at their exact opposite. Since one of its leading characters is explicitly and only identified as the Irish Man, and the other male lead comes from England

and could as readily be described as the English Man, further bearing in his name the suggestion of imperial sway, the play is like *Faith Healer* in being a meditation on national identity and role playing. At the most obvious and explicit level, the Irish Man plays up his Irishness and construes JPW as English when he is at his most defensive, when he wants to shore up his own uncertain self by scoring a few cheap cultural points. He stresses the difference in their critical reaction to Gigli's singing; the Irish Man claims he values the 'instinctual' response over JPW's complaint that he cannot understand the words: 'Oh but the English, the English, what would they know anyway!' (201). And his departing taunt to the Mr King who has been left behind by his organisation is: 'Go home' (238). Nicholas Grene has pointed out the extent to which they invert the racial stereotypes: 'The Irish Man couldn't be less like the Irishman of cultural stereotype, dreamy, impractical, fluent with words; he is brutal and inarticulate, boorishly down to earth. [...] It is JPW who has the gift of the gab traditionally assigned to the Irish, JPW whose defensive defeatism is voiced in a volatile stream of words.'[21] We are of course privy through the drama to the process and nature of this inversion – by seeing those occasions just cited in which the Irish Man deliberately reverts to stereotype. But the play of national identities assumes a fourfold rather than a twofold identity. One of the authors cited near the end is Otto Rank, and his essay 'The Double'[22] serves to gloss the extent to which the two men serve to project unresolved conflicts on to one another, to set up the other as a shadow-figure or double, and then seek to kill them. The Irish Man verbally threatens to 'get' JPW at one stage, and the latter feels himself permanently under threat from what he presumes to be a pistol in the Irish Man's pocket. The doubling is verbally signalled by the line with which JPW blearily greets the dawn: 'Christ, how am I going to get through today?' self-consciously and explicitly turning up a few pages later in the Irish Man's account of his breakdown: 'In the mornings I say Christ how am I going to get through today' (166, 173). The dramatic situation, and its interplay of Irishness and Englishness, is constructed through the medium of the two-in-one characterisation.

The dramatic precedent for this interplay between Irishman and Englishman is Shaw's *John Bull's Other Island*, where one character supplies the other's place, with Tom Broadbent taking over Larry Doyle's fiancée and promised place in Parliament. Shaw's Irishman is hardbitten and acerbic, trying to keep down the innate romanticism and dreamy excess against which he inveighs; while the Englishman can afford to indulge a surface romanticism as a cover for the detachment with which he calculates the odds. Shaw's and Murphy's Irish men share a bitterness.

Doyle is an architect who builds at others' behests while what he really seeks is an integrated achievement of Utopia. He has deliberately and coldly embraced a world of facts to attempt to combat and cure the dreaminess with which his Irishness has endowed him. Larry Doyle's most urgent wish is for 'a country to live in where the facts weren't brutal and the dreaming unreal'.[23] The impulse that sends Doyle back to Ireland for the first time in 18 years is the same that drives the Irish Man into JPW's office, a recoil from the world of facts. As the Irish Man says in refusing to give any personal details, 'There's too many facts in the world! Them houses were built out of facts: corruption, brutality, backhanding, fronthanding, backstabbing, lump labour and a bit of technology' (173). And yet 'this practical man is declaring that the romantic kingdom *is* of this world' (190). Or that at least is how JPW chooses to translate it and, through that act of translation, to move into the Irish Man's place and inherit his obsession, to sing like Gigli.

This, however, is not just the story of an Irish Man and an English Man but of an Italian tenor who serves acoustically and symbolically as the ground of their exchange. I have already examined how this third term serves to deconstruct the supposed cultural authenticity of any one of the Irish Man's life histories, as he stumbles over the words 'our-our culture' (179). And there is the presence of the spirit of Gigli throughout the play in the endless sound of his recordings playing from the very opening, long before they are introduced into the diegetic space of the play. This prepares the way for the closing, in which the sound of Gigli's voice does *not* come from the disconnected record player but from the entire stage space into which at that point the solo King is absorbed. But before that Utopian acoustic space can be temporarily gained, a complex series of situations has to be resolved. I wish to analyse the play's personal politics in terms primarily of triangular rather than doubled relations, or to put it another way, to argue that every doubled relationship in the play necessarily implies a third. Gigli as the shifting object of identification between, initially the Irish Man, and, increasingly, JPW King, is the most insistently foregrounded, the most resonant, and yet the most absent.

For there is a third character in the play, Mona. And while some have felt and argued that she is a supernumerary and could be dropped,[24] I believe Mona's presence to be crucial, as the necessary third without which the play's dynamic would fail. It is tempting to confuse her pointed exclusion *in* the play with the marginality of her treatment *by* the play.[25] She enters at the very end of the first scene, even as Jimmy exits, and her greeting fails to hold him. And at the end of Scene Four, when JPW resolves to attempt the impossible leap of faith, we see Mona arriving at the door and hear her cry for help even as her knocking at the door goes unanswered. But the situation at the beginning of the scene – the pillow talk between JPW and Mona – is that rarest of sites in Irish drama, contemporary or otherwise, a bed scene between a man and a woman in which the most intimate questions of identity can be raised and reconfigured. As with Freud, there are four people in the bed, the two absences here being the Irish Man and Mona's godchild. As Mona speaks of the latter, Jimmy speaks of his relationship with the Irish Man. He does so in terms which Mona's presence now reveals as that of a contracted partnership or marriage. The point is clinched when the Irish Man arrives and Mona gets up to leave, an act of substitution and replacement, to which JPW protests, 'He's not my wife', and Mona retorts, 'You haven't got a wife!' (195).

The debate between Irishness and Englishness, heretofore addressed in exclusively male terms, is now examined in terms of gendered relations, those between an English man and an Irish woman. Additionally, there are the class implications of Jimmy's voice: 'Hear that posh voice of yours' (193). The overarching voice of the play is Italian; but the desire this impulse embodies, to get out of one's own voice, to find one's own voice through adopting another, is here related to the long history in Ireland of the 'appeal' of an English voice and accent as cultured. But Mona does not want to sound like JPW; rather, she wants to hear him talk to her. For in his Englishness he is contrasted with Irish men's attitudes towards women. They are aggressive, physically rough, and will not talk. Mona has the experience of a variety of Irish men to back this up. Her narrative primarily contrasts Jimmy's behaviour with her archetypal Irish husband, Michael Quinn, who grunts in reply to her questions as she tempts him with lumps of meat for the dinner. Mona's account of life with her husband balances and complements the Irish Man's account of *his* home life, showing what his lack of communication must be like from the other side. After Mona's departure there are consequences. JPW leaves the bed unmade for the first time, admits to the Irish Man that he lives there, not somewhere else, and that his story of married life is a fantasy.

In other words, it is the exchange with Mona that allows for and prepares the breakdown in the construction of 'home'. And in the role playing by which she draws Jimmy's attention to her, Mona twice establishes that Jimmy's repeated verbal insistence on 'proving' himself is a displacement of male impotence. She does so first in physical terms by diving under the bedclothes and performing oral sex ('Nothing much happening down here, my friend' (191)), and then by telling the story of her godchild Karen-Marie and the female assistant laughing as the child unzips the flies of the male dummies: 'And says Karen-Marie [...] "They're only dummies", she said, "they have no willies"' (195). Throughout this section, the issues are cued by Mona's declaration to Jimmy, 'I'm a subject' (191). The initial role that she will play is patient to his psychiatrist; she shows she is willing to prepare him for the imminent encounter with the Irish Man. But the line resonates and declares the extent to which the play is concerned with 'woman' as enabling subject of male-centred discourse.

There is no more apotheosising of woman as goddess. This is the necessary complement to the play's rewriting of the compact with the Christian God, of the 'I am who am' into the 'I am who may be' (211).[26] *The Gigli Concert*'s version of the Fall concentrates on Adam and God to the exclusion of Eve. Eve is written out of the traditional version of the Fall of Man, so she is no longer the scapegoat of that patriarchal narrative. While Jimmy is doing his 'God and Adam' monologue, Mona is knocking repeatedly at the door in an effort to be heard and to be admitted. In the play's final act, she and Jimmy are in bed for the last time when Jimmy addresses his ideal lines to a real woman in the here and now. The play makes equally clear that Mona is not to take her/the place. For in Jimmy's relationship with Mona there is simultaneity, mutuality, aliveness, no permanence; hence his resistance to the term 'beloved'. Mona now admits she has cancer, that the 'child' she has discussed has been imaginary, and that she has tried to get pregnant to replace the child she gave up for adoption. Mona's life history emerges at the moment of her acceptance by Jimmy and of her leave-taking. Her valedictory term for him is her 'magician friend' (234). If JPW King is a Faust, Mona sees him as a benevolent one, not as an emissary for a punitive god.

JPW cannot say 'I love you' in return. His monologue between Mona's exit and the Irish Man's entrance is the most anguished and explosive in the play: 'I lo–! (*love*) I love! I – ! I – ! Fuck you! I love! Fu–! Fuck you! I love! – I love! Fuck you – fuck you! I love.' (235) In his divided state, what emerges is a mix of Mona's 'I love you' and the Irish Man's roar of hate, 'Fuck you, fuck you' (185). The desire to possess is opposed by the desire to relinquish. Where the Irish Man appears here at his most satanic, JPW is now at his most human as Faust. The Irish Man's pronounced condescension to the itinerants measures the extent to which he has again taken himself prisoner, claiming he is cured 'by readjusting to the world that is destroying him'.[27] But the sign of the ideology in the name of which he has operated is 'the wife'. As he says: 'Supposing my life depended on it, who would I turn to? I went through mothers, brothers, relations. The wife. It all boils down to the wife for us all in the end' (235) – all us men, presumably. The Irish Man has already banished his demons with some traditional Irish remedies: a tear and a smile, a wink and a nod, a prayer ('please God') and a hand-out. He can now 'live' again and does so by investing in 'the wife', someone to depend on, someone to save him. What Jimmy has relinquished by resisting a claim on Mona, the Irish Man is once more subscribing to.

With woman no longer being exchanged as a romantic token, Jimmy can walk out through that door as Mona has done, acknowledging his and her independence and the voluntariness of their chosen ground. He no longer resembles the Irish Man, despite having sung like Gigli. By seeing through the obsession, he has come out the other side of it. In the final primal cry which precedes Jimmy's singing, 'Mama! Mama! Don't leave me in the dark', the ultimate court of appeal is the female, not the male, in a plea that abandons the myth of male sufficiency and creation. The symbolic order readmits the female and a theatre of the impossible, as Jimmy sings like Gigli. He is now free to leave the stage, and Gigli singing 'on forever' (240) to the audience where each man and woman faces the open vacant zone of their own possible self-transformation. The Irish Man's final, parting taunt has been 'Go home' (238). Murphy's remarkable play shows the arduous process involved in a true homecoming to the self.

The Gigli Concert was followed two years later by *Bailegangaire*, another undisputed high point of Murphy's theatrical career. First staged at Galway's Druid Theatre by founder Garry Hynes on 5 December 1985, the play's distinctive blend of traditionalism and innovation enabled its audience to enter fully its imaginative world and respond to the challenges Murphy was formulating. The most instructive comparison with his earlier works is provided by *A Whistle in the Dark*. They are both family plays, two of the group in Murphy's *oeuvre* that are concerned with the fraught tensions of the family romance, the givens of blood ties rather than the random collisions of his outsiders and misfits. As such, the two plays can hardly but draw on their great Greek and Shakespearean predecessors. I have already discussed the workings and place of tragedy in *Whistle*. *Bailegangaire* is a female *King Lear*, its monstrous mother figure enthroned centre stage in her double bed, scorning the 'good' granddaughter while playing up to the 'bad' one.

The play is set in the west of Ireland, some miles outside Galway, 'in the year 1984',[28] as the text more than once reminds us. One reason it needs to keep doing so is that the setting of a traditional thatched cottage speaks of an earlier time, a different Ireland. The change is pointed by reference to such technological developments as helicopters and computers. But if Ireland's rural past and technologised present are juxtaposed in the play, this is socially all of a piece, as is the tension between the Irish emigrés in *Whistle* and the English milieu to which they are transplanted. What is theatrically more venturesome is the juxtaposition in *Bailegangaire* between the specifics of life in Galway, 1984, with the mythic timelessness of Mommo's storytelling. For much of its running time, the play foregrounds the extended narrative of 'how the place called Bochtán [...] came by its new appellation, Bailegangaire, the place without laughter' (92), Mommo's mock-epic account of how two mighty protagonists drove each other into competitive paroxysms of mirth by laughing at their misfortunes.[29]

In placing the oral recital of a folk narrative at centre stage, Murphy is returning Irish drama to its origins. But the nature of that narrative and its relation to the encircling drama is far from straightforward; and in charting the twists and turns by which the two discrete narratives illuminate one another, by which the traditional tale serves as a psychological displacement for the unspoken fears and resentments of its teller, Murphy propels his play into the company of modernist works by Samuel Beckett, Harold Pinter and Sam Shepard. *Bailegangaire* not only recapitulates the process by which storytelling evolved into the drama of the Irish Literary Revival, but also demonstrates the extent to which in postmodern dramaturgy the phenomenon of storytelling has itself become the action of drama. In turn, storytelling in Murphy's plays feed directly into the work of a younger contemporary like Conor McPherson in *The Weir* (1997).

Bailegangaire marks a sea-change in Murphy's drama, in that all three of its roles are written for women. The most extreme example of the extent to which his stage is dominated by all-male activities, and the greatest contrast with *Bailegangaire*, is provided by *A Whistle in the Dark*, where Michael's wife Betty never stands a verbal or physical chance against the six Carney men's jibes. Though the character of Betty gives a good defence of herself and serves to expose the shortcomings of the 'hard men', Murphy still carries around with him the memory of the play's first night in London in 1962:

> A woman came up to me after the play, and she said it was very good, and so on, but 'if you don't mind, Tom, you know nothing about women'. So I wanted to write a play for three women, not just based on that incident of the first night, but it did contribute to it.

By situating his three women in the west of Ireland, Murphy encountered certain associations. At the abstract level of myth, Mommo suggests the traditional feminine personification of Ireland as the *Sean Bhean Bhocht* or Poor Old Woman, of which Yeats' and Gregory's *Cathleen ni Houlihan* is a variant and dramatic precursor. But at the more local level, as an old woman living on the western edge of Europe who has striven to raise children in the teeth of self-destructive impulses within and harsh natural conditions without, Mommo most resembles Maurya in Synge's *Riders to the Sea*.[30] The affinity is suggested by the absence of male figures in the lives of both women. In Murphy's play one granddaughter's husband may have skipped off to England; but the larger, more overwhelming sense is of the unnatural premature deaths of young males. The narrative focus of this concern is the missing third grandchild, whose absence is accounted for by Mommo's repeated insistence that Tom is in Galway. What the play eventually discloses as the terrible secret shadowing the event of the laughing contest is the inadvertent death by fire of the young male child while awaiting his grandparents' belated return home. The event broadens into tragedy as the admission of the death of the one grandson enables, in turn, Mommo's verbal evocation of the fate of all her male children:

> Her Pat was her eldest, died of consumption, had his pick of the girls an' married the widdy again' all her wishes. [...] An' how Pat, when he came back for the two sheep [that] wor his – An' they wor – An' he was her first-born. But you'll not have them she told him. Soft Willie inside, quiet by the hearth, but she knew he'd be able, the spawgs of hands he had on him. 'Is it goin' fightin' me own brother?' But she told him a brother was one thing, but she was his mother, an' them were the orders. [...] They hurted each other. An' how Pat went back empty to his strap of a widdy. An' was dead within a six months. Hih-hih-hih. [*The 'hih-hih-hih' which punctuate her story sounds more like tears – ingrown sobs – rather than laughter.*] Oh she made great contribution to the rollcall of the dead. [...] An' for the sake of an auld ewe was stuck in the flood was how she lost two of the others, Jimmy and Michael.
>
> (163–4)

The same dramatic condensation is in Murphy's *Bailegangaire* and Synge's *Riders* whereby the foregrounded death serves to implicate and unfurl all the others in the monologues through which the two women outline the fate of each of their sons. Mommo's stance, however, differs from the stoic acceptance of Synge's Maurya in its more spirited defiance of the angels of destruction, kept at bay if not defeated by the howls of laughter: 'driving bellows of refusal at the sky through the roof. Och hona ho gus hah-haa!' (164–5).

In the phonetic notation of Mommo's laughter with which this speech concludes, we have only one example of the extraordinary range of sounds uttered in the play. Again, there is a traditional aspect to all this, the reason why figures like Synge and Lady Gregory were drawn as Irish dramatists writing in English to the west of Ireland at the turn of the century. For it was on the Aran Islands and in the remote pockets of the Gaeltacht that they came in contact with the curious, hybridised speech of Hiberno-English and studied the process by which English syntax was disrupted by the forms of Gaelic which underlay them. The resulting Anglo-Irish speech simultaneously gave a home to direct translations from the Irish as well as older forms of English diction. Mommo's narrative is cast in this idiom, though it takes account of the later parodic possibilities exploited by Flann O'Brien in combining epic scale, a pervasive whimsy and a fussy formality in the telling:

> The size an' the breadth of him, you'd near have to step into the verge to give him sufficient right-of-way. 'Twould be no use him extending the civility 'cause you'd hardly get around him I'm saying.
>
> (101–2)

The great virtue of the oral tradition of storytelling lingering in the West is that it preserved many of the distinguishing features of the early culture, not least the Gaelic language itself. In Murphy's play, Mommo's story bears along in its verbal torrent frequent scraps of Gaelic ('go bhfóire Dia orainn, may God protect us', 'cráite, crestfallen', and the 'seafóid, nonsense' with which Dolly ironically dismisses the story), some of them translated, others not. The play thus serves as a continuous aural reminder of the joint linguistic inheritance in Ireland.

But for Tom Murphy this is as much, if not more, a source of confusion as it is of creative diversity. And he increases the sense of linguistic confusion (as reflecting cultural and individual incertitude), not only by stressing the fragmentariness of Mommo's narrative but by isolating the different elements in it, as we shall see. He also invades the suspect cultural purity of the thatched cottage setting and the old woman telling stories with a variety of anomalous sounds from outside. The radio is a continuous presence throughout, bringing a much broader frame of cultural reference to an isolated community. And we are never far away from the sounds of passing motor cars or the putt-putt-putt of Dolly's motor cycle. And if Mommo speaks in the more traditional tongue, Dolly is as up to date as her name suggests in offering a more recognisable verbal representation of the Irish home of 1984:

> I've rubber-backed lino in all the bedrooms now, the Honda is going like a bomb and the *lounge*, my dear, is carpeted.
>
> (107)

In extending the play's linguistic range from the guttural, unarticulated sounds he scores for Mommo's laughter to the brand names of the thoroughly modern housewife, Murphy allows no one speech pattern to dominate but is constantly pitting one against the other and so showing the limitations of each. When Murphy and I discussed the influences that gave rise to *Bailegangaire*, he did so in terms of the way people talk, resulting in an uncanny mix of the familiar and the bizarre which is characteristic of his drama:

> I'd noticed in my visits to the West the way people talk. You can still hear the sound of the sheep on the hillside and the sound of seabirds in their accents and voices. And yet you see a man with a bag of turf on his back and he's got a Walkman on his head and he's listening to music. And you go on another few hundred yards – I'm talking about the remote wilds of Ireland now – and you'll have two fellas mad drunk at eleven o'clock in the morning and they're speaking in a language that the English or any English-speaking race couldn't possibly recognise. And a few hundred yards further there's a hotel where a woman is speaking in the most posh voice, saying 'Your table is laid over there', and the menu is in French. These extraordinary anomalies abound in this country today – transition doesn't cover it.

But if *Bailegangaire* extends the dramatic range and purpose of Irish speech, it also complicates the function of storytelling in the drama. In so doing it develops a dramatic process begun by Synge. The stories recorded by the latter in *The Aran Islands* tell a great deal about the sympathies of the island community, very little about the individual psychology of the teller. Their stance before their material is, on the one hand, to insist on their personal witnessing of the event narrated (as a vouch of authenticity)[31] and, on the other, to refrain from any direct intervention in the narrative so that its essence may be preserved and transmitted. Synge is careful to imitate this scrupulosity in his prose handling of *The Aran Islands* material. But in the plays he takes increasingly freer rein until, with *The Playboy of the Western World*, the story Christy Mahon tells about his father-slaying becomes highly problematic as to its status. Synge's play shows how the story is increasingly fleshed out in Christy's representation with dialogue, movement, gesture, stage directions, and so on, and increasingly serves as a mediumistic mirror to reflect the transformation he undergoes. Synge's play tests its heroine's claim that 'there's a great gap between a gallous story and a dirty deed'[32] by dramatising the uncertain areas in that gap. In Murphy's play the distance separating the story Mommo tells from the lives of the three characters is even wider, the 'gap' between them precisely that which has to be filled in and made meaningful by a sustained process of dramatic interaction.

What immediately strikes an audience is just such a discrepancy between the rambling Christmas story Mommo is relating and the chaotic present-tense conditions of the play itself. Such a problematic relation, rather than a direct and obvious connection, between the play within the play and the larger work is a characteristic feature of postmodern drama. Kristin Morrison has pointed to the pervasive use of storytelling in the plays of Beckett and Pinter and the apparent paradox by which that supposedly most objective of forms has instead replaced soliloquy as a 'convention for expressing psychological inwardness on stage [...]. Now the telling of a story allows characters that quintessentially "modern", Freudian opportunity to reveal deep and difficult thoughts and feelings while at the same time concealing them as fiction or at least distancing them as narration.'[33] Mommo insists on just such distancing by the casting of the story in third-person terms and her refusal to identify with the 'stranger woman' of the narrative. On those rare occasions when the 'they' slips unintentionally into 'we' (124, for example), the narrative breaks down and Mommo is unable to continue. Like Mouth in Beckett's *Not I* (1972) it becomes clear that Mommo 'has suffered an experience so traumatic that she cannot accept it; she must insist that it has happened to somebody else'.[34] Murphy's Mommo and Beckett's Mouth can be compared as storytellers, especially in their persistent refusal to relinquish the third person. The characteristic movement resulting from such an ambivalence is double: facing up to personal events by talking of them in a story, but fleeing them by recourse to fictional concealment. This double movement is reflected in the Beckettian fragmentation to which Murphy subjects Mommo's narrative. In place of the lengthy seamless narratives of Synge's Aran islanders, or even of the three monologuists in Friel's *Faith Healer*, the audience of *Bailegangaire* only gets the narrative of the night in Bochtán in bits and pieces. Some of the breaks are more clearly formal. Twice, Mommo breaks off at obvious high points in the narrative: the arrival of the stranger and his wife at the pub; the declaration of a laughing contest and the initial squaring off of the antagonists. But even here one senses a personal reluctance to go further vying with the onward drive of the story itself. More often the breaks follow no internal logic of the narrative. Rather, they occur at random and are more the consequence of Mommo's tiredness (feigned or real), senility, bloody-mindedness, above all her reaction to the private grief encoded in the outward proceedings. Although there *are* occasions when the *Thief of a Christmas* story takes over and runs unimpeded for several pages, more often (especially in the revised, published script) we are not given enough at a time to allow direct engagement with the story. Instead, the emphasis shifts to the psychological condition of the teller, highlighting the extent to which the story serves simultaneously to conceal her from her lived experience and to provide her only means of access to it.

Of all the plays considered by Kristin Morrison, that which most resembles *Bailegangaire* is Beckett's *Endgame*. Hamm and Mommo are equally obsessed with their stories, narratives which tease the audience into working out connections, possible identifications 'between this chronicle, this value-laden record of past events, and the words and actions which make up the dramatic present of the play'.[35] Both recounted narratives are set in the dead of winter, near Christmas time, and hinge on the fate of a child. But there is also a resemblance in the immediate dramatic situation of the play's present, in the relationship between Hamm and Clov in *Endgame* and Mommo and Mary in *Bailegangaire*. Two figures are on stage, one immobilised, the other attending upon them. Hamm is blind; Mommo sees little in or of the present surrounding her. Both are physically encumbered, dependent on another person (while scornful of them) for motion, sustenance, chronology ('What time is it?', a frequent refrain), pills and painkillers, the latter underscoring each play's graphic concern with the continuous physical pain of the aged human body. Though entirely dependent, each of the paralysed figures acts in lordly fashion to the other person, substituting a master-servant relationship for the more appropriate familial bond of parent-child. Mommo refuses to acknowledge Mary by name until the very end of the play, referring to her instead as 'Miss' and addressing her as a less-than-satisfactory domestic (cf. Hamm's 'Ah the creatures, the creatures, everything has to be explained to them').[36] Mary and Clov in turn are forced to attend to every whim and carry out menial tasks at the behest of the dominant figure. There is also a sense, however, that they are psychologically more secure inhabiting the role of scurrying servant, that it attests to a need for external order and clarity in their lives.

It is to Mary that much of the dramatic responsibility is delegated for mobilising, not only Mommo, but the intricate interaction of the drama itself. The play begins when the narrative begins; but it soon becomes clear that, though oft-told (the two granddaughters are able to recite it virtually by heart), the story has always broken off at some crucial point, since neither knows the ending. The audience comes to realise that Murphy's play cannot end until Mommo's story is ended, until it has been pushed through to a conclusion. The dramatic suspense is therefore generated out of the struggle for the full story to be told. The agent for this working through is Mary, when she realises the extent to which the complications of this adult fairytale are interwoven with the complications of their own lives. Mommo remains oblivious to the latter and equally determined to resist the former. The technical and emotional resolution of the play will have arrived when these two narratives become one.

Initially, such possibilities seem remote. Clearly, these people need each other and have at this point in their lives no one else to depend on. Equally clearly, they are very far from each other. Mommo sits secure in her double bed, scattering imaginary hens from the kitchen and telling a bedtime haunted narrative to the children. But the 'children' are now 39 and 41 respectively. Dolly is landed with an impending and unwanted baby which she will try to palm off on her unmarried sister or else abandon in a field in the middle of the night. Mary sits motionless at the opening, staring at the 'nothing' of her life, on the verge of hysteria, if not suicide. At this point she is resistant to the blandishments of Mommo's narrative, seeing it as pure evasion, irrelevant to the crisis of feeling afflicting them all. In the opening movement she eschews narrative altogether and instead makes a direct emotional appeal to Mommo:

> Look! [*She holds up an iced cake.*] We never knew your birthday but today is mine and I thought we might share the same birthday together in future.
>
> (93)

But the appeal is rebuffed in Mommo's obstinate refusal to recognise Mary's birthday, denying not only her part in giving birth but ultimately Mary herself as existing. In the 'Heh heh heh heh' with which she answers her granddaughter's appeal, '*there is defiance, hatred, in the sound*'. Instead, Mommo looses a large stream of her narrative which only comes to a halt when the decent man urges his horse on up the icy hill: 'Try again, Pedlar' (96). At that point, Mary turns to the table and does try again to make hard upward progress against Mommo's icy indifference. Simultaneously, the audience begins to sense the symbolic overlap and interrelation between the two narratives, the coded message Mommo is translating. It takes Mary some time longer.

In the meantime, she prefers the more formal consolations of a poem by Thomas Hardy, 'Silences', which not only has reference to her own situation, but also reminds her of the fate of her drowned uncle and father. There are two lengthy quotations from 'Silences', the first to suggest alternatives to the story Mommo is reciting:

> There is the silence of copse or croft [...]
> When the wind sinks dumb.
> And of belfry loft
> When the tenor after tolling stops its hum.
> (121–2)

The Hardy poem is little better than the silence between the two women, serving in its aftermath only to toll Mary back to her sole self. What she only gradually comes to realise, as she idly picks up the threads of Mommo's narrative while the latter sleeps, is that no other of Mommo's stories nor the more literary consolations of Hardy's poetic dirge will do, will be adequate to the task of reintegrating the lives of these three strangers. Their lifelines and the plotlines of the *Bailegangaire* narrative are woven into a single knot of meaning by the fatalities at the centre of both: the fate of the third grandchild Tom on the night the narrated events took place ('then') and the fate of Dolly's pregnancy on the night the narrated events are related in full ('now'). The second quotation from 'Silences' refers to 'the silence of an empty house/Where oneself was born' (122) and argues that 'no power can waken it' since the present is incapable of altering what the past has decreed. Mary concludes her reading of the poem by saying 'no' to Hardy's deterministic pessimism. Instead, she wakens Mommo and offers her own aid and participation to realise the full 'power' of storytelling as the means by which we not only take imaginative possession of our past but shape it in the light of present possibilities:

> [*To the book, and dumping it.*] Is that so? Well, I don't agree with you [...] Wake up *now*, Mommo. Mommo! Because I don't want to wait

> till midnight, or one or two or three o'clock in the morning, for more of your – unfinished symphony. I'm ready *now*.
>
> (122)

She urges 'On with the story' (122), and so inaugurates a Beckettian dialogue, with Mommo moaning that she can't go on and Mary insisting that she must.

More than a Beckett-like dialogue ensues. What we are presented with instead is an increasingly theatrical collaboration between Mary and Mommo in the generation of the narrative, with Mary performing more of a directorial function than just urging the action forward. She provides the psychological equivalent of her physical ministrations when she eases Mommo back into the story:

> MARY: To continue. But that decent man and his decent wife the same did as was proper on entering.
> MOMMO: Sure we weren't meant to be here at all!
> MARY: The customary salutation was given.
> MOMMO: That was one of God's errors.
> MARY: Though silently, for they were shy people, and confused in their quandry. Mommo? And then, without fuss, the man indicated a seat in the most private corner.
> MOMMO: An' they were wrongin' them there again! So they wor.
> MARY: They were.
>
> (123)

When Mommo insists 'we weren't meant to be there at all', her unprecedented use of the first person plural closes the fictional distance into an immediate identification with the characters in her story. The pronominal shift back to 'they' in Mary's 'though silently, for they were shy people, and confused' restores the vital mediation of third-person narration, not only to console Mommo but to hold her grief at bay and enable the story to continue.

Mary goes even further in transforming the storytelling into drama when she takes over not only the storyline but Mommo's speech, gestures and role. The proper young woman who flinched at verbal blasphemy now unleashes a stream of invective that derives its verbal energies from the Gaelic tradition's belief in the word magic of the satirist: 'a venomous pack of jolter-headed gobshites [...], an ill-bred band of amadáns an' óinseachs [...], a low crew of illiterate plebs, drunkards and incestuous bastards [...]. Them were his words' (121). The words were those of her great-grandfather. With them Mary is not only re-establishing contact with Mommo, but through the rituals of story-telling and drama playing a part in maintaining and orally transmitting a family legacy.

A crucial element is added to the process of dramatisation by the raucous entrance of Dolly. For she is the reality principle in the play in more than one respect, a direct link with the real world of contemporary Ireland outside the cottage. Where Mary holds herself off from overt physical contact, Dolly revels in her own sensuality. In the time between her first and second appearances, neither we nor Mary can forget that Dolly is outside having sex in a field. Her plan for Mary to adopt her baby culminates in the casual but thematically significant statement:

> DOLLY: After a year it'll be easy to make up a story.
> MARY: *Another* story! [*She laughs.*]
>
> (146)

Dolly herself has little time or respect for stories. We see that she is all too familiar with Mommo's, readily giving the cue whenever Mommo is stuck. But she does so automatically and compares the recital to a 'gramophone record'.[37] We can infer from her surface reproduction of it the number of times Dolly has heard the story of Bailegangaire and hence the truth of her claim to have looked after Mommo while Mary was away. Without the presence of Dolly, the thrust of the narrative would remain too much centred in the past, the Ireland of the 1950s, and the peasant kitchen drama that went along with it. Mommo would not be drawn into the present and the recognition of her two granddaughters; Mary could use the storytelling too much as an escape from her present dilemma, an aestheticising of her aloneness. But Dolly's raucous vitality and unrepentant hedonism not only give a necessary jolt to Murphy's play; more importantly, her concern over the child and its fate presses more immediate claims on the story-telling and insists that if the never-ending, constantly deferred story is to finish on this night, then its proper resolution can and must embrace the present tense, recording the all-too-real dilemma of an unwanted pregnancy in the Ireland of 1984.[38]

The play's resolution is achieved through the tentative redefinition of the elusive concept of 'home'. As we saw earlier, all Mary can say by way of explanation as to why she abandoned a successful career in England is that she wanted to come home; the prospect she returns to, of a senile, bed-ridden old grandmother who refuses to recognise her, is devastating in its ironies on the theme. Dolly is rueful when the word 'home' is mentioned, since ostensibly it points to the house where her fatherless children await her but (as Mary forces her to admit) actually indicates a few precious minutes of sexual gratification. The full story of Dolly's 'home' situation emerges later when she gives details of the previous Christmas, her estranged husband's return from England, a consoling night of love-making followed the next day by a more protracted round of wife-battering that swelled her face up like a balloon. More brutal ironies.

The word 'home' recurs frequently in Mommo's narrative (its mention spurring many of the above reflections by Dolly). Not only is it the ultimate destination of the decent man and his wife coming from the *margadh mór* but the goal of the entire narrative and, by extension, of the play itself. When they come to a fork in the road, the stranger reflects 'that the road to Bochtán, though of circularity, was another means home' (98). This proves true in more than the literal sense for the other half of the couple making up this home. For the wife is the one who takes the crucial step in the night's events when she intervenes to ensure that her man does not walk away from the contest with laughing Costello. She even admits that 'they could have got home' safely at that point for 'the thaw was settling in' (140). But she reflected at that moment on what her definition and experience of home had been in the rural Ireland of the time.

> No, a woman isn't stick or stone. The forty years an' more in the one bed together an' he to rise in the mornin' (and) not to give her a glance. An' so long it had been he had called her by first name, she'd near forgot it herself ... Brigit ... Hah? ... An' so she thought he hated her ... An' maybe he did. Like everything else ... An'. [*Her head comes up, eyes fierce.*] 'Yis, yis-yis, he's challe'gin' ye, he is!'
>
> (140)

And so Mommo urges her man to take out her frustrations in an act of revenge, much as Dada does with his sons in *Whistle*. The consequence of that act is to lose what homes she has, with the death of her grandson that same night and her husband a short time later.

Having brought her story to completion, Mommo ends her night's talking with a prayer, another formal narrative, whose petitioning appeal is extended from those absent (including Tom, whose death is finally acknowledged) to Mary, who is now turned to and directly addressed as a living presence for the first time: 'And sure a tear isn't such a bad thing, Mary' (169). The tears throughout the play have veered imperceptibly between laughter and grief, an appropriate response for that most distinctive of Irish dramatic genres, the tragicomedy. The story of Bailegangaire began with a laughing contest and ended in tragedy. The story of *Bailegangaire* begins in the fruitless grief of mutual isolation and achieves its healing atonement through moments of shared laughter. Mommo's first recognition, of Dolly, comes about through the latter's playfulness: 'Is it Dolly? Aw is it my Dolly! [...] Sure you were always the joker. [...] Dolly, come 'ere to me' (110). The first moment of reconciliation between the sisters is achieved when they put aside mutual recrimination and apply Mommo's narrative ironically, incongruously, finally joyously to their own situation:

> [MARY *laughs*, DOLLY *joins in the laughter*, DOLLY *flaunting herself, clowning*]: And you're his [the baby's] aunt! [*They laugh louder; the laughter getting out of hand.*] .... [*Uproariously.*] Jesus, misfortunes!
>
> (161)

The tonic of their laughter brings Mommo back to life, enabling her to resume and (for the first time in all its many retellings) bring the family narrative to a conclusion. That it also functions as a resolution has to do, I would suggest, with the shift from the predominantly male company of the original laughing contest, determined that events will not conclude before heads are broken and mortal wounds inflicted, to the company of three women, who are collectively able to follow through on Mommo's 30-year-old suggestion and laugh at their misfortunes. Mary has urged that they 'live out the – story – finish it, move on to a place where, perhaps, we could make some kind of new start' (157). She now offers the play's final words, not of heavenly but of humanist consolation, acknowledging the second chance afforded them through their opening up Mommo's past-fixated narrative into the mutually enacted dramatic present

> in whatever wisdom there is, in the year 1984, it was decided to give that – fambly ... of strangers another chance, and a brand new baby to gladden their home.
>
> (170)

The central image of *Bailegangaire*, dominating the stage as memorably as Mother Courage's wagon does in Brecht's play, is the double bed in which Mommo lies. As Thomas Kilroy has written: 'Murphy's play is built upon the image of the bed, the domestic centre, the place of fertility, the nest but now [...] become the place of decrepitude with the voice of that tormented old woman trying to complete the telling of her story.'[39] At the end of several bruising hours, when the true ending of the young life lost that night long ago is provided by Dolly's baby, the two younger women climb into the bed with their senile grandmother. At the end of our interview, Tom Murphy spoke of himself as in a sense getting into bed with four women. Writing the interlocked drama of Mommo and Mary and Dolly, working with an equally strong-minded woman director in Druid's Garry Hynes, Murphy in *Bailegangaire* went beyond the exclusively male camaraderie of his earlier works into a new and challenging theatrical domain.

Tom Murphy's greatest achievements remain *The Gigli Concert* and *Bailegangaire*, his plays of the 1980s on which this chapter has concentrated. He finished out the decade with *Too Late for Logic* (1989), a return to the dilemmas of a central male protagonist, here explicitly a writer. The play is notable as a rare dramatic exploration in the Irish context of the break-up of a marriage, as Christopher seeks to work out a new relationship with his son and daughter. He is simultaneously working on a television lecture on young Arthur Schopenhauer and, while there is clearly a connection between the pessimism of the philosopher and the existential crisis being experienced by Christopher, the TV lecture becomes a deflection from the heart of the matter as much for the play itself as for its central character.

After almost a decade's silence, Murphy returned to the stage by reworking his 1994 novel *The Seduction of Morality* as a play. *The Wake* (1998) is centred on another homecoming, this time from New York rather than the UK, and is the most openly confrontational of all. Where Mary has buried her resentment towards both Mommo and her sister Dolly, as befits a young woman who has trained as a nurse, Vera O'Toole comes home defiant and determined to claim the inheritance the rest of the O'Tooles would seek to wrest from her. There is some discussion at a clan gathering (in the play's best scene) where various notes of outrage are struck at Vera's chosen profession in New York: 'Oh! And the rumour – maybe you've heard it, maybe you've not – about our Vera's "line of business" in America?'[40] For Vera is a prostitute and has earned her living there by metaphorically – and occasionally literally, she suggests – 'eating shit' (118). The representation of the central female character as a prostitute is inherently problematic in both the novel and the play. In the novel it is too readily taken as a given of modernism, a sign for sexual freedom and romantic hedonism to set in opposition to the bourgeois small-town aspirational morality of the other (married and familied) O'Toole siblings. In the play, there is a greater awareness and self-questioning of the role, notably when Vera takes the terms 'whore' and 'cunt' that have been liberally thrown at her by another character and pushes them to the limit to expose their essential degradation. But the on-stage representation of the female body undergone by Vera in the transition from novel to theatre means that from the start she is visually represented as a whore. This is always done with a self-conscious awareness that she is performing a role in public – '*the dress is makeshift but it creates the effect she wants: the marks of a whore and sexy*' (140) – but I do not share the play's confidence that Vera's '*private self*' is revealed by the mere act of '*tak(ing) off her wig*' (97). In the final scene, she is '*sober and in a simple dress*' (156) but the information that she has been put in a mental home for three days in the interim seriously qualifies the notion that she has been 'cured'.

Murphy has been drawn to the characterizations of men as pimps and women as whores throughout his career, as Nicholas Grene has pointed out,[41] but a pimp is just a man in a suit. In *The House* (2000), this is the rough trade that the returned emigrant Christy has practiced in England. As with Vera, that life is never directly dramatized on stage – it remains foreign, abroad, other. In both plays that sexually mystified life elsewhere creates an occlusion in understanding and renders problematic the motivations which drive the returned emigrant. The action of *The House*, like *The Wake*, centres on a piece of property with strong family associations being put up for sale: the O'Toole family hotel which Vera has inherited; the (big) house of the title belonging to an Anglo-Irish family, the de Burcas. Where the symbolic space of the house is associated with female energies – Mrs de Burca/Mother and her three daughters – the pub is the masculine place to which all of the temporarily returned emigres resort. Christy oscillates uneasily between the two spaces – the Big House by which he was fostered and to which he is attracted, the native Catholic culture into which he was born. Mrs de Burca and her three daughters are Chekhovian figures and Murphy went on in 2004 to write a version of *The Cherry Orchard* for the Abbey. But his touch is less sure in this area than Friel's; there is something pallid and two-dimensional about the Big House women, especially in contrast to the fizzing dialogue and the beautifully etched portraits of the different male returnees in the pub.

Both *The Wake* and *The House* find their important female characters stranded, isolated, by the two plays' dependence on social realism. The female characters may feel that a greater degree of self-expression is possible in the alternative locations they find themselves in – the closed hotel which Vera breaks into and occupies, the de Burcas' Big House – but this is only achieved by an almost total social marginalisation. When they seek to enter the bigger/normative world – as when each of the younger de Burca women comes into the pub – they do so by physically and psychologically asserting themselves but with the concomitant effect of offering themselves as a sexualised commodity to that masculine milieu.

Alice Trilogy (2005), first staged at London's Royal Court with Juliet Stevenson in the lead, marked a triumphant return to form for Tom Murphy. The play follows on very naturally from *Bailegangaire* but also redresses the gender problems I have outlined in *The Wake* and *The House*. It explicitly draws on Beckett's later woman-centred plays like *Rockaby* and *Not I* while deftly rendering the changes across the last three decades in Irish society. What is most radical about the play is its form. *Alice Trilogy* is divided into three sections, each individually titled. The temptation to call them Acts is countered by the fact that there is nothing traditional about them; they represent more of a break with conventional dramaturgy than anything even the restless and experimental Murphy has earlier undertaken. The term 'scene' is preferable, given their relative condensation and brevity; it also acknowledges more deliberately the spatial setting in which they are enacted. The three scenes are each set in a separate decade: 'In the Apiary' in the 1980s; 'By the Gasworks Wall' in the 1990s; and 'At the Aiport' in the 2000s, or '2005', as the text explicitly states, contemporaneous with the date of the play's first production. *Bailegangaire* was set in 1984 and hearkened back to Ireland's mythical past. *Alice Trilogy* is set in Ireland's present – the Celtic Tiger of the millennium years – and could scarcely have found a more appropriate setting for that era than the gleaming, booming airport in which the 2005 section is set. But the play refuses to endorse the cultural amnesia that appeared to accompany the Celtic Tiger phenomenon. By so carefully historicising each of its three scenes in terms of the decades in which they are set, the play is always anticipating the future while attending to the present. Its open ending, so marked a feature of Murphy's play-writing, anticipates further scenes to follow in the sequence of a life.

Where *Bailegangaire* features three separate female characters, linked by family but divided by their personal histories, *Alice Trilogy* features the same woman – the eponymous Alice – in three interlinked plays at three different periods of her life. The actress playing the part has to convincingly cover an age range from late twenties to late forties. Certain constants remain across the span of the three decades: marriage (obviously at a young age) to her husband, Bill, who has a reliable job in the bank; motherhood, with three children already in place by her late twenties, two daughters with whom she has a fraught relationship and her adored youngest son, William. But since the play is not naturalistically staged, these constants account for less than they might by way of providing continuity. The three children never appear, and Bill is only physically present in the last scene. Each of the three sections might well be a different play with a different character: each has a completely different setting and a different character with whom Alice engages; each also gets to grips with the particular (and very different concerns) of each decade in Irish society. But *Alice Trilogy* is not just (or even) three one-act plays loosely strung together as an ostensible unity. It is centred in the psyche of Alice and her constant efforts to make sense of the situations in which she finds herself, but to do so on her own terms. The result of that quest is a necessary fragmentation of form and language for which the outstanding dramatic precursor is Beckett.

Murphy's settings, as has been observed earlier in the chapter, have been either realistic – the pub in *Conversations on a Homecoming* – or overtly symbolic – the church in *The Sanctuary Lamp*. From the beginning of his career, he has never been comfortable in either of these types of settings and an important element of the unfolding dramatic action is an attack on the set and its properties – of Michael's house by his brothers in *Whistle in the Dark* and of the church's confession box and pulpit by the strongman Harry in *Sanctuary Lamp*. In the course of Murphy's career, the realistic/symbolic distinction has increasingly broken down, with JPW King's office at several removes from the everyday and Mommo's house a bizarre version of the peasant cottage. *Alice Trilogy* takes that necessary distancing from the everyday one step further – or alternatively renders the realistic abstract. The stage space is largely bare, with the exception of '*a few objects of broken furniture*'[42] and a mirror, and largely in darkness, cut across by two shafts of light. Alice enters '*soft-shoed; silently*' and moves to the centre of the room; stopping, she returns and closes the door, shutting off one of the two sources of light. Like the pacing May in Beckett's *Footfalls*, she looks neither to right nor left and moves as if by routine, memorably expressed in her creator's words as '*like a rat on a familiar run that believes itself to be unobserved*'. We only faintly register that she is in the attic of her home, retreating there for 'this holy half-hour, she is sitting quiet, here upstairs; in the attic room, my dear, [...] stocktaking her assets, mental and material' (14). For what the stage primarily represents is a space apart, a psychic space, in which the predominant elements are non-specific. The few objects in the room are both stage properties and symbolic objects, none more so than the mirror. The visual area is reduced severely by the prevailing grey and concentrated on the one shaft of light bearing down on Alice.

But it is the play's extraordinary acoustic which most deserves comment. The area of sound design has seen as much development and experiment in stage practice in the past decade as the area of the visual has for over a century. And while sound design has become an important element of contemporary Irish theatre practice, it is rarely called for by the playwrights themselves. Murphy in *Alice Trilogy* is the exception (so far). The first scene is accompanied by a soundtrack of '*continuous sound, not very loud, ascended from below*' (3). Though Murphy identifies it as the sound of a radio, he also indicates that it is '*hard to know what it is*'. It is soon joined by a second continuous sound, with a distinctive '*thump-thump*' rhythm – a washing machine, though like the first the sound is bizarre, muffled, distorted, defamiliarised. The third is '*something that sounds like the chirrup of a bird*' – more a symbolic identifier to Alice in her cage than an avian actuality. Collectively, the noise represents the world outside the room. Psychically, it registers as acoustic pressure on Alice, what Mouth in Beckett's *Not I* calls 'the buzzing... so-called... in the ears' (377).

Alice's first words develop the Beckettian associations of her entrance by referring to how others would view her, as not quite right in the head: 'Anybody home? (*Taps her head.*) No one at home. Good [...] Out to lunch. (*Meaning 'crazy', as earlier with 'No one at home'*) (3–4). These opening lines echo the comments made by W in *Rockaby* late in her monologue about how she is regarded and spoken about by the normative community: 'off her head they said/gone off her head/but harmless/no harm in her' (440). Out of her loneliness and psychic need, W is searching for 'another like herself/a little like' (436) and failing to find another human being for genuine contact and communion has instead become 'her own other/own other living soul' (440). Murphy's Alice has done something similar, conjuring a female alter ego in the shape of Al (the contraction of her name indicates the psychic affinity and the dependence). Appropriately, given that there are several quotations from Lewis Carroll's two Alice books in the play, this version of Alice emerges from the looking-glass: '*a figure emerges gradually from, the darkness behind her, a young woman like herself, to stand framed in the mirror and eventually to step out of it*' (4). A central concern of this study has been the male double-act, so it is only appropriate and fitting that in successive years, 2005 and 2006, Tom Murphy in *Alice Trilogy* and Marina Carr in *Woman and Scarecrow* (to be discussed in the last chapter), should virtually simultaneously come up with the female double-act. While it serves many of the same dramatic and psychic needs as the former, a significant and crucial distinction is that the dialogue between the female character and her alter ego is carried out almost exclusively between the two selves rather than in relation to other characters. In the Murphy play, the alter ego only appears in the first of the three scenes. But her presence lingers on in the other two and reminds us that what we are seeing represented on stage is the central character's psychic reality, that we are simultaneously both inside and out of her head. In that first scene, no other character is present. And the psychic character of the scene is enhanced by Al's emergence from a mirror. For as Elin Diamond has remarked when writing of Beckett's women-centred plays: 'Mirrors are long associated with narcissistic women, objects of masculine disgust, and possibly envy. But the mirror is also crucial in Lacanian psychoanalysis as that which inaugurates the ego as other.'[43] Lacan identifies what is central to the mirror-stage: 'the subject makes himself an object by displaying himself before the mirror'.[44] He indicates that the moment at which the mirror-stage comes to an end is inaugurated by 'the dialectic that will link the I to socially elaborated situations'.[45] But there is a crucial gendered dimension to successful negotiation of this stage if the individual is to gain access to the symbolic order of language. Given the phallogocentric nature of language, that link or transit for a woman is neither straightforward nor unproblematic.

Beckett's Mouth in *Not I* recalls those few 'socially elaborated situations' in which she was required to speak; at the checkout stand in the supermarket or 'that time in court' when she was adjoined by a male interlocutor: 'what had she to say for herself ... guilty or not guilty ... stand up woman ... speak up woman' (381). Alice and Al, the character and her mirror surrogate self, form a compact, a necessary resistance to normative discourse. In that debate between the two Alices, both refer to her in the third person:

AL: It's just that she's upset.
ALICE: At the *moment* she's upset.

(5)

This refusal of the 'I' places Alice in the company of Mouth as she repeatedly insists: 'what? ... who? ... no? ... SHE!' (382). This pronominal shift has already been discussed in relation to Murphy's *Bailegangaire*. But the effect here is almost precisely the reverse. Where Mommo has needed to lessen the distance between herself and her narrative history, to forge a psychological and affective link with the past and its trauma, Alice needs to distance herself from the material history of her own life, to view it objectively in gendered terms. Much of the dialogue between Alice and Al takes an interrogative form. The questions Al asks are linked not to the supermarket or to the courts but to the conventions of the radio talk or 'chat' show, a parody of what is being broadcast downstairs. Such programmes had come to the fore in 1980s Ireland as a medium through which women confined during the day to the house or 'home', and still called housewives, could air their problems to a sympathetic male DJ. But though they broadened the range of subjects which could be discussed in public – particularly in relation to women's reproductive issues – these male-female exchanges were still severely contained and censored. Much of what Alice has to say to Al pertains to her husband Bill and could be seen as fitting in to the radio interview genre with the woman protesting 'my husband doesn't understand me'. But when Alice comes out with 'I do not like the Pope, for instance' (8), the remark is far from random. Pope John Paul II had visited Ireland at the end of the previous decade to that represented in this scene. He was given what can only be described as a rock star's welcome, with over 250,000 people attending the open air mass he celebrated in the Phoenix Park. In Ireland of the 1980s, you could not say a bad word about the Pope, on the public airwaves, at any rate:

> ALICE: Penchant for skull caps and kissing the ground?
> AL: Yet he won't look at a woman or wear a condom.
>
> (9)

The pairing of Pope John Paul II and Alice's husband Bill in these exchanges indicates that what is being targeted, questioned, laughed at, is neither the individual male *per se* but patriarchal discourse. As has been widely recognised in feminist analysis of Beckettt's *Not I*, the disavowal of the first person singular and the insistence on 'she' signals a refusal by the woman – Mouth or Alice – to be aligned seamlessly with that discourse. This chapter has stressed throughout Murphy's attack on the norms of language, his fragmentation of English syntax and insertion of non-standard dialect and Irish language terms. That attack aligns itself with *écriture feminine* as a writing against the grain, and underlines Alice's moving declaration late in the scene about what its verbal critique and visual defamiliarising gesture at:

> There's a strange, savage, beautiful and mysterious country inside me. Otherwise, well, give me – a bucking bronco to deal with, because this is slow death.
>
> (23)

In *The Irish Times* Theatre Awards of 2007, *Alice Trilogy* was awarded the Best New Play of the Year award, for the production at the Peacock with Jane Brennan in the lead and directed by the playwright himself. Like Friel before him, and Kilroy after, Murphy as playwright has responded to the Ireland of recent decades through an explicit engagement with Beckett. If all of Beckett's plays were being staged in Irish productions by Michael Colgan at the Gate Theatre from 1991 on, his dramatic legacy was also being picked up and transmitted by his great contemporary equivalents.

4
Kilroy's Doubles

Thomas Kilroy was born in Callan, Co. Kilkenny, on 23 September 1934. The son of a policeman, he was educated by the Christian Brothers in Callan, at St Kieran's College in Kilkenny, and University College Dublin (UCD), where he read English. After several years as a teacher and headmaster in a Dublin school, he received an MA and began lecturing at UCD; he also began writing critical essays for the Jesuit journal, *Studies*. Much of his teaching concentrated on eighteenth-century Anglo-Irish playwrights, much of his criticism on the state of contemporary Irish theatre. During these years, Kilroy began to achieve success as a dramatist in his own right. His comedy *The Death and Resurrection of Mr Roche* was staged at the Dublin Theatre Festival in 1968; his historical drama *The O'Neill* played at the Peacock in 1969. In 1971, Kilroy's novel *The Big Chapel*, based on his Kilkenny background, was published and won several awards. In 1973, he resigned from university teaching to devote more time to writing. The troubled personal period which followed is reflected in his 1976 Abbey play *Tea and Sex and Shakespeare*, in its absurdist, satiric portrayal of writer's block and a riven marriage. *Talbot's Box*, Kilroy's Beckettian dramatisation of the life of the Dublin working-class mystic Matt Talbot, followed in 1977 and provided an early directing triumph for Patrick Mason at the Peacock.

In the 1980s, Thomas Kilroy's play writing was closely involved with some of the most innovative theatrical companies in Ireland and England. For London's Royal Court Theatre in 1981, he provided a version of Chekhov's *The Seagull* which daringly transposed the characters and setting from Russia to the west of Ireland in the 1890s. In 1986, he offered Derry's Field Day Theatre Company his play *Double Cross*, and actor Stephen Rea a tour-de-force in the double role at its centre.

He contributed to another of Field Day's annual productions in 1991 with *The Madame MacAdam Travelling Theatre*, a play about the touring fit-up companies of the 1940s. Kilroy was appointed a director of Field Day in 1988, the only Southerner on the board, and served for four years. He has also worked closely with younger companies on new productions of his earlier plays, notably with Dublin's Rough Magic in 1988 on *Tea and Sex and Shakespeare*, and with Waterford's Red Kettle on *Talbot's Box* the following year. 1988 also saw the staging of another Kilroy adaptation, this time of Ibsen's *Ghosts* from 1880s Scandinavia to 1980s Dublin and from syphilis to AIDS. During this period, he was also Professor of English at the National University of Ireland, Galway.

After his Field Day sojourn, Thomas Kilroy returned to Ireland's National Theatre with two major plays on the main stage, 1997's *The Secret Fall of Constance Wilde*, directed by then-Artistic Director Patrick Mason for that year's Dublin Theatre Festival, and 2003's *The Shape of Metal*, directed by Lynne Parker. Between the two, he wrote a play entitled *Blake* about the Romantic painter, poet and visionary, which so far remains unstaged. Throughout his career, Kilroy has displayed an obsession with the man of vision, a socially marginalised figure, frequently an artist; and the fertile theatrical experimentation is deeply linked to the challenge of representing those visions on stage. That concern continues in these three plays; but alongside it, and receiving an ever-greater emphasis, is the human cost of such a pursuit. As the title of *The Secret Fall of Constance Wilde* indicates, Oscar is not the centre of this particular drama, however much he would like to be. Wilde and Lord Alfred Douglas ('Bosie') are displaced by the emphasis on the usually overlooked figure of his wife Constance, whose tragedy takes centre stage. Most of *Blake* is set in an asylum, where the incarcerated artist is being scrutinised by the authorities. But the most sustained scrutiny is being directed at him by his wife, Catherine. When I put this to the playwright, he agreed, remarking; 'He is tested by the presence of this woman, who was a remarkable woman – illiterate, but at the same time somebody who has access to his vision in a way that nobody else has. She challenges the righteousness of his religion and the righteousness of his sense of personal virtue and in a way just reduces him down to the human.'[1] In both of these plays, the woman performs a traditional role in relation to the male artist, a combination of helpmeet, mother and muse. But in *The Shape of Metal*, Kilroy crosses the gender divide in attributing artistic agency to a woman, the 82-year-old sculptor Nell Jeffrey. Her fraught relations with her two daughters, Grace and Judith, are brought into fruitful interchange with the artistic process she enacts on stage.

What is so striking and unusual about Thomas Kilroy is that he has operated throughout his writing career as both practising playwright and academic critic. In an essay entitled 'The Irish Writer: Self and Society, 1950–1980', he offered as the paradigm of the writer someone

> who spans several enterprises, the imaginative writer who is also an intellectual, deeply concerned with both the life of the mind and the life as it is lived on the street, in the marketplace, in the institutions of social and political power and in the confrontation with this world through ideas.[2]

Written to describe Sean O'Faolain, these lines serve equally to illuminate Kilroy's concerns as a writer. For what distinguishes his still-evolving career is his own ability to span several enterprises. By bringing together the usually exclusive areas of academic criticism and creative writing, Kilroy extends the boundaries of the creative *and* the critical, admitting the intellectual and the imaginative to mutual influence. Kilroy's writing in both domains is drawn to the medium of drama and to the life of the Irish mind, grappling with a post-colonial Anglo-Irish past and looking towards a European future. Critically he takes on the complex issues of identity and culture raised by Irish achievement in drama. Kilroy's scrutiny extends all the way from the Anglo-Irish playwrights, from Farquhar to Wilde and Shaw, who first expatriated and then reinvented themselves, through the problematic relation to modernism of the Irish dramatic revival, to the isolation of the contemporary Irish playwright from a creative working relationship with live theatre.

This last concern was raised in one of Kilroy's very first essays, written when he was 25 and entitled 'Groundwork for an Irish Theatre': 'we should experiment with a theatre deliberately geared to attract back the writer and provide a workshop for him'.[3] The essay took a trenchant look at the then-contemporary (1959) scene and criticised Irish playwrights for a lack of technical excitement in their writing. But Kilroy went on to link this technical deficiency with their shirking 'the painful, sometimes tragic problem of a modern Ireland which is undergoing considerable social and ideological stress'.[4] The expected reply came, from a director of the National Theatre, for Kilroy to practise what he was preaching. Unlike most critics, he went on to do just that. The results did not show immediately, since writing for Kilroy has always been a slow, painful process requiring a long gestation period and frequent rewriting. But answer there came, first with *The Death and Resurrection of Mr Roche* and then *The O'Neill*. These two plays began the creative career of one of Ireland's most technically adventurous writers, but one whose experimentation is never a mere arid exercise in its own right; rather, Kilroy's restless innovations are always in the service of articulating the troubling vision of a modern Ireland undergoing ever greater social and ideological stress.

Ever since Stephen Dedalus, and by extension James Joyce, resisted a religious vocation and undertook instead to become priests of the eternal imagination, many Irish writers have followed in a tradition of pressing Christian tropes to serve the ends of their secular arts. Kilroy's first play, *The Death and Resurrection of Mr Roche*, points in its own title to the central mystery of Christianity, that final death-defying act by which all of Christ's earlier claims to deity were put to the test. The greatest instance of the death-and-resurrection motif in Irish drama is of course Synge's *The Playboy of the Western World*. And Beckett's plays are increasingly marked by characters who inhabit a zone or condition beyond the grave. Irish playwrights have continually drawn on Christ's death and resurrection as a motif, even in such ostensibly naturalistic plays as Kilroy's *Mr Roche*. Indeed, the return of the dead to haunt the living, their refusal to stay dead, is the means by which the fantastic is admitted to the predominantly naturalistic Irish stage, an influence which, once admitted, works to transform the nature of that space.

In Kilroy's play, the sudden death of the title character proves as unexpected a development to the after-hours revellers in their Dublin basement flat and the apparently naturalistic drama within which they are situated, as it does to the audience. Even more, Mr Roche's unexpected return from the dead in Act Two shatters the characters' and the play's facade of daytime normality. Initially, no more has been intended by the play's central character Kelly and his drinking partners than a little friendly horseplay at the expense of their belated and unwelcome party guest. But all pretence at play is increasingly abandoned and the high jinks turn homicidal when Mr Roche verbally assails their manhood and they collectively thrust him into the claustrophobic darkness of the curiously named 'holy-hole'. Act One concludes with the apparently fatal arrest of Mr Roche's heart, an organ whose motions are much derided by the other characters but on whose progress that of the play ultimately depends:

MEDICAL STUDENT: He's dead [...]
KELLY: What the hell are you talking about? [...]

MEDICAL STUDENT:	I'm telling you. He's a goner [...]
KELLY:	Are you sure?
MEDICAL STUDENT:	Look – of course, I'm sure. I'm looking at these every morning, noon and night.
KELLY:	Oh, God! Oh, God! Oh, God![5]

After a long dark night of the soul for Kelly, pressured by this fatal turn of events to reflect on what he has made of his life, the two characters who have hauled off the corpse to attempt its disposal return as a threesome with the dead character not only very much alive but dominating the proceedings. Kelly's shocked single expletive 'Jesus!' at the sight of a revived and triumphant Roche prompts the reply, 'Not quite, my dear chap, but I am flattered by your mistake' (71). The play filters the quasi-religious themes associated with death and resurrection through just such layers of verbal irony, rarely failing to exploit the doubleness of the colloquial Dublin idiom by converting its most demotic utterances to unwitting spiritual diagnosis: 'get Mr Roche to hell out of here'; 'Sure Dublin is going to hell altogether' and the all-purpose 'Jesus!' which can attach itself to the most likely candidate present, as here.

The rational loophole through which such an outrageous resurrection can be admitted to the play is provided by the Medical Student 'Doc's' questionable expertise. His title, like O'Casey's Captain Boyle, turns out to be purely honorific; as he himself explains he never passed his final examinations and is clearly ill-equipped to adjudge the point at which the merely physical passes over into the metaphysical. It is left to Mr Roche to account for what he has experienced in a soliloquy which the play strains to accommodate and which passes right over the heads of his immediate listeners:

> And – it [the sun] – came! Like the beginning of life again. A great white egg at the foot of the sky. Breaking up into light. Breaking up into life. Consider the mystery of it! [...] Breaking up over the roof-tops into particles of silver and gold. And the streets opened up before it. And each tree yawned and shook, the leaves splintering. I was witness to it. [*Pause. Quietly.*] Then the clock began again.
>
> (74–5)

Like a phoenix, Mr Roche has risen from the ashes of the previous night's debauch to a personal apocalypse of visual splendour, a new dawn which contrasts vividly with the prevailing greyness to which the others awaken.

What has drawn their hostility at the outset, the social crime for which Mr Roche is arraigned by this kangaroo court, is his homosexuality. It is a word which Kelly can scarcely bring himself to name or to own and which drives him repeatedly into such linguistic evasions as the impersonal pronoun and the loose generality of 'perversions':

> You don't have to deal with it. It's easy for you to talk [...] I'm just saying I don't fancy the company. Can't I say that in my own flat, can't I? Bejay I'm as liberal as the next man, but I draw the line at perversions.
>
> (36)

The other party to the baiting is Myles, the aggressively heterosexual used-car salesman whose romantic accomplishments rarely go beyond verbal boasting and take their colouring and contours from the more baroque Hollywood melodramas. Myles plays up and displays the stereotyped gesture and phrase as much in his taunting of Mr Roche as in his own amatory pursuits: 'That's right, the young lad. The fair boy of your dreams. [*In a loud whisper.*] What age d'you like them at, Agatha?' (35).

Kilroy's play charts the ironic process by which the outsider who is the butt of the established group's jibes also serves to expose the shortcomings they are so anxious to conceal. Mr Roche presents them with the threat, not just of homosexuality but of sexuality *per se*, notably absent from the lives of a group of bachelors pushing forty, and with a sexuality of mutual concern rather than the battering-ram bluster of Myles.

But the character which is laid bare and anatomised by the presence of Mr Roche is that of Kelly. Kilroy's play took a resolutely contemporary look at the Ireland of the 1960s, not only through tackling the then taboo subject of homosexuality, but through its analysis of the character of Kelly. Come from the country to be a civil servant in Dublin, Kelly is renting a space in the basement of a Georgian house. As this series of symbolic displacements reveals, he is trying to live up to a variety of conflicting images and is uncertain in all of them. The past he has suppressed is not just personal but social and is addressed in terms of sexuality. Kelly is the one who relies most completely for a sense of self on their Saturday night/Sunday morning ritual of drinking themselves blind to their condition. His wildly disproportionate reaction to the newcomer makes it clear that Kelly has something to hide, never more so than when his tirade breaks off before it can reach its announced climax:

> Get him away, away to hell. Dirty, filthy pervert. If you won't do anything about it, I'm telling you I will. I'm telling you that now.

> [*He exits to bedroom where at first he stands indecisive and nervous. Later, he bends to look through and listen at the keyhole or walks about the room.*]
>
> (28)

Having made the suggestion to thrust Roche into the holy-hole, Kelly's guilt at what ensues brings to light the associated guilt of his former knowledge of Mr Roche in the full, biblical sense. With his reluctant admission to his married friend Seamus of a homosexual encounter, the final frame in Kelly's elaborately constructed social facade gives way and he undergoes a personal annihilation akin to death:

> KELLY [*Intoning. Dead voice*]: Something to do – with the way the ground steals the heat outta the air. Walking through a field now at this hour you'd feel the clay taking the heat outta your feet.
>
> (64)

But at roughly the same time, as we learn later, out in those fields the rain is pouring down on the exposed corpse of Mr Roche and reviving it. As Kelly emerges from the closet, Mr Roche, representing that aspect of Kelly which the latter has always sought to deny, simultaneously emerges from a deathlike state with his *alter ego's* confession. He returns as the embodiment of that which Kelly has repressed and ends the play firmly seated in the centre of Kelly's flat, answering his telephone. This is a double act which would be treated more fully in Kilroy's later *Double Cross*. Here, Kelly is still trying to evade the truth about himself by integrating with the group, to derive a sense of identity by running with the pack. In terms of either personal or national politics, this has been less and less a credible option in the years since the play was first produced. Although for much of its action Mr Roche has been verbally and physically abused, it is finally the straight bourgeois characters who stand accused by him in the closing lines as 'the living dead' (80): the Doc, so long an alcoholic morgue attendant he can no longer distinguish the living from the dead; Seamus, who engages as little with the evening's events as he does with his wife and the marriage he is temporarily fleeing; Myles, the self-proclaimed stud, who rushes home in time to appear before his deaf mother at the breakfast table; and Kelly, who paradoxically seeks to escape the implications of latent homosexuality in the arms of boisterous male camaraderie.

Thomas Kilroy has declared one of his obsessions as a writer to be (like Beckett) with failure. But in Kilroy's case there is a historical and cultural concern with what he perceives as 'a failure to achieve a wholeness of community in the Irish experience'.[6] Accordingly, in *The O'Neill*, he goes back to the sixteenth century – as Brian Friel was later to do in *Making History* – to get at the historical roots of this dividedness in the Irish experience. The play examines those conditions that contribute to a lack of community through the emblematic experience of Hugh O'Neill. The question that animates Kilroy's play is how this figure, the Earl of Tyrone 'called The O'Neill by the Irish in their ancient fashion',[7] as the Secretary of State remarks, could have led a rebellious Irish Army to victory against the English at the Battle of the Yellow Ford in 1598 when he had been expatriated and bred up to a civilised ideal at the court of Queen Elizabeth. The play instigates a retrospective enquiry into what caused the change of loyalties. It focuses on the complex nature of Hugh O'Neill as Anglicised native, as the locus of two conflicting ideologies of Irishness and Englishness.[8] As in so much of Kilroy's work the vision is double, shifting from the more obvious external conflict between the individual and the demands of his society, represented here by Hugh being back among his family and his tribe, to the conflicts within that individual.

In Act One the tension in O'Neill between the old Gaelic world into which he was born and the new English world in which he was raised hinges on issues of gender and sexuality. The dramatic emphasis is on Mabel Bagenal, young, beautiful and English, whom he desires. What follows between them is more than a conventional love scene. The strength of desire that Hugh articulates not only breaks with propriety but is the spur to drive him beyond all inherited political and family loyalties: 'I love her with a violence that is like constant anger' (28). The attraction is reciprocal: O'Neill is longing for the 'good order and civil security' (25) of English society, Mabel for the 'romance' of Ireland. But his decision to make her his third wife and to impose her on his people has the opposite effect from integrating the disparate halves of his identity; rather, it drives them apart.

In Act Two, which follows the victory of the Yellow Ford with the overwhelming defeat at Kinsale, the conflict shifts to the political and the military. The dramatic interplay of Anglo-Irishness is developed between Hugh O'Neill and Lord Mountjoy, the man sent over by the English to deal with the situation. O'Neill has come to the recognition that he is his people's servant and that they own him now. Accordingly, he leads the Irish into a first encounter which sees him defeating the Bagenals, his wife's people, and gradually realising that he is sowing the seeds of war, helping to prepare a political situation that will be passed on for centuries. It is the double perspective Hugh O'Neill is able to take, a detached view of historical circumstance, which alienates and estranges him from those he leads. It also unites him with Mountjoy. The *doppelgänger* or double motif between the Irish chieftain and the English lord emerges in Act Two through its symmetries of scene and speech. Hugh strives to delay the headlong Irish impetuousness and to deflect the inevitable bloodbath, but without success. The concepts of victory and defeat have no real meaning for either Mountjoy or O'Neill. Each of them on their opposing sides is progressively isolated on the stage, stalking back and forth, trapped in a tragic recognition of the forces released by the conflict between the Irish and the English: 'savagery and fear and greed' (61). After the Battle of Kinsale, Mountjoy is anxious to be away from 'this diseased climate' (66). O'Neill is confronted with the headless corpse of his own son, which he fails to recognise.

The defeat at Kinsale sees the old Gaelic laws replaced and the Catholic religion outlawed. The effects are dramatised in a third Anglo-Irish relationship which succeeds those with Mabel Bagenal and Lord Mountjoy. It is between O'Neill and Queen Elizabeth I. His exchange with Elizabeth is the personal shadow-play in which Hugh has been engaged all along, 'a kind of fierce kinship' (60) through which he has sought to realise himself, a dynamic of Yeatsian opposition. The true sign of O'Neill's defeat occurs when the relinquishing of his power is accompanied by the news that the Queen is dead. He retreats into a dream state and the play ends with the personal and national tragedy of a split between abject political victimisation and the consoling nostalgia of a romanticised past.

Thomas Kilroy won an initial following with these two plays of the late 1960s. Their full radical implications were somewhat concealed by the efforts of both protagonists, Kelly in *Mr Roche* and Hugh O'Neill, to identify with the group. The characters' ultimate failure to do so, and the harrowing existential crisis they subsequently undergo, points the way to the developments of the later plays. There, the central figure is already an outsider, an eccentric from the point of view of those around him. The dramatic emphasis falls on those isolated figures and their interrogative impact on the norms of the surrounding society. The concern is twofold in terms of dramatic conflict: on the defensive violence provoked by such a character in the society to which he has become unassimilable; and on the fragmented identity which such a character owns, the series of masks by which he conceals the hollowness at the centre of his being.

Kilroy wanted to treat of the conflict between individual and society, and the essentially artificial nature of the self. To do so required a much greater degree of theatrical freedom than the conventions of naturalism would allow. In *Tea and Sex and Shakespeare*, the explicit subject of writer's block paradoxically generated the greatest range of theatrical effects in Kilroy's work so far and paved the way for *Talbot's Box* in the following year. In their open theatricality, both plays stand in marked contrast to the narrow, stifling realism of much Irish theatre. They provide a significant bridge between the early ground-breaking plays of Brian Friel (Kilroy acknowledged the impact of Friel's *Philadelphia, Here I Come!*, as we have seen) and the 1980s collaboration on an Irish 'theatre of images' between playwright Tom MacIntyre, director Patrick Mason and actor Tom Hickey in such works as *The Great Hunger*. In Kilroy's two 1960s plays, there is a sense of both the central characters and the playwright struggling to free themselves from naturalistic conventions, a sense of these forms as neither relevant nor adequate to convey the experience of a fragmented, chaotic existence. After a break of seven years, Kilroy returned with *Tea and Sex and Shakespeare*, a declaration of theatrical independence proclaimed through the subversive slogan of the title, which follows up the reassuring offer of tea with the explicit offer of sex before bringing on Shakespeare to mediate an impossible situation.

From its opening image of a mock-hanging, *Tea and Sex and Shakespeare* declares open season on the norms of naturalism and proceeds instead to pull rabbits out of hats – or surrealistically dressed characters out of a cupboard or wall. Declan Hughes directed the play in 1988 for Dublin's Rough Magic Theatre Company, one of the leading independents (of which he was an artistic director). He said that the company was tempted to go much further visually with the playfulness, but they realised that 'we had to give the audience a place to return to and one rooted individual that we try and follow through this storm of craziness'.[9] We do, therefore, have a base in reality through the character of Brien and the play's central situation: what Hugh Kenner in relation to Beckett has described as the impasse of 'a man alone in a room, writing'.[10] Brien has attempted to shut his door firmly against a pressing world and the social forces threatening his freedom of expression: the nine-to-five world of the office in which wife Elmina works; the bourgeois phalanx of Daddy's rugby, Mummy's niceness, and the tea-drinking ritual. But Brien only slams shut one door to open another: the Pandora's (theatrical) box of a large on-stage cupboard out of which pop surrealistic versions of all the forces against which the beleaguered writer is trying to secure himself, not only from without but from within. As Brien and the play ask: 'Am I inside or outside or what?'[11] An early example of this is the theatrical illusion which has neighbour Sylvester going into his room on the right only to emerge from the cupboard on the left dressed as a secret agent, finally to re-emerge from his room in even more outrageously colourful 'realistic' gear. While the audience is held by this theatrical triple take, Brien shouts frantically for his misplaced biro to write it all down. The act he is seeking to perform requires not mere writing but, like the play itself, a whole range of theatrical representation: mime, costume, music and stylised imagery.

As always in Kilroy, the crucial issue of identity is bound up with the struggle to find an appropriate theatrical form. Brien in comic desperation tries to deal with his married life with Elmina and the various guilts it has engendered, with the intimate connection between his literary and sexual impotence. His creator strives to realise on-stage the phantasmagoria of one man's experience in a dramatically arresting way. Brien finds out he cannot do it all on his own. The landlady Mrs O and her nubile daughter Deirdre are summoned repeatedly, against his better instincts and the kind of high-toned play he would prefer to write, to drag Brien back down to the demotic. A faulty telephone is employed to try to contact God, who is not returning calls, and an indignant Shakespeare, who is.

Part of Brien's neurosis, and a reason for Shakespeare's presence in the title, is the attempt to keep rewriting his own life as a version of *Othello*. He casts himself in the role of the jealous husband, with Elmina as his Desdemona, and the Anglo-Irish Sylvester as a psychologically and culturally intimidating Iago. But if Brien convinces himself he is inhabiting *Othello*, the play itself pushes in the direction of a late 'dark' comedy like *The Winter's Tale*, where the self-generated tragedy of a Leontes is reworked by confessedly artificial means into a comic resolution. The late Shakespearean play that seems even more relevant is the 'rough magic' of *The Tempest* with Brien as a Prospero who keeps losing his grip. The spirits of the island, the ghostly presences Brien has theatrically conjured up and which haunt him throughout the play, take increasing control of his psyche. Finally they force from him the secret hurt at the heart of his imaginings and can only be exorcised in a final gust of purgative laughter. The ending of *Tea and Sex and Shakespeare* draws audaciously on Wilde's *The Importance of Being Earnest* and its missing handbag to discover a baby over which Brien weeps. The baby is transparently an artificial one, a dummy.

If the figure of the blocked writer was rather too close for comfort, Kilroy's choice of Matt Talbot in his other great theatrical experiment of the 1970s, *Talbot's Box*, provided a figure with whom it was difficult for him to identify: a Dublin working-class mystic who burdened himself with the trappings of devout Catholicism – chains, scapulars, daily Mass and devotions, fasting – to the point of eccentric extremism. In dramatising the life of such an antipathetic character, Kilroy challenged his own sympathies and simultaneously chose a public figure who already held a recognised place in the affections of his Irish audience.

Talbot's Box begins in a morgue with the title character's corpse already laid out for burial, the rest of the cast assembled to perform the last rites. The audience may well wonder whether Matt Talbot, and the unfortunate actor condemned to play him, will be forced to maintain that posture all night and whether therefore Talbot's role in the drama is akin to that of Beckett's Godot, a vacant centre around whom hypotheses are spun. But the stage form adopted by the play is not narrowly mimetic. Rather, Kilroy frames the action with '*a huge box, occupying virtually the whole stage*',[12] and develops this as a resonant, flexible metaphor. The box suggests in turn a coffin, a confession box, a witness stand, a wooden bulwark constructed by Talbot the carpenter against the encroaching chaos and Tom Kilroy's own box of theatrical tricks, the props and stratagems of the playwright's trade openly on display.

Beyond the symbolic setting, a further indication that Matt Talbot will not be confined exclusively to his role as corpse is given when an actress representing a statue of the Virgin Mary breaks up her 'pious pose' to ask: 'How long do I have to stand like this?' (11). The answer is not long in coming, since each actor apart from Talbot is called on to play many parts. The doubling up of roles, usually a fact of economic necessity, is acknowledged and utilised within the stage narrative itself for several reasons: metamorphosis is one of the chief properties of drama and so it is entirely 'natural' within its privileged frame that people should assume a number of identities; there is also a social truth to this endless role playing and role changing about which the play has some satiric points to make; and finally it admits to the deep kinship between the stage and the act of dreaming whereby the normal boundaries that separate and define human beings no longer operate and one identity blurs readily into another.

The play from the start is Talbot's dream, in his long sleep of death, and the figures who gather about him are in part the fevered, accusatory fragments of his past. Prepared as we are both by the innovative staging, and by the amusing but pertinent gag of the protesting statue, we realise it can only be a matter of time before we witness Talbot's resurrection, an existential impossibility overcome and represented by the physical act of standing up. When Talbot eventually does so, struggling against the gravitational weight of the chains encumbering him, the phoenix-like quality of the event is suggested by the light he generates, lacerating and painful for the spectators to contemplate:

> *With a sudden, startling energy, he rises on the trolley and flings both arms out in the shape of crucifixion. As he does so, blinding beams of light shoot through the walls of the box, pooling about him and leaving the rest of the stage in darkness.*
>
> (19)

This apotheosis does not last long. The four figures on-stage, unable to endure the illumination, block the apertures of the box. With the dying of his light, Talbot '*sinks to the trolley, and kneels upon it*'.

In a series of retrospective scenes Talbot struggles first with the lethal personal legacy of his father's drunkenness (reminiscent of O'Casey, but more hard-edged, less sentimental), then with the more equivocal family claims of brother and sister, and finally with a young chambermaid who has left her position to occupy another that Talbot could never be brought to share:

> WOMAN: D'ya never think of ... of having children of yer own? D'ya never think of having a home? [...]
>
> TALBOT: I think of them day 'n' night.
>
> WOMAN: Well, why don't you do something, so? Instead of always stuck in that bit of a room. How do you stick it in there?
>
> TALBOT: I've measured it. The length and the breadth of it. I fit into it.
>
> (51–2)

Like Emily Dickinson, Talbot's life, informed by a perpetual sense of death, is 'shaven, / and fitted to a frame'.[13]

Matt Talbot is subject not just to familial but to social pressures. In the trial-like situation which emerges to place him in the witness box, the charge most frequently levelled against him is that of strike-breaking. For one of the several specific time-periods with which the play aligns itself is 1913, the time of the Dublin Lock Out from which the Irish Labour Movement dates its origins. In the other great dramatic treatment of this event, O'Casey's *Red Roses For Me* (1943), the protagonist

Ayamonn Breydon apparently experiences no difficulties in transferring his Promethean private visions of the first half of the play to the arena of public dispute and his role as strike-leader in the second. But Matt Talbot's private mysticism cannot be so readily assimilated to the social protests of his fellow-workers. His stance is not so much radically opposed to the strike as occupying another plane entirely, so that the personal vision he follows inevitably brings him into conflict with the law operating between workers and employers. This conflict is memorably crystallised in a scene where, as the severely debilitated Talbot and another worker carry a plank between them in an endless circle, the foreman engages the latter in an argument over ideological differences while both contrive to overlook the plight of the individual before them – one man's comrade, the other man's employee – who is so evidently on the verge of physical collapse.

Talbot remains on his knees for most of the play while the stage is taken by a succession of admonishing voices. None of them owns a distinct personality since each is the mouthpiece of some group with a vested interest in co-opting Talbot to its own ends. What occurs is a series of physical assaults: first, the littering of the dazed Talbot with a pile of religious texts, then a full-scale assault on the box-set itself, as we sense but do not see the collective force of a large crowd pressing on its (and Talbot's) outer limits, warping and buckling the frame: '*there is pushing, scratching, beating against walls* [...] *then a great uproar and beating which threatens to demolish the great box*' (35). The assault is one of the most succinct paradigms of the pressures exerted on the solitary man of vision, be he mystic, artist or homosexual, by the force of external historical events – in this case, the Great Strike of 1913 – and an equally determined assault on radical individuality by society's various pressure groups – ecclesiastical, political, social – determined to convert that organic integrity into the inert specimen or example.

When the culprit becomes the scapegoat, he or she is literally singled out from the rest of society. What marks and marks out the central characters in Kilroy's plays is their isolation. This has its visual, emblematic dimension, as in the stage-description of Matt Talbot: '*The body is that of a frail old man* [...] *naked with the torso, arms, shoulders, and legs painted garishly with stripes of red and blue*' (13). Although he later dons his ordinary working clothes, nothing effaces these distinguishing marks from Talbot in the course of the play. All the other people on stage, as noted earlier, indulge in an orgy of role-changing, so that an existential morgue attendant becomes a captain of industry as a mere change of clothes suggests the relative ease with which social roles can be acquired or changed. The same facility occurs at the level of language. The drama makes great play of the inter-changeability of slogans, as in the following litany: 'He was a tool of the Church against the workers!', 'He was a scab! He was a scab!', 'He was irrelevant!', 'He was a saint!' (36). The net effect is of self-cancellation, not of Talbot's integrity, but of those who make the claims. In a garrulous play, Talbot says relatively little, often a mere monosyllabic yes or no. When he does launch into a rare climactic testimony to his sense of things, the language breaks out of a cage of single-minded slogans into a Blakean dialectic of reconciled opposites that levels all social distinctions:

> Blessed be the body,
> For its pain is the message o' the spirit.
> Blessed be the starvin' peoples of the earth,
> For they bring down the castles of the mighty.
> Blessed be the dung o' the world,
> For on it is built the City on the Hill!
>
> (37)

None of the nets of explanation or inquisition flung at Matt Talbot succeeds finally in capturing or defining him. This is a suitably Pyrrhic outcome for a play which examines its own processes and so must defend the integrity of the singular vision repeatedly put at risk by the necessary efforts to render it public as drama. He does not seek to fly from the various nets flung at him, as Stephen Dedalus had done. Talbot's movement towards self-realisation is not upward and outward, but inward and down, the fall into a darkness which yields the fortunate consummation. He sees clearly the violence in himself and others and sees it as proceeding from a 'terrible hunger' for what others might have. And so, in turning to look for what is missing inside himself, he learns to embrace the darkness: 'Tis because I wanta meet the darkness as meself [...] I think meself the darkness is Gawd' (47).

Kilroy's earlier plays ultimately endorse the singular vision of a Roche or a Talbot by setting it against the claims of a conformist society. Such radical individuality is a good deal more suspect, and its social implications gauged in a more political way, in *Double Cross*, Kilroy's best play to date. It was first produced in Derry on 13 February 1986 as the year's offering from Field Day before touring the country and travelling to the Royal Court in May. Of the play's two roughly contemporary Irish protagonists, Brendan Bracken and William Joyce, the first became Churchill's Minister of Information in World War II while Joyce adopted the persona of Lord Haw Haw in his notorious radio broadcasts from Nazi Germany. In Kilroy's theatrical presentation, both rely uniquely on the power and projection of their voices as a means of self-realisation. This lends an ironic appropriateness to their fates, one killed by cancer of the throat, the other hanged as a traitor. The issues raised by the play are at the centre of Field Day's debates, which my next chapter will develop: the relation between language and identity, the hazardous crossing-over of established boundaries, and the inevitability of betraying self and others.

Double Cross also lines up revealingly with Kilroy's previous plays. All engage with the notion of personal freedom, of resisting the pressures of social conformity and marking out a space of existential possibility. Kilroy has dramatised these issues through such socially displaced figures as the homosexual, the writer and the Catholic mystic. Part of that endeavour has been the choosing of an objective other identity to mediate the struggle for self-realisation. The process in *Double Cross*, with its two Anglo-Irish outsiders as British and German insiders, is even more theatrical. As Kilroy makes clear in his notes to the play, he perceives Bracken as an actor, Joyce as a writer, 'a creator of fictions'.[14]

Brendan Bracken's story occupies the first half. The venue for Bracken's self-presentation is 'the world of English politics and what used to be called Society' (6). The reference to Society recalls Lady Bracknell's memorable line in Wilde's *Earnest:* 'Never speak disrespectfully of Society, Algernon. Only people who can't get into it do that.'[15] English – specifically London – society has always proved, in Kilroy's words, 'susceptible to the charm of a master Thespian' (6). And so Bracken is represented as in the line of Anglo-Irishmen from Farquhar to Wilde who made their way by means of on- and off-stage theatricalisation to a position of influence in English society. The transfer involves the shedding of Irish identity and a complete re-invention of the self. As a result, the role is an actor's delight, since Bracken responds with a different face or rather voice to each encounter. In one long telephone scene he runs a gamut of identities, blithely switching tones, accents and biographies as a different speaker comes on the line. The question dramatised by the scenes with his lover Popsie is whether there is a stable, enduring identity behind all the protean impersonations. Bracken breaks down at the threshold of intimacy and instead seeks concealment in theatrics. When confronted with his Irish background, he asserts that 'it must be totally suppressed' (37). Only during a rooftop bombing does that suppressed Irish identity emerge: 'Me father was wan of the lads, so he was, wan of the hillside men. He took the oath. He was out in the tenants war of eighty-nine.

Bejasus I was' (43). At this stage we can see the larger point for Ireland as a post-colonial society that, because Bracken cannot live with or admit his past, its legacy of secrecy and betrayal, he can never escape it.

While Brendan Bracken plays the Anglo-Irish role of court jester to English society, William Joyce in Act Two (the same actor is to play both roles) seizes on his more absolute displacement and the medium of radio to delight in the possibilities of subversion at every level. As Lord Haw Haw, he takes pleasure in detonating verbal bombs in the primed imaginations of the British public. His private relations with his wife Margaret are more revealing than Bracken's with Popsie. The scene in which he alternately pleads with her for forgiveness and verbally abuses her for being 'intimate' with another man recalls the letters of the other Joyce to Nora. In both cases there is a master plot in which the women have their allotted ambivalent places. This Joyce finally scripts his own end by handing himself over. Preferring to submit to trial and execution, he perversely confirms his fabricated identity as a subject of the British crown.

Both characters dramatise a response to two historical crises. The first was post-revolutionary Ireland in which the ambivalences of Anglo-Irishness no longer remained possible. Yeats responded with the creation of an eighteenth-century Protestant Ascendancy; Bracken and Joyce moved to England and set about re-inventing themselves as Englishmen. The equivalent ambiguous strains of England in the 1930s resolved themselves with the outbreak of World War II into patriotic Toryism and German Fascism, with Bracken opting for one extreme and Joyce for the other. W. J. McCormack has written of how the year 1939 brings to crisis the concern of writers like Beckett and Francis Stuart with 'the possibilities of self-betrayal' and the broader issue of 'Ireland's identity and integrity *vis-à-vis* Europe at war'.[16]

Through Bracken and Joyce, Kilroy demonstrates the inevitability of self-betrayal in the course of any movement outwards towards a world of possibilities. Bracken speaks eloquently at one point about distance and space as freedom (28); but the play notes that the results may be either criminal or artistic. Even as art, the aestheticisation of that choice is a denial of history, of socially determining forces. And so each character becomes not a free individual who has shed his Irish past, but someone who has traded in the role of historical victim for the mirror-image of oppressor and placed all his faith in the symbols of the culturally dominant race. He has mastered those symbols, as signified in the fluency of his language, and become their perfect embodiment for demonstration through the media.

Kilroy's stage play began life on the radio as *That Man Bracken* (1986). At first, I thought it might best be served by radio, where it could become the play of contending voices it aspires to be. But to do so would be to capitulate to the mystification, the historical appeal of voice as presence, which Kilroy expends a full battery of theatrical effects on deconstructing and subjecting to critique. What the stage offers *Double Cross* through its inherent carnality is the possibility of a true doubleness which admits the other, the presence which both Bracken's and Joyce's endless subjective monologues seek to crowd out and deny. Kilroy's play was a major contribution to Field Day's agenda in urging the drama and politics of creative contradiction.

Five years later, Kilroy offered in *The Madame MacAdam Travelling Theatre* a play that was not so much a sequel to *Double Cross* as its parallel or counterpart. Also set during World War II, this play returned the scene to Ireland and more explicitly used theatrics and transvestism as a means of interrogating questions of identity. This is the Ireland discussed in my opening chapter, an Ireland which had reacted to a world war by declaring a state of emergency and, in so doing, had itself quarantined from what was happening in the rest of the world. This viewpoint is expressed by the Sergeant in Kilroy's play when addressing the travelling players from England:

> SERGEANT: [*explosion*] 'Course it's an emergency. What else would you call it with the whole country upside down with that carry-on over the water? Isn't Churchill trying to starve us out? [...] Anyways. It's yer war. If that's so, why aren't ye fighting it?[17]

But the play will not permit the kind of absolute separation the Sergeant is insisting on here, between Ireland and the rest of the world, and between an Ireland in a state of neutrality and an England at war. The division is undercut from the start by the unlocalised space of a dark stage and particularly by the acoustic '*Distant sound of a bomber approaching, passing overhead, passing away*' (1). Because if, as Madame MacAdam confidently asserts, they have arrived in independent Ireland ('a town called, I believe, Mullingar'), then what would bombers be doing flying over a neutral zone? Either the players are still very close to the border partitioning the South and the North of Ireland, or an intimidating military presence need pay scant attention to such distinctions and can fly over them with impunity. The sound that seems so historically precise serves to set free the play from one fixed place and moves it into a border zone of threat and uncertainty, as much a feature of the present as of the past There is also the ground recently traversed by the players. Madame MacAdam and her consort Lyle Jones are English, sufficiently so to give the Sergeant warrant to address and dismiss them collectively as 'ye'. Lyle gets to play all the lead parts in the Shakespearean or Romantic scenarios they perform, to wear the tinsel crown; Madame accompanies on the piano or as the woman wooed or fills in as the working-class serving girl. But off-stage it is Madame MacAdam who gives Lyle his directions, whose name fronts the troupe and who maintains its equilibrium.

During their career, they have crossed the Irish Sea, to Belfast, where they have acquired Sally, whose Northern accent and perspective are used to disturb any absolute distinction between the English players and the Irish locals. As she says to the Sergeant: 'Look. This lot came over from Liverpool to Belfast. We crossed the border. [...] I only wish I could get away home again' (29). Sally's fondness for Belfast as a 'home' itself puts in question received views of the place, and undermines the sense that a greater security is gained by crossing the border into the South. The other key member of the troupe is Rabe, from England, but Jewish, whose abiding memory is the burning of his father's shop by Blackshirts. The apocalyptic imagination that drives Rabe offers him no respite or hiding place in Ireland, and he has only found temporary refuge with the players.

The visit of the theatricals has a disturbing effect upon the local community. This is the central theatrical conceit of the play, which deploys overt theatrics to expose the codes constructing everyday social discourse. The most serious example of this cross-over between theatre and life is the manoeuvrings throughout of the Local Defence Force and of Bun Bourke, town baker and squad leader. When he is addressed as Bun, the reply is savage: 'Don't call me Bun. D'ya hear! In this uniform I'm not Bun Bourke. I'm Squad Leader Bourke. Be day I may be a baker but be night I'm officer in charge of this Squad' (2). In a play of sometimes varying effectiveness, there is a chillingly effective theatrical coup at the end of Part One, as we see projected across the back '*in silhouette, the figures of Bun Bourke and his LDF men but now in Nazi uniforms*' (49). The full context of the scene needs to be considered. The shadows of war, of Fascism and of Nazism, which the florid on-stage theatrics and the Irish setting keep in the background, here dominate the foreground and reverse the relationship. The shadows fall on Ireland, which is not immune. They fall into this scene on the naked, terrified figure of Rabe, who is pointing to the shadows and who is the only one who can see them, exaggerated and grotesque like the figures from his nightmare. When they appear, Rabe is in bed with the young local girl Jo in their travelling caravan; the 'real' Bun lurks outside, ready to expose them.

Jo is the most overlooked character in the play and, despite the fact that she wears heavy glasses, the most clear-sighted. This quirky, determined, comically serious young woman, teetering over from adolescence into adulthood, is the most distinctive creation in Kilroy's play. Jo's ability to intuit the future, to register the sense of impending apocalypse, links her with Rabe, sexually, and with Madame MacAdam, psychically. When the two young girls approach him, Rabe is attracted to the intense, independent Jo, rather than the conventional Marie Therese. The curtain of the van lifts later to show the couple '*lying naked on a makeshift bed*' (47) and Jo remarks: 'If they were to open up the curtain there now we could perform for all and sundry. Couldn't you see their faces!'

While the audience refers to its own views on theatrical nudity, the lovers have one on-stage spectator, Bun Bourke, who reflects:

> Pure people don't carry on, putting on the act, trying to be what they're not. [...] Paint and powder, mincy-mancy, shaping and dressing and stripping and putting on the act.
>
> (46)

These are the terms in which he couches his opposition to the players and their disruptive presence. But we have seen Bun and his men putting on and off military costumes, and the change of behaviour that has accompanied their act of transvestism. In her book on cross-dressing, Marjorie Garber argues that 'dress codes' in society 'had as their apparent motivation the imposition of discipline' and 'also a sense of hierarchy', ordaining 'a certain set of behaviours' which she terms 'vestimentary'.[18] The behaviour of Bun and his men in uniform is marked by a rigid impersonalised set of behaviours, of dominance and control. What proves threatening about theatre is that it both raises this vestimentary aspect of power in society and then effaces the distinctions which society strives to keep in place. It does so by encouraging the act of cross-dressing in its on-stage practices. This mechanism of substitution, Garber asserts, is of 'the very essence of theatre: role playing, improvisation, costume and disguise'.[19] The Sergeant in Kilroy's play feels like a sham, wearing his costume while behaving like a criminal. His fear is that the costume is seeping into and becoming him; but his final efforts to change go for nothing, as we realise when he jeers at the costumes they have stolen from the departing players:

> SERGEANT: [*handling costumes*] Lord save us but aren't they poor looking enough in the broad daylight! Would you credit it that people could be so fooled now? What'll they do now I wonder without their bits of covering and their hokery pokery? The misfortunates!
>
> (78)

For all that Kilroy's play talks of a theatre of transformation, it equally casts doubt on such a possibility. And too much of the talk is interchangeable, part of an ongoing disquisition on drama. Both in reading and performance, *The Madame MacAdam Travelling Theatre* remained continuously interesting; but there is a disjunction between its theoretic concerns and the form in which it is expressed. Madame MacAdam is frequently more of a mouthpiece than a character; Jo manages to be a real live character while having a great deal to say. Perhaps the play could have been more fully entrusted to her direction?

Madame MacAdam states in her opening how the troupe came to 'a crossroads somewhere in Ulster' (2). As an offering from the Field Day Company in 1991, Kilroy's play brought the theatrical affairs of that Ulster ensemble to something of a crossroads. The next chapter will investigate the ten years of annual productions premiered by Field Day in Derry. After a week's run of each play, the company took to the roads of Ireland, North and South, and toured in the best tradition of the travelling theatre companies.

The highlight of 1997's Dublin Theatre Festival was a stunningly theatrical production by Patrick Mason of Kilroy's *The Secret Fall of Constance Wilde*. Where Madame McAdam seemed unintentionally partitioned from the rest of the play which bore her name, Constance Wilde (nee Lloyd) has been consistently sidelined and erased in almost every account of Oscar Wilde. As Richard Pine points out: 'It is a cruel reflection on their personal fates that [Oscar's] wife Constance and their two sons, Cyril and Vyvyan, lost their identities in the aftermath of the scandal, and that oblivion cast a retrospective shadow over the life which they had led before that fall.'[20] Kilroy has undertaken in his play consciously to redress that balance, to rescue Constance Wilde and the two boys from oblivion by dramatising the private tragedy which accompanied the spectacular public fall of Oscar Wilde, to place the passionate and probing intelligence of his Constance at the dramatic centre while Oscar for once is sidelined.

In reviewing *An Ideal Husband,* George Bernard Shaw wrote that 'Mr Wilde is to me our only thorough playwright. He plays with everything: with wit, with philosophy, with drama, with actors and audience, with the whole theatre.'[21] This chapter has been arguing as much of Thomas Kilroy in relation to contemporary Irish theatre. He is certainly our most thorough playwright, not only placing memorable characters on the stage and engaging them in absolutely honed verbal exchanges but drawing on a much wider range of theatrical techniques than has until recently been the norm in Irish theatre: mime, choreography, abstract and symbolic design. In this play the three human figures represented by the actors playing Constance and Oscar Wilde and Lord Alfred Douglas are joined by puppets – small white puppets to represent the two boys and '*a gigantic puppet [of] a Victorian gentleman, red cheeks, black moustache, bowler hat, umbrella [and] frock-coat*'[22] to represent the play's oppressive patriarchs. These puppets are manipulated on stage by six attendant figures wearing '*white, faceless masks*' (11) who remain present throughout and (it is suggested) are also 'Figures of Fate' manipulating the lives of the three human protagonists. Kilroy was inspired by the Japanese puppet-theatre of Bunraku. But, as Nicholas Grene has pointed out, where the onstage puppeteers of the Bunraku are dressed all in black 'and consequently by convention are invisible', the white blank faces of Kilroy's sextet 'draw our attention to them'[23] as they set each scene, cue the action and stand by as a public dimension of even the most private scene (which they frequently applaud).

The Secret Fall of Constance Wilde begins with Constance and Oscar both seriously ill and near the end of their comparatively short lives. He is broken by the years in prison, she by the mysterious 'fall' down the stairs in their beautiful house in Tite Street and by progressive paralysis and degeneration of the spine. Their two falls are aligned by the dates: Oscar's release from prison in May 1897, with a bare three years to live, and Constance's death at forty on 7 April 1898. Husband and wife confront one another over an abyss, of the graves they are about to enter, of the past which they must seek to enter once again. Oscar is reluctant, but Constance insists that their story, and hence her story, must be told. The play's overt theatricality does not pretend that the events we are witnessing on stage are occurring for the first time. They have all already happened and are being re-presented to the audience. But the protagonists have to relive the experience also and come to a fuller understanding of what they have shared.

The earliest memory is set in Dublin – the place or origin to which they returned to declare their intention to marry. They sit in Merrion Square and are watched from Number One (his parents' house) by Lady Wilde/Speranza. The absent father, Sir William Wilde, is only directly referred to in a single line: 'I despise my father' (22). Elsewhere, he alludes to his father's renowned sexual appetites by dubbing him a goat. But the reference draws from Constance the disclosure that scandal attaches to her own father – arrested for exposing himself in public. Oscar's jesting response is that they are being brought together 'out of a mutual interest in patricide' (23); but there is more truth in what he says than he may be aware. Kilroy orchestrates and doubles the lives of his two characters in this way, foregrounding the sexually scandalous Victorian fathers whom they are seeking to flee and who have psychologically damaged them. The arrival of Lord Alfred Douglas/Bosie triangulates this aspect of the play also, shadowed as he is by his father, the Marquess of Queensbury, and the libel case Wilde subsequently brought against him. In seeking to escape their pasts, as represented by their fathers, the three characters are doomed to repeat them and carry them forward into their subsequent relationships. In their wooing scene, Oscar tells a moving fairy tale where three young princesses die and then admits that they are the three sisters Wilde lost in life, one legitimate, two illegitimate:

> CONSTANCE: [...] So your father is the king?
>
> OSCAR: Yes. (*Pause.*) King Billy Goat who sent his three daughters away to die. Everything I write is autobiographical. With the facts changed, of course.
>
> (25)

Constance is offered a role as sister by Oscar but insists on that of wife. As she acknowledges from the play's outset, she has been cast – by her husband and by society – in the role of the 'good' woman, 'the good woman who ran away with her children, away from the horror, the filth; the good wife who kept him in money throughout even while he betrayed her' (13). As she comes increasingly to realise, it is a process in which she herself has colluded and which she now seeks to redress. There was much of the 'new' woman in the Constance Lloyd whom Oscar Wilde met and fell in love with. She had attended art school, had aesthetic views on dress which influenced her husband, edited and contributed to various journals, addressed conferences on women's support for world peace and brought Oscar along to support striking dockers. She was also, in relation to his own writing, his first and best critic; and it is a combined compliment and criticism made by Kilroy's Oscar that she is always trying to correct him.

The exchanges between the young couple are attended to by an onstage audience of one: Lord Alfred Douglas. Oscar moves from his wife's embrace to that of Bosie's while Constance remains present as the mute witness of what transpires between them. The play brings wife and lover together in a series of charged encounters. Each wants to be all-in-all to Oscar but has increasingly to concede that each only has access to separated areas of his partitioned psyche:

CONSTANCE: I know an Oscar that no one else can know!
DOUGLAS: Precisely. There are many Oscars. That is what makes him so seductive.

(59)

In the Richard Ellmann biography of Wilde, the marriage is largely seen as a front, a façade of respectability to protect Oscar from the possible scandal his increasingly overt homosexuality might cause.[24] In the Kilroy play, there is a genuine ambivalence; Constance and Bosie are the two sexual and psychic opposites between which Oscar oscillates. For Kilroy's Oscar, his home life in the beautiful all-white house in Tite Street – personified by Constance and the children – offers a refuge from the fleshly reality he pursues in the streets of London: 'You're a long line to the shore when I'm far out at sea' (40). But it is a perfection that Constance must finally refuse. Her 'fall' is as much sexual as actual: into the knowledge of the sexual variations her husband practices with other men and into an admission of the sexual abuse she has suffered in her own family household.

In the closing speech, Constance addresses her children on the subject of their father: 'He had this terrible, strange vision. He sacrificed everything to reach out to that vision [...]. You see what he did was to try to release the soul from his body, even when his body was still alive' (68). *The Secret Fall of Constance Wilde* offers a brilliant meditation on the contrasts between role-playing in life and in the theatre. But it also allows for an ethical probing of guilt and of the conflict between the body and the soul which Wilde had to draw on all the genres – poetry, essays, drama, fiction, parables, dialogues, epigrams – to convey and which Kilroy has distilled into this ambitious work. What the play does is to deny Wilde the one-man show he so often sought to represent his life as. It restores a dimension of his creativity which has too often been left in the shadows, perhaps because that dimension most fully depends on the human cost at which his creativity was achieved.

Kilroy's plays have long been fascinated with the figure of the artist – from the writer Brien in *Tea and Sex and Shakespeare* through to Oscar Wilde and William Blake in the 1990s. Women have come to the fore in his recent plays, but all of them – Madame McAdam, Constance Wilde, Catherine Blake – have been helpmeets to the male artists. Anna McMullan has remarked that, even as Kilroy has explored 'the masks of gender and authority [...] it is primarily the male figures who are irresponsible and creative, leaving the women to "pick up the pieces"'.[25] The result is a tendency 'to produce a binary opposition between (male) creativity and (female) reality, (male) invention and (female) responsibility'.[26] But the plays increasingly dramatise a process of the female usurping on the male, on protesting at the absolute identification of the artist with the masculine principle. That Constance Wilde's life and its tragedy should be eclipsed by Oscar's is one of the motivating forces behind *The Secret Fall*. In *Blake*, the man of vision is challenged and forced to contend with human reality by his wife Catherine. This process goes one crucial transformative step further in 2003's *The Shape of Metal* with Nell Jeffrey claiming the autonomy so long accorded male artists. Nell is an 82-year-old sculptor and the setting throughout her studio, cleared now of all but a single piece, which is shrouded. In the play's extended flashback to 1972, we see the sculptor actively at work as a dynamic 52-year-old. But even in her decrepitude, with her hands too weak to work, Nell is still trying to shape some meaning, not only out of metal but out of the mess which is her life.

Like *Bailegangaire, The Shape of Metal* is a play of three women, Nell and her two daughters. The youngest, Judith, now in her late forties, is alive and present throughout, an active force driving and goading her wayward mother to confront and tackle issues she has shirked. One is the issue of her paternity. Nell Jeffrey has never married, has had multiple sexual relations throughout her life and more than one father for her two children. At one level, she has claimed for herself the same freedom from traditional sexual mores as the male modernist artist; at another, she has left tangled emotional wreckage in her wake for her children to deal with. Judith has been attending the deathbed of the man she has always believed is her father, Eddie, who has now denied that this is the case. Prompted by Judith, Nell consults *'a heap of old diaries/journals'*[27] and, in a conscious evocation and parody of Beckett's *Krapp's Last Tape*, finds when she reads out the crucial details from her past that she is scarcely able to remember them:

> (*Goes to chair, looks at diary.*) Hmm. 1956. (*Reading.*) What's that word? 'Bedded'! Yes! Date? 'March 12th, '56 [...] Finally bedded the

divine ~ Charlie in Room 217 looking out over Stephen's Green.' (*Looks up.*) Divine Charlie? Who the fuck was that?

(40)

When Nell finally locates the correct dates in the diary, they confirm that Eddie was after all Judith's father. His denial of this is interpreted by his daughter as an effort to set her free from the burdens of patriarchy; but Judith has to wrest this interpretation free from her mother's insistence on placing herself at the centre of all emotional conflicts involving the family.

This is also a play about artistic fathers. Nell's speeches are punctuated with long reminiscences of her meeting with Swiss sculptor Alberto Giacometti in Paris of the 1930s: 'Year before I met Giacometti on the Rue Hippolyte-Maindron' (15). One can see both the artistic and biographical point of this young aspiring sculptor seeking out the great man, who broke with the surrealists in his desire to sculpt a head from the life as Nell is determined to sculpt a lifelike head of her other daughter, Grace.[28] The play's second Giacometti reference reveals that the introduction was effected by an Irish family friend, Sam. The third indicates that Sam 'was childishly pleased to be able to show his two Dublin gels [...] the artistic sights. All the Becketts are like that. Always immensely attentive to others, the Becketts' (29). Beckett is, therefore, explicitly named and referenced throughout *The Shape of Metal*. His emergence and transformation in the text are interesting. At first Giacometti is presented solo, with Beckett first emerging as an acquaintance who hails from the same Dublin Anglo-Irish Protestant circle as Nell Jeffreys. By the end of Nell's story, Beckett has become 'Beckett', the artistic brand name by which he is known throughout the world. In this way, Kilroy ingeniously creates life tissue for Nell Jeffrey, a great artistic figure who never existed (unlike Wilde or Blake). But the paired presence of Beckett and Giacometti do more than that. They share a number of late modernist ideals: the importance of seeing things 'from a distance, at a remove', in order to see them 'whole' (28); the necessity of failure being embodied in the work of art.[29]

The final set-piece of Nell's stories about them hinges on artistic difference, rather than shared ideals. Giacometti, capering like a dancer, extols natural contact between the foot and the ground. Beckett, '*storm(ing)* at this' (55), pulls off his boots and holds them 'aloft above his head like a trophy. [...] "Regard!" he yelled and, as always when he was in a fit, the Irish brogue came out. "Regard!" he cried [...] "These! They carry us everywhere!"' (55–6). Boots are man-made, constructed, something to put between our feet and the earth. Direct contact with (and imitation) of nature is not Beckett's presiding aesthetic, nor is it Kilroy's. Their theatre is constructed out of empty space which is filled with human sounds and man-made objects. The artist's studio or its bare-staged equivalent, not the rural cottage, is their chosen setting. This is the place in which theatre is made, the vacuum into which props are summoned: 'I'm back again, with the glass,' as Clov remarks in *Endgame* when he returns with the telescope.[30] Adrian Frazier has rightly identified it as 'a laboratory for the examination of human life and identity'.[31]

In an essay written over 30 years ago on 'Yeats and Beckett', Kilroy focused on the idea of 'shape' as central to both playwrights:

> When we consider what Beckett means by 'shape' in the plays, we become involved [in the idea] that stage-space is given meaning by what is enacted there, hence becoming part of the meaning of the action itself.[32]

Beckett once defined his artistic imperative as finding a form to accommodate the mess. Nell Jeffreys speaks in *The Shape of Metal* of 'a lifetime trying to create perfect form. The finished, rounded, perfect form' (53). Thomas Kilroy has always striven with form but there has always been a central human conflict at the core of his drama. Rarely have the emotions been as raw and fierce as those in this play; rarely has the concern with form been as urgent or necessary; rarely have the two been as ineluctably intertwined. In *The Shape of Metal*, form and feeling converge in the tragic figure of Nell's other daughter, Grace. The crucial term 'shape' is explicitly associated with her by Nell: 'You must shape a statue of him, Mummy. That's what she [Grace] said. I remember because of her use of that word. Shape' (48). Grace is here talking of the young man in the village to whom she has become attracted, a relationship which her mother is determined to bring to an end. The process by which she does so is initiated when she 'asked him to pose' naked (47). But the consequences of her direct artistic and personal engagement with her daughter are even more catastrophic. Grace never appears directly in the present of the play. She has disappeared to England, like the two sisters in *Dancing at Lughnasa*, and has been missing for decades. Grace appears as a 25-year old in the play's 1972 flashback, when the evidence of her chronic depression is already all too evident. As the stage directions indicate, Nell seeks to soothe her daughter's torment, '[*using*] *her fingers delicately, almost shaping the face, eyes, ears, hair*' (31). Grace initially '[*goes*] *limp, drawn into her mother's body for support and the two of them sway together, almost in dance'* until her whole bodily movement alters and, as if 'escaping immense danger', she leaps away and cries out: 'You are killing me! [...] Hands – fingers – pushing'. If this is Grace encountered through the filter of her sister Judith's memories of that day in 1972, the play opens with Nell calling out Grace's name in her sleep while she dreams of her daughter pleading with her to 'sculpt my head, Mummy, as promised [...] stone or metal to be transformed into Grace finally at peace, head still and quiet, no terrible dread anymore, no mad panics'. The Grace who speaks onstage is represented as an illuminated speaking head, a live onstage sculpture, as it were. This is even more the case when Grace puts in her final appearance, '*this time as a bronze death head on a plinth, bronze head which speaks, the mouth moving but the eyes closed over, metallic*'(51). The echo is strong of the three talking heads installed in the urns in Beckett's *Play*. Here, Nell is confronted with the perfection of art as that which she has always sought: something finished, complete, dead. Grace's words confirm this even as they mock Nell Jeffrey for her failure to address her daughter's suffering in life rather than remove it through art: 'Mummy shaped Gracie's head into metal. Mummy's fingers moulding eyes, ears, nose. Mouth, head. Cold. Cold metal. Peace. Silence. All finished. Nothingness' (51).

The paradox at the centre of Kilroy's art is that he has employed an unprecedented range of theatrical artifice, not at the expense of feeling but in the service of it. Nicholas Grene has noted this in his response to how the two puppets are deployed to represent the young Cyril and Vyvyan Wilde: 'Nothing was more moving in the stage experience of *Secret Fall* than the puppet children.'[33] The image and voice of Nell's missing daughter are the most abstract and yet the most moving elements of *The Shape of Metal*, a play which is haunted as much by the ghost of Samuel Beckett as it is by that of Grace Jeffreys.

5
Northern Irish Drama: Imagining Alternatives

Since conflict is the essence of all drama, it should be no surprise that the situation in Northern Ireland across 30 years from the late 1960s to the late 1990s has generated a considerable number of stage plays. Catholic versus Protestant, British versus Irish, republican versus loyalist, the gun versus the ballot box: to live in the North is to inherit and inhabit a drama of conflict whose contradictions often resulted in lethal consequences. Far from offering too little material to an aspiring playwright, events in Northern Ireland posed a threat to all our notions of aesthetic form and dramatic coherence. As a political entity, the North has an inherently unstable structure. Any play dealing with the situation there has to acknowledge that instability in its own structure to some degree. Language in the North is also a double-edged weapon. It offers the playwright a rich and varied dramatic speech, one saturated in the cadences and imagery of the King James Bible if the speakers are Frank McGuinness's sons of Ulster, one drawing on a distinctive Anglo-Irish idiom and the subterranean influence of the Irish language if the locale is Brian Friel's Ballybeg. But, especially in the arena of public speech, the North's linguistic inheritance is a divided one, since occupying the same geographic space are two diametrically opposed political communities. The spokespersons for each side, even after the Belfast Agreement, are still prone to totalising claims which not only contradict each other but are mutually exclusive. To utter both on the same stage is to have not so much dissent as dissonance.

From the early 1970s, drama from and about Northern Ireland develops in significance. There is an important two-way process. The 'Troubles' play (as it came to be known) is a reflex of the intensely politicised and frequently violent events in the North that developed across three decades. On the other hand, the playwrights and all those involved in the act of theatre have attempted to create a space in which some kind of meaningful dialogue can be enacted, in which indeed the limitations of language can be both demonstrated and gone beyond. The Northern Irish play has marked dramaturgic features which may collectively constitute a new genre. Its particular stage procedures have developed in relation to the complex political and social order out of which they have emerged. Like most of the works so far considered, Northern Irish plays are anti-hierarchical; for the most part, no single character dominates. Instead, characters are displaced from their original affiliations into potentially new groupings and proto-communities. Structurally, Northern Irish drama moves in opposition to the well-made play, emphasising instead discontinuity, fragmentation and juxtaposition.

The Northern Irish play is extremely sceptical with regard to language and language's implicit claim to validate reality. Repeatedly in these works the verbal norms of everyday experience are called into question by the full range of theatrical language – gesture, mime, dance, song, music, symbolism and stage imagery. On its stage, the living are confronted by the dead; speech is challenged by silence. The terms are Beckettian, recapitulated in a key of heightened realism. This chapter will take some measure of the range of Northern Irish drama by examining, first, the late Stewart Parker's *Pentecost* as a token of his insufficiently acknowledged achievement, then the rich contribution made in this area by women playwrights like Anne Devlin and Christina Reid. This will establish a context in which the brave experiment of Derry's Field Day Theatre Company will be assessed. The chapter will particularly focus on two major plays by Frank McGuinness, the most outstanding of the next generation of contemporary Irish playwrights, and conclude with the work of Gary Mitchell, which came to prominence in the late 1990s and raised serious questions about the extent to which the North had left its past behind.

Stewart Parker was born in East Belfast in 1941 and attended Queen's University there. Having spent some time studying and teaching in the United States, Parker returned to his native Belfast shortly after the outbreak of the Troubles; he subsequently lived in Edinburgh and London. Tragically, he died of cancer on 2 November 1988. Like Synge and Behan before him, here was an Irish playwright cut off in his prime; but the sense was greatest with Parker that he still had much more to contribute. Nevertheless, between 1975 and 1988, Stewart Parker produced an impressive and wide-ranging body of work: eight stage plays, nine radio plays, seven television plays and a six-part television series, *Lost Belongings*, which updated the legend of Deirdre and the Sons of Usnach to 1980s Belfast.

Only three of his stage plays were published during his lifetime, despite their evident quality. It was a subject that exercised him, and one we discussed when I phoned him in London in March 1988 to request a script of *Pentecost* (which he immediately and generously supplied). Ironically, his major work of the 1980s, *Three Plays For Ireland: Northern Star, Heavenly Bodies, Pentecost*, was in the press when Parker died, and appeared posthumously. He at least had the satisfaction of seeing 1987's *Pentecost* acclaimed in a superb Field Day production by Patrick Mason and a cast of five, headed by Stephen Rea. When it played the Dublin Theatre Festival in that year, *Pentecost* was given the John Player Theatre in the South Circular Road, a sizeable enough venue, but one away from the mainstream Dublin theatres. It was precisely the same venue where Parker had made his theatrical debut with *Spokesong* in 1975. That play had been the hit of the Dublin Theatre Festival and had gone on in 1976 to win London's *Evening Standard* award for Best Newcomer. In the years between, Parker's plays had met with a mixed reception. And if Belfast's Lyric Theatre made some amends for their rejection of a native son's first play by commissioning *Northern Star* (1984), that play had been too tricky in its use of pastiche to win a popular audience and had played Dublin's Olympia Theatre to half-empty houses. So, for his next theatre festival slot, Parker was back in the John Player where he had begun 12 years earlier, no longer a promising newcomer but a seasoned dramatist. *Pentecost* went on from that stage to win the Harvey's Best Play of the Year award and to gain Parker a whole new audience. It seems his fate was to be repeatedly rediscovered. And now his death adds a further complication to that fate. One tangible legacy is the Stewart Parker Trust, set up shortly afterwards to encourage new playwriting in Ireland; another is posthumous productions of *Spokesong* by the Lyric and by Dublin's Rough Magic Company, directed by its co-founder (and Parker's niece) Lynne Parker. But more productions and more critical writing on the man and his considerable body of work are required to make clear his importance in the landscape of contemporary Irish drama.

I would single out two of many possible contributions. Stewart Parker's unique sense of play enabled him to treat of that most deadly and potentially deadening of subjects, the crisis in Northern Ireland, with an unrivalled lightness of touch. Of the multiple facets of that distinctive sense of play, I would highlight music – itself a recurrent feature of contemporary Irish drama but nowhere more so than in Parker. Indeed, I first encountered his writing in the early 1970s when he contributed the first rock column, 'High Pop', to an Irish daily national newspaper. His first play, *Spokesong*, was virtually a musical, with lyrics by Parker himself and music by fellow Northerner Jimmy Kennedy, who composed 'Red Sails in the Sunset' and 'South of the Border (Down Mexico Way)'. In his second play,

Catchpenny Twist (1977), the two male leads are archetypal Parker dreamers, a pair of would-be song-writers hoping for success in the Eurovision Song Contest while earning a living writing ballads for opposed sets of paramilitaries – with disastrous consequences. The presence of music is more muted and traditional in *Pentecost*, but Stephen Rea is as wedded to his trombone in Parker's play as he is to his saxophone in Neil Jordan's film *Angel* (1982); and the play ends with Lenny's version of 'Just a Closer Walk with Thee'.

And that brings me to Stewart Parker's second contribution. After the pioneers Sam Thompson and John Boyd,[1] he was the first among the younger generation of the Northern Protestant community to write and have plays produced for the contemporary stage; other fine playwrights were to follow, including J. Graham Reid and Christina Reid (no relation). Parker broke the mould by bringing a colourful imagination to bear on a community often characterised as dour and grey, by showing that it had a culture worth celebrating, and by bringing an openness of response to key historical events in the life of that community: the 1798 Rebellion and Henry Joy McCracken in *Northern Star*; the 1974 Ulster Protestant Workers' Strike in *Pentecost*. In his John Malone Memorial Lecture at Queen's University in 1986, Parker came closest to declaring his artistic intent:

> If ever a time and place cried out for the solace and rigour and passionate rejoinder of great drama, it is here and now. There is a whole culture to be achieved. The politicians, visionless almost to a man, are withdrawing into their sectarian stockades. It falls to the artists to construct a working model of wholeness by means of which the society can begin to hold up its head in the world.[2]

Parker's words and works remain as a challenge to achieve that model of wholeness for a fractured society. In his own writings, *Pentecost* comes closest to achieving that ideal and, in concentrating on it as the single play by Stewart Parker that will be analysed here, I would have it stand for and draw attention to the sum of his achievements.[3] When I first saw and subsequently read the play, Stewart Parker was still alive; now, the reading of *Pentecost* which follows is haunted by the unexpected ghost of its author, his whimsical, humane presence.

In his Introduction, Parker observes that '[p]lays and ghosts have a lot in common. The energy which flows from some intense moment of conflict in a particular time and place seems to activate them both'.[4] The dead figure in *Pentecost* who returns to haunt the living is not a victim of sectarian assassination, as one might assume from the play's setting of Belfast in 1974.

Rather, Lily Matthews, respectable widow of Alfred George Matthews, has died of natural causes at the respectable age of 74. This makes her as old as the century and in many ways a representative of the history of the Northern Protestant community over that period. The lives which are under threat are those of the still-living characters in the play at one of the numerous crisis points in the ongoing turbulence of the North. Or as Lenny the trombone player ruefully remarks: 'Sure, every bloody day in the week's historic, in this place.'[5]

But even so a particularly historic turning point was the Ulster Workers' Strike of 1974, given their determination to force the hand of the British Government and bring down the recently established power-sharing Executive, Northern Ireland's first concerted effort at a form of democratic self-government. With the army and police standing idly by and the streets full of gun-toting civilians, nobody out there is safe, as we discover when Lenny's estranged wife Marian returns in Act Two from searching for her car. Marian enters and is revealed '*in the light of the torch, mud-spattered with her coat ripped, and scratch marks on her face*' (222). Given the life-threatening reality of what is just on the other side of that door, the on-stage space in *Pentecost* functions as it does in so many Northern Irish plays as a kind of stay or refuge, an asylum or temporary holding-ground, one step removed from the (war) zone of historical circumstance. Marian identifies the house in which she has chosen to live and which provides the play's setting in just these terms, when she complains to Lenny: 'You've been living with me again, here in this house, the very place I chose as a refuge' (225). The larger point Marian misses, but the play does not, is that her estranged husband, and the two friends of theirs who follow in his wake to make up the play's ill-assorted quartet, are all equally in need of a refuge, a site for healing in which their psychological even more than their physical wounds can be tended.

Lenny's relationship with Marian – they cannot live together, cannot live apart – has certain features in common with Northern Ireland itself; but husband and wife are both reluctant to declare their marriage a failed entity. Ruth, a friend of Marian from ten years earlier, is on the run from her truncheon-wielding policeman husband, not for the first time and not for any crime she has committed beyond that of marrying him. The fourth stray who finds his way to the house Marian has chosen for splendid isolation is not Ruth's pursuing husband, as we might expect and as someone prophesies, but a new, belated character called Peter, an ex-college friend of Lenny's. Peter has returned to Belfast after spending the intervening years in Birmingham, and so brings the perspective of the outside world to bear on the local situation. Since the view is that of mainland

Britain and the place it is trained on is the place he has left, Peter's attitude is complex, both in its own divided allegiances and the response it provokes in Lenny.

Pentecost is, however, more crowded than this cast of four would suggest. For the house into which Marian moves at the beginning of the play, '*a respectable working-class "parlour" house, built in the early years of this century*' (171), is a haunted site and not as empty as it strikes the pragmatic Lenny. He has come to inherit it by way of a great-aunt, and when Marian declares a wish to take it over in return for finally granting him a divorce, Lenny protests: 'Marian, you can't possibly live in this gaff, it's the last house on the road left inhabited! – the very road itself is scheduled to vanish off the map' (179). Here, Marian's declared concern is to preserve the house as an example of the lived culture of a Belfast working-class Protestant family from the early years of the century. But that act is finally seen by Marian as a fossilisation of the past, and resisted in a change of heart. She finally decides that what the house needs is air and light to be lived in.

The agent of this change of heart is the ghost of Lily Matthews, the house's last occupant, who refuses to be evicted, even by death. Marian conjures up the ghost of Lily at least partly out of her own loneliness and need. Her motives for so doing, like those for moving into the house in the first place, remain initially unclear, even to herself. When she and Lenny first enter the house, it is within hours of Lily vacating it, and the space is still very much suffused with her presence. Entering the area where a woman has just died, the couple are both paradoxically made aware of her presence, evidenced through a half-finished cup of tea. Marian begins the process of identification which is central to the play by trying to imagine the moment of Lily's death. Since they never knew her and do not remember her, Marian is consciously using her imagination to reconstruct a past, to create a memory where none had previously existed. She cannot do it all by herself; Marian is going to require the active presence and participation of the ghost of Lily Matthews.

The process by which a dead person is evoked in Irish drama usually depends on a theatrical, symbolic and personal prop, like the half-drunk cup of tea, a physical synecdoche through which a more complete reality is summoned on to the stage. In Act One, Scene Two, Marian lifts a piece of unfinished knitting out of a basket. The knitting signifies a severed lifeline and the distinctive fabric of one person's existence. In Marian's suggestion that she 'might just finish it off for you' (180), there is also a sense of the larger overall pattern beginning to emerge, the dramatic weave of the ensemble. The sceptical and the visionary are fused in Marian's response. Her superficially ironic yet emotionally charged dialogue with the absent figure of Lily now brings the old woman on-stage, dressed as she exited two months earlier:[6]

> LILY MATHEWS, *in Sunday coat and hat and best handbag, appears in the shadowy doorway leading from the pantry.*
>
> LILY: I don't want you in my house.
>
> MARIAN *keeps her eyes on the knitting pattern: on guard but not entirely frightened, aware that her mind is playing tricks on her.*
>
> MARIAN: You needn't try to scare me, Lily.
>
> LILY: Don't you 'Lily' me. I don't want you in here, breathing strong drink and profanity.
>
> (181)

As in Hugh Leonard's *Da* (1973), where Charley returns from his father's funeral to find the old man sitting in an armchair, the body in question is dead but refuses to lie down – or to stop talking. Parker's *Pentecost* is in a tradition of Irish drama which recognises that, far from having to keep to narrowly realistic boundaries, theatre is a means of bringing the dead back to life. And in the objective and ineluctable materialisation which is the theatrical act, those ghosts cannot be read entirely in psychological terms once the playwright has decided to bring them on stage. Clearly, as even Marian herself realises, her own isolation, the strong drink, and her confession that she does not like herself much are all psychologically sufficient to call up an imagined *alter ego*. Friel did no less with his youthful protagonist in *Philadelphia, Here I Come!*, as we have seen, with Private Gar invisible to everybody else. The same is true of Lily Matthews in relation to the other three characters in *Pentecost*. As far as Lenny, Ruth and Peter are concerned, Lily is invisible, non-existent, a sign merely of Marian's cracking up: 'You've been talking to yourself, you've been counting spoons, you've been babbling in tongues in the middle of the night!' (226). But in theatrical terms, once an actor has been assigned to embody an *alter ego*, a *doppelgänger* or a dead person, there is nothing to choose between the real and the ghostly. Nor is the 'real' figure given precedence or consistently placed in the foreground. Friel preserves a fine equilibrium between his split persona(e) in a dramatic double-act as balanced as that of Beckett's Vladimir and Estragon. This doubling not only erodes any distance between the real and the uncanny but redefines the nature of the surrounding drama. For the inhabitants of Ballybeg in Friel's play, there is only one Gareth O'Donnell to see and commune with; but for the audience there are the two Gars who become more real to us than the two-dimensional characters of the fictional rather than the theatrical space.

The same holds true for Lily Matthews in Parker's *Pentecost*. Her ghostly manifestation not only challenges Marian's reality and her grip on it, but undermines the reality the play is representing. Ironically, while Lily appears to urge strict segregation into Protestant and Catholic, her presence on-stage succeeds in crossing boundaries and established lines of demarcation between the living and the dead. When the ghost invokes her dead husband by saying, 'You'd be singing on the other side of your face if my Alfie was here' (181), Marian, closing her eyes, insists: 'There's nobody here. Nobody.' The audience at this point must register a double-take and apply Marian's words reflexively to herself no less than to Lily's ghost since who, in the theatre, on the stage, is *really* there? Parker has fun with this phenomenon of theatre all the way through, by having the ghost insist that she is more real than the intruder. Lily reverses the 'real' situation by treating Marian as the ghost who needs to be exorcised and singing hymns whenever she appears to keep the Antichrist at bay. But Marian plays along with the inverted terms of the identification by stressing the interdependence between living and dead:

> I need you, we have got to make this work, you and me. [...] You think you're haunting me, don't you. But you see it's me that's actually haunting you. I'm not going to go away. There's no curse or hymn that can exorcise me. So you might just as well give me your blessing and make your peace with me, Lily.
>
> (210)

As Marian examines this revenant on the details of her history, she gradually uncovers the intimate hidden life which lies behind the imperturbable respectability of Lily's facade. In the process, she increasingly takes over the role of Lily as she comes to understand and identify with it.

The appearance of one ghost in this haunted house soon draws others in its train. Lily's husband Alfie is a ghost twice over. He may have died and left her widowed only fifteen years earlier, in 1959. But it turns out that Lily was wed as an 18-year-old virgin to this man who, having endured the trenches of World War I and having been one of only two on his street to survive, returned as a living corpse: 'Alfie had come back, that's why. Back from Passchendaele. Hellfire Corner. Back from the dead. [...] All in the one week, married and moved in, he couldn't wait ... not after what he'd seen ... this house was his life, same as mine' (182).

Marian confronts Lily, however, with incriminating objects that suggest a very different story. A rent book signifies the ghost of a third party, Alan Ferris, an English airman; and in a photograph of husband, wife and lodger, the outlines of a triangular relationship may be glimpsed. In regard to her and Alfie's childlessness, and the sterility the house ultimately represents, Lily is confronted by Marian with a child's christening gown, which gives the lie to their married relationship. Lily's husband was sexually maimed in World War I and her one brief experience of passion has been with the airman. In her most intimate exchange with Marian, Lily evokes a visionary landscape of water and air redolent of her sexual need: 'All we did was stand and look, across the water' (229). This scene develops in ways reminiscent of Beckett's *Rockaby* as Marian sits in the rocking chair and begins speaking Lily's innermost thoughts for her, based on a reading of her diary: 'That was the moment when it hit you, though [...] the moment you realised that you were going to give yourself.' The fruit of Lily's relationship with the airman was the baby she could not bring herself to acknowledge and so abandoned, the sign of her betrayal of a whole way of life as she saw it.

There is more than one buried child being exhumed in this exchange. Earlier in the play, Lenny has let slip to Peter that he and Marian had a baby, something which is never mentioned when they are on stage together discussing their marriage. The irony is that Lenny and Marian only married because she was pregnant and then, in having the baby, discovered that they loved each other: 'For five months. That was how long it lasted ... that was how long the sprog lasted. At that point he checked out, he'd seen enough' (205). Ruth, too, has had more than one miscarriage, not least because her policeman husband repeatedly beats her, and has been told she can have no more pregnancies. At the close of the first half, the play's three women are linked across the generations, across the sectarian divide, and across life and death itself as the revelation of Lily's abandoned child leads Marian and Ruth to acknowledge their own. When Lily finishes telling her suppressed story of the affair with Alan that resulted in her child, Marian responds with an eloquent gesture, reaching out the hand of the living to that of the dead and holding it against her heart in an act of restoration:

> MARIAN *takes* LILY'S *hand and holds it against her own heart.*
> MARIAN: Forgive me, Lily.
> *Lights fade to blackout.*
>
> (232)

Since Alfie was impotent, neither 'the first nor the last to come back from the dead in that condition' (231), the house was sole witness to Lily's labour, birth, and what she suffered in giving up her child. This denial of life and her sense of abandonment are the emotional truths which her Unionist cry of 'no surrender' tries to deny and silence.

Lily does not appear after this, in the play's fifth and final scene. When Marian speaks at the close there is no longer any need to present them as two separate or distinct selves on the stage. This last scene has generally been criticised, in the context of the overall praise *Pentecost* has attracted, as Parker's imposing an overly religious and didactic conclusion. Marian's speeches are dramatically convincing, more so than the other three, as she follows the Pentecostal injunction to 'speak with other tongues, as the Spirit gave them utterance' (240). She does so, first, by speaking for and as Lily Matthews in the culmination of the dramatic process we have witnessed throughout the play. She imagines a scene during a World War II bombing raid on the Belfast house, with Lily lying alone on the parlour sofa on which she and Alan had made love. In a direct evocation of Ibsen's *Ghosts*, Peter and Ruth earlier in the play had entered that same parlour and made love there. The past repeats itself in the present and the hidden sexual history of these interconnected lives is brought to the surface. But through Marian's subsequent storytelling, the past is not allowed to dominate and determine the present, as it does in Ibsen's drama. Rather, it is reshaped in the light of present possibilities, as in Murphy's *Bailegangaire*. Marian as actress/playwright speaks out of her own understanding of Lily Matthews's twilight existence, imagining a unique moment of sexual and spiritual ecstasy in which Lily wished to die, 'waiting for the chosen bomb to fall on her and cleanse her terrible sinfulness and shame' (237). Instead of the oblivion and release she craves, Lily is suffered to survive, 'condemned to life. A life sentence' (237), in words that recall the mordant Belfast graffito, 'Is there a life before death?' Marian realises that her own making present of that half-lived life, her selfless articulation of what lay hidden in Lily's journal, has released the petrified ghost from the bonds of hypocrisy.

Marian proceeds to speak for and in the tongue of yet another denied ghost, her dead infant son Christopher. In so doing, she speaks up effectively for the Christ in each of the four living characters that is the tenor of the play's closing: 'I denied [...] the ghost of him that I do still carry, as I carried his little body' (244). There is a ghost trio at this point, made tangible through Marian, since her talk of Christopher is also Lily speaking out on behalf of the love-child whose existence she has denied. As *Pentecost* ends, Lenny reaches out and touches Marian's hand '*privately and unobtrusively*' (244), as she had earlier touched Lily's, in a reciprocal gesture of support and restoration.

The considerable impact of *Pentecost* has to do with the fact that the personal relationships between the two couples do not tell the whole story.

There is the extraordinary presence of the ghost of Lily Matthews, as we have seen, bringing to bear an earlier generation's experience. And the characters of Lenny, Marian, Peter and Ruth themselves resonate in mythic and historic terms. Ruth speaks most directly from the Gospels, as a voice of compassionate prophecy; Peter voices the words of his apostolic namesake; Lenny fuses spirituality and sexuality in the music he plays; and Marian discovers her buried self through the ghost of her dead child. But that dead child in turn evokes all the recent dead of Belfast and the North, 'our innocent dead' (245), who call on the living to redeem them. The characters' personal politics speak to the politics of the province, reminding everyone who cares about the North that 'there is a whole culture to be achieved'. *Pentecost* is Parker's most eloquently humane play, the one in which irony and whimsy give way to a passionate dramatising of personal and cultural renewal.

Through its questioning of inherited norms of identity and relationships, the Northern situation has brought several women playwrights to the fore. This development is all the more noteworthy in contrast to the prevailing situation in contemporary theatre in the Republic of Ireland, especially in Dublin, where women playwrights have been markedly absent from the main stages. All the evidence suggests that women in the Republic have been writing plays but that those plays have not been staged.[7] *The Irish Times* in 1982 held a woman playwrights' competition in connection with the Dublin Theatre Festival, in which the impressive number of 188 plays were submitted for a prize of £1000. The award went to Mary Halpin's *Semi-Private*, which was staged at that year's theatre festival; the runner-up was Christina Reid's *Tea in a China Cup*, the next play I wish to consider. In the overall context of Dublin theatre practices, that 1982 competition appears like an aberration and has never been repeated; the other plays on the short list I have never subsequently heard of; and Mary Halpin is still mainly known for *Semi-Private*, a play which has had several successful revivals.

Of them all, Christina Reid, from a Protestant working-class background in Belfast, is the only one who has since had a continuous career in theatre. If I concentrate on *Tea in a China Cup* as her best play to date and as one which covers the same terrain as Stewart Parker's *Pentecost*, albeit from a very different perspective, I would wish to stress that Reid has so far written ten plays, that six of them have been published, and that she has had periods as writer-in-residence both at the Lyric Theatre, Belfast, and at London's Young Vic. Since politically it still remains part of the United Kingdom, the North retains close cultural and economic ties with London; and so a Northern woman playwright like Anne Devlin can look to the

Royal Court Theatre there, and its fostering of feminist playwrights like Caryl Churchill, as a testing ground for her emergence as a dramatist. But the contemporary Northern writer benefits also from the flourishing of an indigenous Irish theatre movement in being encouraged to write out of her own background and experience, to represent directly the conditions that go to make up the current situation in the North.

The dramatic emphasis of Christina Reid's *Tea in a China Cup* is on the supportive interaction of working-class Protestant women, and is laced with sardonic humour. The story is told through the mother-daughter relationship of Sarah and Beth, heightened by the fact that the mother is dying and that the daughter is struggling to come to terms with her own identity. The present of the play is Belfast in the early 1970s, and, like Parker's *Pentecost*, it reaches back to 1939 and beyond to represent the experiences of its Protestant family in the two world wars and in the contemporary Troubles. Unlike the Parker play, however, Reid's is not told continuously from and in the present tense, with the past brought on as a walking ghost. Rather, it resembles Friel's *Philadelphia, Here I Come!* as a memory play in which the central character more directly re-experiences the past through a series of flashbacks. In Reid's play Beth becomes a narrator standing apart from events and a participant in the very events she describes. But where Reid goes further than Friel's practice in *Philadelphia* and *Dancing at Lughnasa* is that the memories enacted by her play extend beyond those of her protagonist. As Beth remarks at the outset of one of the flashbacks:

> I couldn't possibly remember it, I was only an infant, but I've heard that story and all the other family stories so often that I can remember and see clearly things that happened even before I was born.[8]

The memories dramatised in *Tea in a China Cup* are more than merely personal: they are neither absolutely identified with nor restricted to a central bourgeois individual. These dramatised memories extend beyond their teller into the lives and material conditions of those by and among whom she was reared.

The first key memory, long preceding Beth's birth, is 'the day my mother's brother Samuel went off to fight for King and Country' (10). That day is brought before us on-stage through a lighting change which reveals the characters and setting of 1939; but the present remains continuous through the watching, narrating presence of Beth. The scene, however, is one from which she is absent, one to which she bears the relationship of a ghost, of someone who has not yet come into being. Within the flashback there is a further reference back to World War I through the '*sepia photo of the grandfather in First World War uniform*' (10), the previous occasion on which the sons of Ulster went off to fight for King and Country. The grandfather is still present, having survived that great war, and urges his son to emulate his behaviour in the second by (as one of the women puts it) 'filling his head full of nonsense about the great times you had with the lads in France during the First World War' (11). The grandfather has been one of the few lucky enough to survive, albeit with shrapnel in his leg, and has validated the slaughter through the stories he has told. Present throughout the scene is the women's critique of the war, as they counter the claim that the war will 'make a man' of Samuel, by charging that a real man would stay at home and stand by his family.

In dramatising the women's reaction to the men going off to war, the play structurally embodies two kinds of history. In 'Women's Time' Julia Kristeva has characterised the time of history as 'linear time' and has examined the problem of 'reconciling maternal time (motherhood) with linear (political and historical) time'.[9] The project of history has traditionally constructed a patriarchal narrative along the lines of linear time. In *Tea in a China Cup*, this narrative is figured as a succession of sons marching off to fight in great wars and earning their place in the newspaper chronicles: 'four of his cousins are at present serving their country overseas ... his father served with the Royal Ulster Rifles in the last great war' (20). If male linear time constitutes a history, it is a written history and one that writes reproduction out of the record. Kristeva posits an alternative women's time, in which there are 'cycles, gestation, the eternal recurrence of a biological rhythm which conforms to that of nature'.[10] From the women's point of view, the experience of war is that of being left; and the play fixes its attention on those who are left behind. Where the men stress the glory, the women stress the threat to life, the numbing repetition of death. Their own cycles of reproduction are subsumed and made to serve the ends of history, rearing sons for the slaughter. If the sepia photo hints at a previous war, the scene of Samuel's departure for World War II focuses a double concern, extending to that of Sarah for her 'wee Sammy' (11), a three-year-old son with the same name as his uncle. The search for the missing child in the 1939 scene, for a character we realise must be Beth's older brother, raises the question of where he is now, his absence in the present of the play. His name links him to Uncle Samuel going off to war and a possible death. And these prophetic intimations are realised in Act Two when he has grown up and there is an identical replay of the earlier scene, a repetition that is only interrupted when they realise the possible fate of the younger Sammy by looking at the photo of his dead namesake.

The news of Samuel's wounding and death in World War II is conveyed in two ways. The women come on and read a series of terse impersonal communiqués from his commanding officer, recording the fate of Gunner Samuel Bell, Number '1473529' (15). But in the uncertain period between his hospitalisation and his death, fear for Samuel's recovery sends the grandmother, great-aunt Maisie and Sarah into the company of a fortune-teller. Maisie the believer has brought them there, the grandmother is neutral in her beliefs, but the sceptical Sarah voices every rational objection to the superstition of fortune-telling. The prophecy is that Samuel will come home; but the fortune-teller holds back Sarah to reveal that he will be coming home in a box. Her idiom shifts from her profession's hocus-pocus to a realistic frankness as she talks woman to woman. In the face of Sarah's anger and disbelief, the fortune-teller makes good her prophetic claims by revealing the baby that is growing in Sarah's womb and about whom nobody else yet knows. Her visionary seeing is related to the act of theatre, especially drama written by a woman since Beth as narrator is 'seeing' what she has never witnessed directly and we the audience are seeing what isn't there. In the newspaper account of Samuel's death, he is ranked among the legion of the dead; in the fortune-teller's announcement, she counters his death with the no less fateful news of Beth's pregnancy, the two hearts she sees beating. In this juxtaposition, women's experience is asserted against history, against a narrative of absence and death.

The personal is extended into the communal through the use of ritual, and the action of *Tea in a China Cup* foregrounds a series of rituals in which its women characters participate. There is, firstly, the ritual announced in the play's title, that of drinking tea in a china cup. The verbal effect of its first appearance is to lift everyday speech into a formalised, repetitive exchange in the manner of Didi and Gogo's discussion of waiting for Godot:

> BETH: Would you like a cup of tea?
> SARAH: I would love a cup of tea.
> BETH and SARAH (*together*): In a china cup.
>
> (8)

The tea-drinking ritual punctuates and scores the play. At first glance, it would appear to be one that Beth and Sarah share equally. But in the above exchange, Beth is suppressing her differences with her mother in the light of Sarah's impending death. For as the play demonstrates, she seriously questions the value her mother places on a china set, what it represents to her as the height of respectability. The service which Beth acquires through her marriage is too good ever to be used. Her judgment on her feckless husband Stephen is that he too much resembles the fine china set, 'for looking at, for ... show, not for everyday use' (37). The fiercest exchange between mother and daughter occurs at the height of the Troubles when Sarah has been ordered to leave her house but will not go without the china:

> BETH: Mum, they're only...things...bits and pieces...they can all be replaced...
> SARAH: They're my life!!
>
> (57)

In their final exchange, Beth persuades Sarah to let her serve tea in the very finest china cup, but as she is off-stage preparing it, Sarah slips away into unconsciousness. The single teacup and saucer are kept by Beth as a personal memento of her mother, placing a personal value on what the auctioneer in the final scene regrets as an incomplete Belleek teaset, whose monetary value is thereby much reduced.

Another key ritual in the play, one that is much more public but involves a similar transvaluation, is the Orange parade with which *Tea in a China Cup* opens. At first we register the music and singing as off-stage noises; our attention is primarily focused on the pleasure experienced by a patently ill woman as she sits on the sofa and listens. Sarah takes up the words of the marching song after the parade has passed and continues singing them. The more usual critical view of the Orange marches as sectarian is given by Beth when she enters, since mother and daughter differ a great deal in their view of religion. But what Sarah's carnivalesque enjoyment conveys is the personal and communal view rather than the overtly political or demonised perspective on the Twelfth of July. The history of her experience of previous memorable Twelfths is passed on from mother to daughter as an oral legacy, a transmission of storytelling through the generations. For the play's key ritual, its dominant dramatic technique, is the act of storytelling. Christina Reid has spoken of how her 'mother and grandmother were storytellers' and of how she was 'constantly in the company of women and children. It made me realise how ageless women are – they can talk like young girls at any age and on their own are tremendously uninhibited and bawdy – a side that they would never show to men.'[11]

An exception might be made if the man were dead. For Beth as a child has accompanied her grandmother and great-aunt Maisie to attend another ritual, the laying out of the dead. In this form of alternative women's drama, another version of which is the fortune-telling scene, Maisie is the chief performer, covering the mirror to ward off evil spirits and, in sharp contrast to the hypocritical pieties of the official obsequies, taking the corpse by the hair and telling Grandfather Jamison to his face that he was a 'vindictive oul bastard' (35). More often, the men are figured as completely absent. There are those who have gone off to war who, even if they return, do so as shadows of their former selves. But there are also the men in the mother's and daughter's lives who are characterised by their absence and (it is implied) by neglect of their wives. The climax of this tendency has Beth on her wedding night waiting alone in the hotel bedroom while her husband talks to some clients in the bar. After a telephone conversation with him, she phones her childhood friend Theresa for comfort and to thank her, ironically, for the gift of a book entitled *The Invisible Man*. But the telephone scene, if it re-enforces the groom's distance, serves to summon up Theresa's presence, since we have seen her with Beth at various stages throughout the play. On those occasions, too, their conversations have often been about men – not individuals but the myths they have been fed about the sex. This adolescent chat is inflected by the political divisions, since Theresa is Catholic; and what emerges in their talk about masculine nuns and rapacious landlords is the extent to which political fears are expressed in terms of sexual deviancy. Beth's friendship with Theresa persists throughout the play, despite the disapproval of their elders, and is a key bonding to offset their experiences with men.

Paramount is the relationship with the mother and especially the process through which, in helping her mother face her death, Beth is led to tell the story of her own life, to re-evaluate what she has become. And that analysis leads to a crisis, Beth's growing conviction that she has never lived her own life. The movement of the play correlates with what has been articulated by Kristeva as the second phase of the women's movement (the first is the demand for the same rights as men). This second stage dramatises a 'female society' 'as a sort of *alter ego* of the official society, in which all real or fantasised possibilities for *jouissance* [play, pleasure] take refuge'.[12] But Kristeva goes on to warn, in terms that are directly applicable to Beth's dilemma and complex, that 'there is also the connivance of the young girl with her mother, her greater difficulty than the boy in detaching herself from the mother in order to accede to the order of signs'.[13] Beth confesses to Theresa late in the play: 'I'm scared, Theresa… my mother's dying and very soon for the first time in my life I am going to be alone… and I'm scared… my head is full of other people's memories [and] I don't know who *I* am… or what *I* am' (61).

The third generation of feminism, according to Kristeva, is a mixture of the demand for women's '*insertion* into history', a demand which Beth now insists on as her equal right and the legacy of the first phase of the women's movement; but in the third phase this is now accompanied by a 'radical *refusal* of the subjective limitations imposed by this history's time on an experiment carried out in the name of the irreducible difference'.[14] Nowhere does *Tea in a China Cup* mark this radical refusal more than in its structure, where it celebrates women's time by an undermining and displacement of linear history. The opening of Act Two recapitulates the opening, with Sarah lying on the sofa listening to the Orange march. This is back to *Godot* in one sense, but different in another, since the cyclical recurrence of the play is now dramatised in terms of life, not death, of biological continuity, of remembering rather than forgetting. It foregrounds a gallery of women, some of them Beth at different ages, some of them her mother, grandmother and great-aunt, to dramatise 'how ageless women are, [how] they can talk like young girls at any age'.

Beth's new life, like that of Nora in Ibsen's *A Doll's House*, exists outside and after the play, even though the play has suggested a way in which that life might be structured. Anne Devlin's *Ourselves Alone*, by contrast, is a play which more directly juxtaposes and shows the insertion of women's experience into the project of male (nationalist) history. *Ourselves Alone* was first produced in 1985 at the Liverpool Playhouse and at London's Royal Court Theatre. Subsequent productions have been staged at Washington's Arena Stage (in 1987) by Les Waters, and in Los Angeles (in 1989) by Dona Dietz. These have clarified the dramatic strengths of the play which were somewhat muffled in Simon Curtis's original production.

Ourselves Alone does not flinch before the intimidating male-dominated prospect of Irish politics, but rather rises to meet and challenge it. For its title is a direct translation of 'sinn féin amháin', one of the hoariest phrases in the country's political lexicon. Devlin's strategy becomes clear when we realise the ironic, increasingly ambiguous purpose of her title in the play it denominates. The phrase 'Sinn Féin' is still very much alive in contemporary Northern politics, describing the political party representing those who wish for a 32-county Irish Republic, still six counties short of completion. Sinn Féin decided early in the 1980s to enter the electoral process and pursue their political goal of a united Ireland with, as their spokesman Danny Morrison said, an armalite (gun) in one hand and a ballot box in the other. In 2007, after urging their fellow Republicans to renounce violence, they entered into a power-sharing Executive with the Democratic Unionist Party, who no longer insisted on addressing them as 'Sinn Féin/IRA'. In *Ourselves Alone,* Frieda describes how involvement in Sinn Féin has been part of her brother Liam's political evolution: 'My brother's changed his political line three times at least since sixty-nine.

He joined the Officials when they split with the Provos, then the INLA when they split from the Officials; the last time he was out on parole he was impersonating votes for the Sinn Féin election.'[15]

But the political party which has discomfited Northern Irish elections is *not* the primary focus of the play. For what Anne Devlin has done is to take the slogan most traditionally associated with militant Irish republicanism and reassign it to women. Thus revised, 'ourselves alone' comes to apply to a group of women in the play, two sisters Frieda and Josie and their 'sister' Donna. The dramatic *and* political struggle of *Ourselves Alone* is between those (mostly men engaged in various forms of nationalist politics) who speak an ideologically determined line and those (mainly women) who attempt to speak out of a sense of their own experience. By doing so, the women become living contradictions to the political agenda and are frequently silenced, either by being shouted down, ignored or physically struck in the face. The creativity for which they speak takes different forms (Josie's decision to have her baby, Frieda's desire to be a singer/song-writer). Overall, the play dramatises in a predominantly feminist key the resistance to political sloganeering which has characterised Irish theatre since the turn of the century. One key predecessor is O'Casey's *The Plough and the Stars*, especially the scene where Nora Clitheroe runs to find her commandant husband and returns from the battle lines to report what she saw: the look of fear in the men's eyes as they are martyred to an abstraction. Devlin builds on this when she has Frieda cry out in desperation: 'We are the dying. Why are we mourning them! (*She points at the portraits of the dead hunger strikers. She exits.*)' (39–40). Anne Devlin's play is like O'Casey's in its revisionist approach to recent (Northern) Irish history, but goes much further in its dramatic representation of women's experience as an explicit rebuke and questioning of male heroics.

Ourselves Alone finds its centre not in the hierarchical organising and mobilising by the men, but in the collective psychic and personal space established by the three women. In this regard, the second scene of Act One and the concluding scene of Act Two are crucial; they are the only two occasions throughout on which all three women are together. Devlin reveals in an Author's Note that the first scene was germinal in the play's conception: 'I began this play with two women's voices – one funny [Frieda] and one serious [Josie] – and then I found I had a third – the voice of a woman listening [Donna]. And all the women were in some ways living without men. And then the father and a stranger came into the room. And I found myself wondering who the stranger was and what he was doing there' (7). The stranger is Joe Conran, the Englishman who will become Josie's lover. But before the arrival of the men, the acoustic space of the stage is filled with the sound of women talking, whereas in the rest of the play the dialectic is male/female. This feminist emphasis does not mean that the scene is without conflict. The arrival of Frieda soon establishes the friction between the two sisters, one concentrating on looks, the other on education. One insists that her only loyalties are personal; the other counters that 'there are no personal differences between one person and another that are not political' (23). But they are arguing for the same goals from different viewpoints as they (and we) can see from the moderating third presence of Donna, the woman whose biological separateness can make her function as another kind of sister.

Their three-way exchange, whatever the personal differences aired, establishes an even-handedness in the dramatic representation that is central to feminist theatre. It resists the patriarchal tendency to focus on one character (the leading actor or 'star') and to subordinate all other concerns of character and plot in a strict hierarchy. This dramatic decentring of emphasis, first introduced by Chekhov, has been politicised by Brecht and taken on by women's theatre as a precondition of performance style and often of creation through collaboration.[16] No one female character has any social or political role which enables her to wield authority over another. In this respect, Josie's position within the republican organisation, which allows her to stall Joe Conran's sexual advances by reminding him who's doing the interrogating, has no place within this feminist environment. The hierarchy reasserts itself with the intrusion into the scene of their father Malachy, who immediately sets about the process of differentiation by criticising Frieda and hugging Josie as a 'mate' (to which she replies, 'I'm not your mate. I'm your daughter.') (25). The arrival of the other man arouses Frieda's competitive instincts and she makes a play to draw Conran away from a likely alliance with Josie.

The play's final scene reunites the three women temporarily after their intervening liaisons with a man. Josie has been with Conran, whose baby she is now carrying but which she declares her own, finally its own. Frieda has had a passionate fling with John McDermot and with the Workers' Party until recognising that the physical pleasure finally does not compensate for the loss of selfhood and self-determination exacted by the relationship. Donna, whose man spurns her on his return from prison, stoically yields to each inevitability in turn, including their ultimate reconciliation. What deepens the final scene, as it has at key moments throughout the play, is the way these meetings in the present resonate off shared moments in the undifferentiated solidarity of their youth. Frieda recalls 'a long time ago, a moonlit night' (90) when the three of them 'slipped off from the campfire to swim leaving the men arguing on the beach'. Merging in the archetypal imagery of sea, moon and stars, the three women achieve a collective epiphany before the male voices intrude once more:

> And we sank down into the calm water and tried to catch the phosphorescence on the surface of the waves – it was the first time I'd ever seen it – and the moon was reflected on the sea that night. It was as though we swam in the night sky and cupped the stars between our cool fingers. And then they saw us. First Liam and then John, and my father in a temper because we'd left our swimsuits on the beach. And the shouting and the slapping and the waves breaking over us. (90)

These images cast their radiance retrospectively on the entire play, lighting up the intimate connections which bind the women together beneath the fragmentary and divisive experience of their adult lives.[17]

If the women are regarded as intruders into the Republican club, with its 12 male hunger strikers pictured on the wall, the asylum in *Ourselves Alone*, the area in which the contradictions of sexual and nationalist politics most potently interact in Devlin's drama, is in bed. In shedding their clothes, lovers at least temporarily shed their constraining and divisive social roles, those imposed by history and tradition: 'that bed is like a raft and that room is all the world to us' (16). The result may be a renewed sense for the woman of her own desirability, the integral plenitude of the body that so much energy in Irish politics and religion is devoted to denying. It may issue in a child, a creative act which lessens the woman's dependence on the man and enables her to declare, as Josie does: 'You don't have to worry about me, Joe. I've got two hearts' (79). The bed is the scene of hauntings, as in Donna's remarkable speech about the devil lying with her and Josie's recurrent fear of the shadowy figure at the end of the bed, 'priest or police' (78), in addition to her sleepwalking. Earlier, Josie admits that, in freeing her from the restrictive grip of a single overdetermined identity, the act of lovemaking gives her the protean freedom to take on other identities at will: 'Sometimes when we make love I pretend I'm somebody else. [...] Sometimes I'm not even a woman. Sometimes I'm a man – his warrior lover, fighting side by side to the death. Sometimes we're not even on the same side' (17). By embodying on-stage a range of cross-questioning, *Ourselves Alone* displays a positive deconstruction to counter the purely negative theatrics of the politicians and the bombers.

A major development in women's theatre in Northern Ireland was the founding of the Charabanc Theatre Company in 1983.[18] It was a collaborative venture proceeding from five actresses who were stung by the absence of strong roles for women in Irish theatre. As one of them remarked: 'I played nothing but Noras and Cathleens.'[19] Three went on to form the nucleus of the company: Marie (Sarah) Jones, Carol Scanlon Moore and Eleanor Methven. The writing and the performing (but not the directing) were collaborative. For their first project, the company decided to research the lives of Belfast women working in the linen mills early in the century, and to focus a play around the mill workers' strike of 1911. Belfast playwright Martin Lynch, who they initially approached, encouraged them to write it themselves 'and I'll help' and the result was *Lay Up Your Ends*, which established Charabanc's combination of social acuity and satiric comedy. In the subsequent years of the 1980s, they divided each year into six months of researching and writing the plays followed by six months of extensive touring in Northern and Southern Ireland. Their touring venues were designed to cross the notorious theatrical class divides, in venues ranging from conventional theatres to community centres and church halls. Each year saw a new play: *Oul Delf and False Teeth* (1984), *Now You're Talkin* (1985), *Gold in the Streets* (1986), *The Girls in the Big Picture* (1987) and *Somewhere Over the Balcony* (1988). Their topics ranged from discrimination to emigration, not to the more glamorous USA (as in Friel) but to the more overlooked UK, their settings from a reconciliation centre to Belfast's Divis Flats. The latter was the setting of *Somewhere Over the Balcony*, which I saw in Dundalk Town Hall and which characterised many of the Charabanc Theatre Company's strengths. The play's time was the eve of the anniversary of the introduction of internment, and so we were shown the women who had undergone another kind of internment after the seizure of their husbands/men. The women left behind are potentially more subversive, since they remain on the outside. As such they are continuously under surveillance. We become aware as audience of (and are implicated in) the presence of off-stage anonymous forces watching the women on the balcony of their flats. This Beckettian set-up provokes great resistance in terms of the verbal defiance the women hurl at their observers and even more by the surrealistic verve with which they re-imagine the conditions of their daily lives. When these verbal resources and the women's mutual support fail, there are moments of extreme isolation and an underlying sadness. *Somewhere Over the Balcony* was the last collaborative Charabanc venture. In 1990, Marie Jones, who had increasingly become prominent in the play-writing process, singly authored *The Hamster Wheel* and differences arose during the subsequent production. She left the group to pursue a solo writing career and Charabanc disbanded in 1995. Marie Jones broke through in the 1990s to an unprecedented level of success for a contemporary Irish woman playwright with *Women on the Verge of HRT, A Night in November* and *Stones in His Pockets* all enjoying long runs in London and New York.

Charabanc had been preceded in the Northern Irish context by the Field Day Theatre Company, whose inaugurating act was a production of Friel's play *Translations*, premiered in Derry's Guildhall. Co-founded by playwright Friel and actor Stephen Rea, Field Day was soon augmented by a board comprising writer and academic Seamus Deane, poet Seamus Heaney, musicologist David Hammond, poet and critic Tom Paulin and (in 1988) by playwright Thomas Kilroy, the only Southerner in the group. In terms of theatre, Field Day arose from Rea's wish to act in productions in his own place (his career was largely London-based) and from Friel's to have his plays staged initially in the place where they had been written and which they represented. Through this act, Rea and Friel did much to reverse the trend whereby the sources and settings of Irish drama were often in remote rural areas but were staged in the metropolitan centre, whether Dublin, Belfast or London. Field Day did not intend for the productions to remain local. From their Derry base, the Company toured their annual productions around Northern and Southern Ireland, setting an influential pattern for cross-border cultural activity. They also made a frontal assault on London theatre. *Translations* was the first Irish play to enter the repertoire of the British National Theatre, and later Field Day productions like *Making History* transferred directly there after their Irish run. Field Day, therefore, forged the way for the increased hospitality of the London stage from the early 1990s on to other contemporary Irish playwrights such as Frank McGuinness, Billy Roche and Conor McPherson.

For Field Day their enterprise was cultural and political in its scope and energies. Even the choice of venue in a town like Derry which until recently had no proper theatre was symbolically apt: the mayoral Guildhall with its legacy for the Catholics of gerrymandering and second-class citizenship. Since then, all these features have changed; but the need for such change was highlighted by Field Day's activities. The first nights in the Guildhall saw a complete political spectrum of Northern Ireland – from Sinn Fein and the SDLP on the Catholic side through Unionists of various shades on the other – sharing the same space[20] and anticipating by several decades what was put in place by the power-sharing Executive. Under Deane's direction, Field Day also undertook a series of pamphlets on cultural, political and religious issues relating to Irish writing in order to 'contribute to the solution of the present crisis by producing analyses of the established opinions, myths and stereotypes which [have] become both a symptom and a case of the current situation'.[21] This developed into the mammoth *Field Day Anthology of Irish Writing* project which sought to recuperate works by Irish writers from the English canon (like the plays of Shaw and Wilde) or which had enjoyed very little prominence; it also extended the definition of Irish writing beyond the literary genres into such areas as political speeches. What it singularly failed to do was find much room for women (all 12 of the associate editors were men) and the subsequent furore led Deane to commission a subsequent two volumes focused on women's writing. In the political vacuum that was Northern Ireland in the 1980s, Field Day wished to make the same kind of cultural intervention, to conjure a 'fifth province' as they put it, as the founders of the Irish Literary Revival had a century earlier. The enduring importance and continuing influence of the work they did justifies the comparison.[22] For the purposes of this book, the plays it commissioned remain Field Day's most important contribution. Three have already been discussed – Kilroy's *Double Cross* and *The Madame McAdam Travelling Theatre* and Parker's *Pentecost* – and I propose at this point to discuss two more: Friel's classic *Translations* as a rereading and rewriting of historical drama, and Tom Paulin's *The Riot Act* as the translation of Sophocles' *Antigone* into a gritty Northern Irish vernacular. Between them, these two plays illustrate how the Field Day enterprise deployed history and myth in response to an inherited cultural and political impasse.

Much of the critical commentary generated by *Translations* has concentrated on the historical incidents reflected in the play, notably the shift from an Irish- to an English-speaking native populace, or on the novel way in which Friel draws on George Steiner's philosophical meditations on language and its origins in *After Babel*. Too many analyses of the play have overlooked the terms, gestures and syntax of the theatrical language in which it is embedded and transmitted. The two dramatic images I took away from the original Field Day production of *Translations* were of a man on his hands and knees scrutinising a large map which takes up much of the available floor space; and a pair of lovers speaking words which neither of them can understand as they struggle in physical mime towards some kind of mutual understanding.

The two scenes not only play off each other but resonate off culturally central prior texts in drama of the English language. I am thinking particularly of Act III Scene I of *Henry IV, Part One* in Shakespeare's second historical tetralogy, where the disaffected English rebels gather in the Welsh camp. The central act on which all the male parties present are engaged is the reading of a map, a map of the lands they hope to conquer and divide. This act is followed by a parting between the Englishman Mortimer and his bride, the Welsh chieftain Glendower's daughter, one-sided because both husband and audience are unable to translate the woman's Welsh speech into English. Among other things Friel's *Translations* is a response – linguistic, political, cultural and dramatic – to issues raised in and by Shakespeare's history plays. Another way of putting this is to view *Translations* as a modernist rewriting, a revision of Shakespeare along the lines of Tom Stoppard in *Rosencrantz and Guildenstern Are Dead*. From the modernist perspective, the tragic hero or military ruler is deemed marginal or irrelevant. The original Shakespearean text is decentred, dramatically in favour of off-stage or minor characters, in social terms from the nobles and their power struggles to the people who suffer at whatever remove the consequences of those decisions. Where Stoppard translates his protagonists to the aesthetic domain of a Beckett-like limbo, Friel keeps the terms political while moving the scene from the centre of power and rule, London (or its local equivalent, Dublin), to the remote fringes of Donegal, the island's northernmost county, furthest from the political and cultural capital.

Where Welsh people are denied a transmissible voice in Shakespeare's drama, their Irish equivalents now struggle to speak and make themselves understood. In that scene from *Henry IV, Part One* the contentious young Englishman Hotspur makes the one reference to the language with which we are concerned when he says that, rather than Welsh singing, he would prefer to hear his dog 'howl in Irish'.[23] This bears on *Translations* and Manus's opening efforts to teach Sarah to speak. Without language, capable only of inarticulate and unintelligible sounds, Sarah is scarcely distinguishable from the dumb beasts of the field, a point re-enforced by the play's setting of an abandoned stable which now doubles as a classroom. The first words she is coaxed into uttering are 'My name is Sarah',[24] the first act of naming in a play obsessed with the theme and a crucial step in Sarah's sense of her own identity. Manus's cry of jubilation – 'Soon you'll be telling me all the secrets that have been in that head of yours all these years' (385) – testifies to language as the key to memory, the means by which identity is not only formulated in the present but accumulates across time. This theme is taken beyond the purely personal span of a life, Sarah's or anyone else's, and is retrospectively extended to the beginnings of Western culture through the sounds simultaneously emanating from the other side of the stage. There Jimmy Jack is reciting Homer in the original Greek. Friel is subtly attuning his audience's ear to apprehend more than one language within a single stage space and so preparing us to receive two more, the Latin words and phrases that are the schoolmaster Hugh's set-pieces, and the Irish language which all but two of the play's characters speak.

Since the setting of *Translations* is a schoolroom and the dramatic occasion a class, the pupils and the audience have a right to an English translation of Homer and Virgil. But the case presented by the fact that the majority of the characters are Irish-speaking is a different and a difficult one. Two possible solutions to this linguistic dilemma have been advanced by Shakespeare in the *Henry IV/Henry V* plays. The first, as in the scene in Glendower's Welsh camp, is to leave the language in its original form. His stage direction there reads: '*Glendower speaks to her* [his daughter] *in Welsh, and she answers him in the same.*' This practice differs from a similar scene at the close of *Henry V* with which Friel's love scene *has* been compared (the King's wooing of a French bride in her tongue). There, Shakespeare writes the basic French and presumes on the understanding of his courtly audience. In that play, too, with its first appearance of a Stage Irishman, he gives Macmorris, Fluellen and Jemy not their native Celtic languages but the appropriate English dialects of Irish, Welsh and Scottish. In the case of the original Welsh in the earlier play, directions are given for dialogue which he himself does not provide – a scene surely without precedent in Shakespearean dramatic practice. The result of using the original Celtic language from the viewpoint of a dramatist committed to writing in English is a series of unintelligible sounds or, effectively, silence. This possibility emerges briefly in the opening of *Translations* where, since what we hear Friel's characters speak is English, any international audience could only assume that Sarah is being taught to speak in that language and that her 'grunting' represents Gaelic. The second alternative, adopted by Shakespeare in *Henry V*, is to have the Irishman, the Welshman and the Scotsman speak in dialect. Friel is wary on this score. There is now too long a tradition in Irish writing of phonetically transcribing Irish speech into English; too often it brings with it the impression of quaintness and condescension. Dialect is marked by strangeness and deviation from the pattern of standard English and so, while careful to preserve certain features of Northern Irish syntax and speech, Friel writes in the main a dramatic prose scrupulously free of phonetic variation, one in which phrases from Steiner's *After Babel* can be inserted without disruption, and which at times approaches Beckett's in the suggestion that it has itself been translated through the filter of another language to achieve a more neutral style. But an Anglo-Irish dialect such as Shakespeare's Macmorris or an O'Casey character speaks isn't really the issue. For, as Friel noted in a diary kept during the writing of *Translations*, the people 'would have been Irish-speaking in 1833. So a theatrical conceit will have to be devised by which – even though the actors speak English – the audience will assume or accept that they are speaking Irish. Could that work?'[25] This proposes a third solution, more radical than the two Shakespeare experimented with, whereby the same language connotes two separate linguistic realities.

It is not immediately apparent that the English we are hearing is, in fact, Irish. But Jimmy Jack's Greek prepares us for a situation where two languages which neither side adequately comprehends occupy the same acoustic space. The pupils gathering for their class and trying out their Latin provide this unwitting disclosure to the audience:

> MAIRE: *Sum fatigatissima* [...] That's the height of my Latin. Fit me better if I had even that much English.
>
> JIMMY: English? I thought you had some English.
>
> MAIRE: Three words. Wait – there was a spake I used to have off by heart. What's this it was? (*Her accent is strange because she is speaking a foreign language and because she does not understand what she is saying.*) 'In Norfolk we besport ourselves around the maypoll.' What about that!
>
> (388)

It is significant that it is Maire who raises the issue, since she will later challenge the schoolmaster on the score of not teaching them English. She wishes to learn it in order to emigrate to New York, to make space in her large family and to send money home. To the economic forces operating on her is added a historical dimension in her reference to Daniel O'Connell, committed to pursuing social progress for Irish Catholics through the adoption of English. But Maire's interest in acquiring English has been piqued by the arrival of British soldiers in the area to carry out the first ordnance survey map of Ireland. They are accompanied by the schoolteacher's prodigal son, Owen, who returns outwardly successful but paid by the English as translator, a go-between to mediate in a variety of ways between the army and the natives. Friel's 'theatrical conceit' reaches its height at the end of Act One, with Owen translating Captain Lancey's instructions for the benefit of the locals and softening their political edge in the process. This act of translation is all done through the medium of English, however, which gives a farcical tone from the audience's point of view to the mutual incomprehension of both sides.

But there *are* some Irish words spoken in *Translations*, most of them concentrated in Act Two. This middle act is divided, unlike the others, into two scenes, one showing Owen and his friend Lieutenant George Yolland engaged in their task of translating and map-making; the second is a love scene between Yolland and Maire. These two scenes jointly probe the play's deepest concerns. Act Two has scarcely begun when Yolland urges 'Say the Irish name again' and Owen replies 'Bun na hAbhann' (409–10). We now have a parallel of the opening, the private language lesson between Manus and Sarah, with the twist that the Englishman wishes to become fluent in the Gaelic language that his map-making is helping to destroy. The cultural boundaries are not as clearly drawn as they were in Shakespeare in the antagonism between Hotspur and Glendower. For it is one of Friel's revisionist ironies, following Shaw's Doyle and Broadbent in *John Bull's Other Island,* that the Irishman is now the hard-headed pragmatist, the Englishman the romantic in love with the Celtic temperament and landscape. So it is Owen, the native, who severs the intimate identity between Bun na hAbhann and the place it denominates as 'the foot of the river' by substituting for it the new name 'Burnfoot' (410),[26] while the outsider Yolland insists on keeping some link with the memory of the dead by retaining the name *Tobair Vree* or 'the well of Brian'. This prompts Owen to an exasperated outburst against the mythologising tendencies of those with whom he has to deal:

> Back to the romance again. All right! Fine! Fine! Look where we've got to. (*He drops on his hands and knees and stabs a finger at the map.*) We've come to this crossroads. [...] And we call that crossroads Tobair Vree. And why do we call it Tobair Vree? I'll tell you why. Tobair means a well. But what does Vree mean? It's a corruption of Brian – (*Gaelic pronunciation*) Brian – an erosion of Tobair Bhriain. Because a hundred-and-fifty years ago, there used to be a well there, not at the crossroads, mind you – that would be too simple – but in a field close to the crossroads. And an old man called Brian, whose face was disfigured by an enormous growth, got it into his head that the water in that well was blessed; and every day for seven months he went there and bathed his face in it. But the growth didn't go away; and one morning Brian was found drowned in that well. And ever since that crossroads is known as Tobair Vree – even though that well has long since dried up.
>
> (420)

As Yolland is quick to point out, however, Owen is attempting to deny part of himself, his memory of the name and what it connotes, passed on by his grandfather and now relinquished as worthless. Yolland has spoken of the inhabitants of Ballybeg bringing him into direct contact with a living mythology: 'And when I heard Jimmy Jack and your father swapping stories about Apollo and Cuchulain and Paris and Ferdia – as if they lived down the road – it was then that I thought – I knew – perhaps I could live here' (416). But the imaginative belief spoken of by Yolland is only adhered to by the older members of the community. And in the course of the play we watch Jimmy Jack slip over into absolute credulity in the bodily existence of Pallas Athene, abolishing the fictional distinction implicit in 'stories' and ending up wedded to chimaeras. Hugh O'Donnell the schoolmaster has a self-consciousness that can value the mythology perpetuated in and through the Irish language even as he realises the physical deprivation for which it compensates: 'Yes, it is a rich language, Lieutenant, full of the mythologies of fantasy and hope and self-deception – a syntax opulent with tomorrows. It is our response to mud cabins and a diet of potatoes; our only method of replying to... inevitabilities' (418–19).

If the Celtic side of things is more complex and self-conscious in Friel than in Shakespeare, more self-critical of the myth-making tendency, then the dramatic representation of the English side is remarkably consistent, arguing perhaps that their attitude to the bordering tribes has altered little over the centuries. The sappers are led by Captain Lancey, as hard-edged and determined as his name suggests in the advancing of British interests. Lancey speaks throughout in a public idiom as a representative of 'His Majesty's government' (406), a fact which his initial air of forced *bonhomie* and Owen's best efforts only partially conceal. Similarly, in *Henry IV*, Hotspur voices sentiments in the Welsh camp that the other Englishmen present may share but which, for the sake of a political alliance, they would have him suppress. In both, the drama largely derives from the tension between a determined air of friendly co-operation and the underlying allegiances of those present, which acknowledge a differing set of cultural and political beliefs. In both scenes, one character is markedly anti-social, standing apart and striving to point up how opposed the interests of the two sides are to each other. Hotspur's counterpart in this is Manus, motivated by the ambiguity of his brother's role and jealousy at Maire's attentions to Yolland. When Owen is through translating Lancey's speech, Manus angrily points out what the verbal exchange was most designed to cover up: 'It's a bloody military operation' (408). Owen silences him with a 'Shhhh', just as Hotspur's uncle over-rules his nephew in an aside. But the profound differences between the two sides, which are only signified later in *Henry IV* by the non-appearance of the Welsh forces to back up the rebels in battle, are central to Friel's play and are dramatised in the figure and fate of Yolland.

Where Lancey epitomises eighteenth-century Lockean empiricism, fired by industrial and colonial expansionism, Yolland is an example of the culturally compensating Romanticism that followed it. Though he invokes William Wordsworth as a neighbour, Yolland is closer to the second generation of English romantic poets in translating his yearning for the landscape, language and spirit of the place into a feminine figure, Maire Chatach, or Maire the curly-haired. Her appeal for him as a beautiful, remote muse is only heightened by the fact that she speaks a different language and, therefore, cannot directly answer his entreaties. From the first, Yolland has striven to set himself apart from his company and its leader by refusing to speak in Lancey's official idiom; when called upon to add some words to the captain's speech, the lieutenant can only stammer out a few sentences. In private speech throughout Act Two with Owen and then with Maire, he is eloquent on the score of what he has come in contact with, 'a consciousness that wasn't striving nor agitated, but at its ease and with its own conviction and assurance' (416). But that new-found eloquence embodies and springs from a deep division. For it seeks to deny what he is by birth and breeding and to conjure up another life, another identity, by the magic power of words – in this instance the rhapsodic litany of place-names that makes intimate speech between himself and Maire possible:

(YOLLAND *extends his hand to* MAIRE. *She turns away from him and moves slowly across the stage.*)
YOLLAND: Maire.
(*She still moves away.*)
YOLLAND: Maire Chatach.
(*She still moves away.*)
YOLLAND: Bun na hAbhann? (*He says the name softly, almost privately, very tentatively, as if he were searching for a sound she might respond to. He tries again.*) Druim Dubh?
(MAIRE *stops. She is listening.* YOLLAND *is encouraged.*)
Poll na gCaorach. Lis Maol.
(MAIRE *turns towards him.*)
Lis na nGall.
MAIRE: Lis na nGradh.
(*They are now facing each other and begin moving – almost imperceptibly – towards one another.*)
MAIRE: Carraig an Phoill.
YOLLAND: Carraig na Ri. Loch na nEan.
MAIRE: Loch an Iubhair. Machaire Buidhe.

YOLLAND: Machaire Mor. Cnoc na Mona.
MAIRE: Cnoc na nGabhar.
YOLLAND: Mullach.
MAIRE: Port.
YOLLAND: Tor.
MAIRE: Lag. (*She holds out her hands to* YOLLAND. *He takes them. Each now speaks almost to himself/herself.*)
YOLLAND: I wish to God you could understand me.
MAIRE: Soft hands; a gentleman's hands.

(428–9)

In the previous scene, Owen builds to a tentative union between languages, fusing into 'Oland' the different names by which the two cultures address and identify him – Owen and Roland.[27] He and Yolland become giddy at the prospect of a new beginning, a sense of Edenic possibilities opened up by this creative act of naming. But their unitary imaginings are fuelled by the poteen from Anna na mBreags, or Anna of the lies, and Yolland and Maire's dream of a life together suffers from a similar intoxicated oversight. Act Two of *Translations* is a deceptive interlude in that it leaves two characters alone on stage for the most part – Owen and Yolland, Yolland and Maire – filling the space with their hopes and the current of feeling they have generated through the Gaelic placenames. But this is done in the absence of the other characters and the historic and political contexts they represent. As soon as his brother Manus re-enters the scene, 'Oland' reverts into the Owen/Roland dichotomy. And the consummation of the love scene between the British Army sapper and the Irish woman is witnessed by Sarah, who will betray them intentionally to Manus and unintentionally to the community at large.

The scene between Yolland and Maire climaxes in a kiss, as does that between Mortimer and his Welsh wife: 'I understand thy kisses, and thou mine, / And that's a feeling disputation' (III, i, 204–5). The latter's private scene is played out before an on-stage audience and reminds us that, however much genuine feeling Mortimer appears to display on the occasion, his marriage with Glendower's daughter is a political move designed to secure his release from captivity, convert enemies to allies, and bolster the rebellion. The embrace between Maire and Yolland has an unintentional, uninvited audience of one (they have also been seen leaving the dance together). The political consequences of the private act Sarah witnesses follow later that night and throughout Act Three of the following day. The primary difference is that Mortimer's marrying into the Welsh tribe has been sanctioned by both sides. In Friel's play, the opposite is the case:

> Do you know the Greek word *endogamein*? It means to marry within the tribe. And the word *exogamein* means to marry outside the tribe. And you don't cross those borders casually – both sides get very angry.
>
> (446)

In Act Three of *Translations*, the personal desires expressed in the previous two scenes are swept aside or, more accurately, are swallowed up by the retributive logic of tribal impulses. The disappearance (and presumed murder) of Yolland on the one side is answered by a military sweep and the threat of eviction on the other. Where there is relative equality of status between the Welsh and English leaders, here there is only one formally constituted army with one set of acknowledged leaders. Apart from the strike-by-night guerilla tactics of the Donnelly twins, the Ballybeg natives are helpless in the face of forces which threaten to appropriate their native ground.

Friel has had Yolland refer earlier to the map-making enterprise as an 'eviction of sorts' (420), revealing the extent to which he views this characteristic nineteenth-century event as occurring also on the linguistic and psychological planes. The culmination of the process is not only the threat of literal evictions but the dramatic expression of the idea as a series of exits or displacements from the stage. Yolland disappears between the acts; Manus departs abruptly, with no fixed destination in mind, early in Act Three; Maire turns up in a distracted state but without Yolland has nowhere to go; Owen, the man who sought to straddle the divide where Maire has sought to cross over it, subsides into subservience; and Jimmy Jack and Hugh retreat from the chaos of the immediate situation into the consoling lies of Anna na mBreag's poteen and the increasingly less familiar lines of Virgil's *Aeneid*. Where the climax of Shakespeare's *Henry IV* overrides the failure of the Welsh forces to appear and the slaying of Hotspur with the glory of Prince Hal's public apotheosis, no such triumphalist ending is possible for an Irish dramatist writing of historical themes. Friel's muted, deliberately anticlimactic ending concentrates on registering the losses, as the brief linguistic and personal reconciliation of Act Two is decisively overborne. The one note of hope resides in Hugh's change of heart about learning the new place names and his agreement to teach Maire English, but that note equally sounds a warning: 'But don't expect too much. I will provide you with the available words and the available grammar. But will that help you to interpret between privacies?' (446). Any Irish audience brings with it the historical knowledge that 12 years in the future lay the devastations of the Great Famine, already presciently present in the sweet smell of potato blight that hovers in the play's atmosphere. And the location of *Translations* bears on a twentieth-century exercise in cartographic division and a city which still mobilises its loyalties around naming itself 'Derry' or prefixing 'London' first. Within the play, the overall sense is one of loss. For the characteristic music is not that which to Mortimer's English ears 'makes Welsh as sweet as ditties highly penned, / Sung by a fair queen in a summer's bower, / With ravishing division, to her lute' (III, i, 208–10). In Ireland, the 'ravishing division' cuts deeper in the music we have learned of ourselves; and Friel's play, reaching for tentative harmony in Act Two, modulates finally into a solo note of brooding desolation.

George Steiner provides a link between Friel's revising of the history play in *Translations* and Tom Paulin's treatment of the Antigone myth. The former acknowledged the influence of Steiner's *After Babel* and its study of the complex relations between language and the consciousness of a culture.[28] In 1984, Steiner published *Antigones*, which addresses the recursion of the Greek dramatic myth in nineteenth- and twentieth-century thought and speculates on the reasons for its persistence:

> Whenever, wherever, in the western legacy, we have found ourselves engaged in the confrontation of justice and of law, of the aura of the dead and the claims of the living, whenever, wherever, the hungry dreams of the young have collided with the 'realism' of the ageing, we have found ourselves turning to words, images, sinews of argument, synecdoches, tropes, metaphors, out of the grammar of Antigone and of Creon.[29]

That same year, without prior knowledge of Steiner's magisterial survey or – they claim – of each other, three Irish poets independently moved into the theatrical medium by writing versions of Sophocles' *Antigone*: Aidan Carl Mathews, Brendan Kennelly and – the version I wish to consider – Tom Paulin, whose *Antigone* is surprisingly the only one of the three to issue from a Northern Irish perspective. *The Riot Act* was first presented by Field Day at the Guildhall, Derry, in September 1984.

Tom Paulin himself does not slot conveniently into the dividing categories so favoured in the North. He was born in England and teaches at Oxford University; he was raised in Northern Ireland as a Protestant Unionist when his parents returned to Belfast; and in the late 1970s he changed sides, switching allegiances from his Unionist heritage, not so much to its Catholic counter-image, as to a Utopian vision of nationalist identity that would reconcile Protestant Dissenter and Catholic Republican. The zig-zag movement Paulin deliberately traces and the mixed heritage he embodies enlarges the perspective of the 'Northern problem' to set it in the dual context of the Republic of Ireland (the 'South') and Britain ('Westminster').

The first person to draw a comparison between recent events in the North of Ireland and Sophocles' *Antigone* was diplomat-critic Conor Cruise O'Brien. He did so very early on, in October 1968, within the same month as a Civil Rights march was set upon by the Ulster police. In *The Listener*'s reprinting of his Belfast lecture, O'Brien defined the action of Sophocles' heroine as 'nonviolent civil disobedience',[30] terms deliberately meant to echo the Civil Rights Movement in the US, especially Martin Luther King's doctrine of passive resistance and his argument that civil laws are just or unjust when viewed in the light of a divine or higher law. In his very next sentence, however, O'Brien lays the blame for all the deaths in the play – Haemon, Eurydice and her own – squarely on Antigone and her provocative defiance of legitimate civil authority. But what of the play's final judgment (via the Chorus and the prophet Tiresias) that Creon is to blame for unleashing this cycle of death by denying the dead Polyneices to the earth and by forcing the living Antigone into the grave? O'Brien concedes that Creon's

> decision to forbid the burial of Polyneices was rash, but it was also rash to disobey his decision. [...] Creon's authority, after all, was legitimate, *even if he had abused it* [emphasis mine], and the life of the city would become intolerable if citizens should disobey any law that irked their conscience.

Increasingly throughout the piece, Conor Cruise O'Brien becomes the apologist for Creon and his practices. In so doing, he ignores his own finer perception that 'Creon and Antigone are both part of our nature, inaccessible to advice, and incapable of living at peace in the city'. His remark in the same paragraph, 'the play is still performed', suggests the intimate *intersection* between theatre and politics by wittily transferring the play and its cast from the stage to the realm of political action.

Translated into the local conditions of Northern Ireland in 1968, *Antigone* collectively represents the Catholics marching to protest the inequities of a state which had consistently discriminated against them; individually, Bernadette Devlin (as she then was) steps into a role which could have been written for her, a strikingly young and impassioned woman standing up against the oppressive, patriarchal institutions of Stormont and Westminster. If in O'Brien's overall argument Creon wavers between being an individual ruler in whom power is vested or an institution which preserves the fabric of the status quo, then in the political allegory he represents both the local power of Stormont and the longer arm of the Westminster law which took over direct rule in 1972. At the close of his piece O'Brien rounds again on his rational pragmatism to weigh the gaining of peace against 'the price of [the] soul', an unintentional anticipation (as he later noted)[31] of the title of Bernadette Devlin's autobiography, *The Price of My Soul*.

In 1980, Tom Paulin displayed his first interest in the *Antigone* when he used the occasion of a *Times Literary Supplement* review to focus on the writings of Conor Cruise O'Brien.[32] In particular, Paulin notes the revisions *The Listener* lecture underwent when reprinted in O'Brien's 1972 book, *States of Ireland*, and the increasing shift its revision registers 'from the instinctive and intuitive [stance of an Antigone] towards the rational [Creon]'.[33] Four years later, O'Brien himself had declared:

> I find myself no longer in sympathy with the conclusion. Antigone is very fine on the stage, or in retrospect or a long way off, or even in real life for a single, splendid epiphany. But after four years of Antigone and her under-studies and all those funerals [...] you begin to feel that Ismene's commonsense and feeling for the living are preferable.[34]

The final issue for O'Brien is one of obedience to authority, allegiance to the body politic, the rationally codified and transmitted civic law. But if Conor Cruise O'Brien is denying the Antigone within him, Tom Paulin in his essay denies the Creon he once was by attempting to exorcise and castigate his earlier affiliations:

> Until about 1980 [...] [I] reacted like most members of the Unionist middle class and believed that Conor Cruise O'Brien was putting 'our case'. But there was something different in the air as the decade ended. I started reading Irish history again and found myself drawn to John Hume's eloquence, his humane and constitutional politics. As a result, O'Brien's articles in *The Observer* began to seem sloppy and unconvincing and I felt angered by them.[35]

Tom Paulin takes his argument a crucial, transformative step further when, in *The Riot Act*, he writes a version of the Sophoclean tragedy that has figured so largely as a mythic paradigm for politics in the North. The debate with O'Brien was clearly an incentive, and Paulin's dramatic articulation of the issues gives them a range that transcends mere personal polemic. Where the figure of Creon in the prose debate was subordinated to a gloss on the development of O'Brien's conservatism, the dramatic persona now absorbs O'Brien as only one of several possible, partial reflections. Cruise O'Brien can be heard most clearly through those lines of Creon where he aligns his own motives with the pragmatic defence of the state. But when Paulin's Creon speaks in more distinctive tones, we hear accents that come from the mouths of Northern, not Southern, Irish politicians. More precisely, we hear a verbal medley of the two reigning powers in Northern Ireland, Westminster and Unionism. Creon comes on stage shortly after assuming power and his remarks are very much those of the new man on the job. His 'I shall be doing a very great deal of listening'[36] sounds, as Fintan O'Toole remarked, like a 'parody of a Northern Ireland Office political functionary appealing for public support',[37] and Mitchell Harris finds the sentence itself echoing 'the opening remarks of Northern Ireland's incoming Secretary of State',[38] Douglas Hurd, in the summer of 1984.

But if Paulin's Creon starts out by sounding like a Westminster functionary, other identifications soon emerge. He is introduced by the Chorus as 'The big man', an unmistakable reference to the Reverend Ian Paisley, demagogic leader of the Democratic Unionist Party and subsequent First Minister in the 2007 Executive. Creon's speech, therefore, is not only that of the practised public official from 'the mainland' but also of someone from Ulster, a Unionist anxious to reassure those he represents by sounding the code words of the tribe: 'law', 'order' and 'loyalty'. Creon's speech progressively mutates into distinctively Ulster, rather than Oxbridge, tones and dialect, as in his reference to Ismene: 'And this one here – / the sneaky, sleaked one – / she lived in my house too' (30). In Sophocles' play, the character of Creon under the pressure of events abandons his appeal to the larger forces of stability and civilisation, in whose name he acts, and soon has no higher argument than the brute assertion of will and violence. Similarly, Paulin's Creon ends up sounding like nothing more than the hard man of violence, self-appointed leader of a gang of bullyboys deciding who shall live or die: 'Bring out the dirty bitch / and let's be rid of her' (42).

Paulin takes fewer liberties in presenting Antigone herself and is all the more taxed to have her make a distinct dramatic impression. He finds his means by concentrating on her speech. Since he is looking for a distinctive Anglo-Irish or Hiberno-English usage in his writing, the issue of speech is as central to Paulin's translation of Sophocles as it was to Friel's *Translations*. Drawing on Richard Jebb's nineteenth-century translation, Paulin pares away the florid Victorian embellishments to arrive at a short verse line, lean, terse, understated. The play is written in a pared, minimal style, conversational yet urgent, whose Anglo-Irish speech and syntax find a home and context for such lexical outcasts as 'screggy', 'sleg', 'pobby'. The characterising epithet of his *Antigone* is introduced in the first scene by Ismene when she responds to her sister's declaration of filial allegiance by calling her, or more precisely her manner of speaking, 'wild':

ANTIGONE: He's my own brother,
and he's yours too.
I can't betray him
ISMENE: You're talking wild –
it's Creon's order.

(11)

This brings us back to Paulin's debate with Cruise O'Brien. The closest point of *rapprochement* between them is reached when Paulin seizes on a play written by O'Brien, *Salomé and the Wild Man*, and Salomé's admission to the sophist, Philo, that they are both 'lonely for the wild man'.[39] Paulin goes on to define 'wild' as 'a word with a distinctive usage in Ireland' and to cite as a characteristic example 'Yeats remembering the pre-revolutionary Constance Gore-Booth "With all youth's lonely wildness stirred"'. This wildness is not a barbarism to be set over against civilisation since the terms of these polarities have been too long co-opted by the English, casting themselves in the light of the bearers of civilisation, order, rule and moderation, and the Irish as the unkempt barbarians who will not be tamed but Caliban-like insist on wallowing in the mud. The 'wildness' may be transvalued as exuberance, primitive earthiness, an integrity of body and soul that resists social integration or confinement within limits. This lonely 'wildness' is something Antigone claims and it is something Creon will have to share or understand to some degree, as Paulin uses it to find unlikely agreement with W. B. Yeats and Conor Cruise O'Brien.

From the perspective of Creon and the state, Antigone's behaviour is 'wild' only in the sense that it is unruly, potentially anarchic, threatening the stability of civic order and the rational principles it incorporates. But Antigone, in Paulin as in Sophocles, claims she is acting in accordance with another set of laws, one that is not officially inscribed in the edicts of the state. Her loyalty is to her own, kith, kin, family, tribe; and in insisting on proper burial rites for her brother, she insists she is exonerating herself before another court of appeal than the state's tribunals. Although Paulin has cut back on Antigone's references to the gods, and resisted giving them a Christian gloss, those he retains are the references in Sophocles where the gods are not located on high but down among the dead, the instinctual aboriginal forces to whose submission Antigone owes her 'wildness':

> ANTIGONE: It was never Zeus
> made that law.
> Down in the dark earth
> there's no law says,
> 'Break with your own kin,
> go lick the state.'
> We're bound to the dead:
> We must be loyal to them.
> I had to bury him.
>
> (27)

This notion in the Greek original of two sets of laws at variance with one another transfers best to those contemporary societies where some inequity or variance is widely perceived to exist in the man-made laws, where a portion of its people suffer under such inequities that a gap is opened into which other laws may be summoned. It is no surprise, therefore, to learn of productions of *Antigone* in Poland or to find two black prisoners in South Africa's Robben Island, jointly performing the play in Athol Fugard, John Kani and Winston Ntshona's *The Island*. The long interposition of Westminster in the making of Irish laws North and South has encouraged an equivocal, ambivalent adherence when laws are widely perceived as externally imposed and alien. A revolutionary opposition emerges that derives its own laws from a sense of extended family, the sibling relations of fraternity and sorority. It is just such a 'familial' political gesture that Antigone offers Ismene at the beginning of the play: 'Will you go in with me or not?' (10). When Ismene rejects the offer, she effectively rejects this pact of sisterhood and is no longer regarded as such by Antigone. As she goes to face her own death, Antigone addresses in turn each of the members of the House of Laius who have preceded her – Oedipus, Jocasta, Eteocles and Polyneices – and in so doing, invokes their invisible presences as more palpably real than those on stage. Implicit in the play's mythos, Antigone's dual allegiance to the dead and to her family also resonates off the Irish hunger strike. Richard Kearney writes of this tradition of sacrificial martyrdom:

> In 1980, a Maze prisoner reiterated this sentiment when he wrote on the wall of his cell: 'I am one of many who die for my country ... if death is the only way I am prepared to die.' The *many* here refers to a long litany of martyrs whose sacrificial death for Ireland has been translated into the 'sacred debt' of the 'freedom struggle'.[40]

Finding no point of sustained identification with the formal body of laws encoded in the North, Republican prisoners apprehended under those laws turn in a self-consciously ritualised and dramatic way to the extended family of those who have preceded them. The only mediators that a baffled state can resort to in such an impasse are members of the immediate family (most powerfully and persuasively, the mother) and a priest, the two chthonic forces released by such a supra-rational gesture.

Kearney continues by remarking that 'one of the most popular responses to this sacrificial attitude has been the emergence of ballads, snatches or rhymes which, like myths, are often authored by nobody yet known to everybody'. Before she is led away to her cell, Paulin's Antigone sings her own dirge with snatches of an anonymous folk-song:

> ANTIGONE: [*Sings to herself*]
> I heard her cry
> as I climbed the track –
> my friends are cold
> though my bairns are dead.
>
> (46)

This touches upon Antigone's declaration in Sophocles just before she is led off that she is condemned to have 'no part in the bridal-song, the bridal-bed, denied all joy of marriage, raising children'.[41]

What is dramatically so surprising about this final wrenched admission is the lack of emotion previously displayed. Nowhere prior to this does she refer to the fact that she is affianced to Haemon; Ismene brings it up. Antigone's subordination of her own personal life to the cause that compels her is so absolute that Jean Anouilh softened it in his version by offsetting the martyr-to-be with a nervous, romantic schoolgirl. Paulin only makes several brief additions, not to sentimentalise his tragic heroine, but to show the 'excess of love' from which her actions spring: when she breaks off in her repudiation of Ismene to admit 'it tears my heart, though'

(32); and when the term 'love' is added to both her sister's and fiancé's names in the direct address of intimate endearment. 'Excess of love' comes, of course, from Yeats's 'Easter 1916': 'And what if excess of love / Bewildered them till they died?'[42] My reference is not gratuitous; for Paulin has Creon react to the transformed perspective that the death of wife and son have wrought in him with the same poem's most famous line:

> CHORUS: It was too late
> you changed your mind.
> CREON: I changed it, but.
> (*Touching his son*): Aye, changed it utterly.
>
> (60)

Yeats's 'Easter 1916' is part of Paulin's design, therefore, in its fusion of personal sacrifice and political event. His Creon undergoes a profound change of mind, unlike the annihilation of Sophocles' ruler to a state of nothingness. Both Creons are denied what they most seek, to follow after wife and son, barred access to the realms of the dead they have all along denied. Paulin makes clear that what Creon suffers, under the weight of that concentrated catastrophe he has just borne, is a living re-enactment of Antigone's fate.

What registers the transformation is, once more, the language. Creon now speaks Antigone's tongue, addressing his (dead) son for the first time in terms of loving kinship as 'my own wee man' and 'bairn' (60). The penultimate Choral ode places this individual transformation in a larger context by locating the gods neither in Olympus nor in Hades but in 'the quick of tongues plunging' (55), conferring the power to heal wounds not only on speech but on such other rites of celebration as dancing, music, prayer and poetry. Creon 'shared power with no man' and, in setting his face against power sharing, has paid a bloody price. When he finally capitulates and performs the burial rites of Polyneices, he takes 'olive branches / and green laurel leaves / to crown and lap him in' (58), the 'green' deftly highlighted among the classical emblems of peace. In Paulin's poem 'Under Creon', the questing narrator responds to a 'free voice (singing) / dissenting green';[43] he has written elsewhere that he and his work assume the existence 'of a non-sectarian, republican state which comprises the whole island of Ireland'.[44]

Though green of hue, this state would seek to represent the traditions on both sides which have been driven underground. For Paulin repeatedly points, not just to a proscribed Catholic republicanism, but to a genuinely Protestant 'dissenting tradition in Ulster [which] created a distinctive and notable culture in the closing decades of the eighteenth century, [but which] went underground after the Act of Union and has still not been given the attention it deserves'.[45] In 'Under Creon' it is this alternative dissenting tradition which the speaker searches for in the gaps in the imperial shrub:

> The daylight gods were never in this place and
> I had pressed beyond my usual dusk
> to find a cadence for the dead: McCracken,
> Hope, the northern starlight, a death mask
> and the levelled grave that Biggar traced;
>
> like an epic arming in an olive grove
> this was a stringent grief and a form of love.
> Maybe one day I'll get the hang of it
> and find joy, not justice, in a snapped connection,
> that Jacobin oath on the black mountain.[46]

When Creon at the close of *The Riot Act* turns his back on the daylight gods by which he has lived his life and says, 'All I want's the dark' (62), he is not only seeking annihilation (as in Sophocles) but acknowledging the instinctual forces he has all along denied. This places him in the company of Beckett's Krapp, who turns from seeking illumination in the light to a deliberate embrace of the dark; politically, it belatedly recognises the claims of the dead; and culturally it at least marks out the space through which the Field Day Theatre Company looked for common cultural ground between two dissenting traditions.

If there was one play that got away from Field Day, it must surely have been Frank McGuinness's *Observe the Sons of Ulster Marching Towards the Somme* (1985). Field Day did much to illuminate the historical conditions of the Catholic Nationalist population in the North, but little to redress the perception that the Protestant Unionist community is dour, humourless, and generally without culture. Protestant writers like Derek Mahon, Tom Paulin and Stewart Parker who contributed plays to Field Day were generally in reaction against that perceived character of Ulster Unionism. So perhaps it is appropriate that it took a Catholic from Buncrana, Co. Donegal, to write a powerful and moving drama about eight representatives of the six thousand Ulster-men of the 36th Ulster Division who were slaughtered at the Battle of the Somme on 1 July 1916. In so doing, McGuinness not only managed to challenge and confront his 'own bigotry', as he put it in an interview,[47] but to raise unsettling questions about the extent to which Catholic Nationalism has exclusively appropriated the concept of 'Irishness' for the past century. As he reveals later in the same interview, the idea for the play came to McGuinness when he saw the names of dead soldiers on two war memorials in Northern Ireland, and sought to imagine their lives back into existence, to remember them 'as friends, sons and lovers'. There is a debt here to Friel's *Translations* in the litany of names as a means of finding a connection with the past, but the names invoked are not those of Gaelic Ireland, but of a more recent Ulster past it is less easy to romanticise.

Sons of Ulster is not presented as a documentary narrative or a traditional history lesson. The first of its four sections, 'Remembrance', is set in the present and is a monologue by the one man to have survived the holocaust, condemned to a living death while all his comrades perished. References to the more recent past remain deliberately imprecise. What the aged Pyper registers is more the sense of increasing loneliness, isolation and betrayal, which assails not only him but the Ulster Protestants whose reluctant spokesman he has become: 'Sinn Féin? Ourselves alone. It is we, the Protestant people, who have always stood alone. We have stood alone and triumphed, for we are God's chosen.'[48]

In dramatic terms, what we are confronted with is an old man sitting alone on stage, asleep when we first see him, in a completely bare space, a void. He is woken by the sound of a low, persistent drumming, to protest 'Again. As always, again. Why does this persist?' (97). The situation is familiar from Beckett, of Malone being prodded by unseen forces into writing, of Hamm in *Endgame* wishing for it to end while compelled to continue. Pyper does not even have the wretched ministrations of a Clov; and yet in his absolute isolation he uses the first person plural. This occurs three times in his opening monologue. The first is the most defiant and accusatory: 'What more have we to tell each other?' Given that he is alone, this address must be seen as at least in part a dialogue with his creator or God, here imaged in the Beckettian form of an off-stage, invisible yet insistent presence who torments the speaker into remembering. The first sound heard in the play is a voice of defiance raised against the creator, not only for giving birth to the protagonist but for orchestrating slaughter on a mass scale, all carried out in his name and for his greater glory.

But the play is also something of an anti-drama, resenting the intrusive presence of the audience who have come along with their expectations and are goading the reticent Pyper into addressing them. He and McGuinness are well aware of the challenge that dramatising a world war poses to notions of representation and form. Pyper will not play the role of military historian, nor will he offer a narrative that will give 'that slaughter shape' (97). And there will be no comforting present, no stable illusory fixed point from which to view safely these events of the distant past.

The second reference for the 'we' is addressed to himself, to the voice or voices in his head that will not be silent or let him be. Before this voice, it is less easy for Pyper to defend himself. The self-accusation is directed against the fact that he has survived, when it was he who most wished to die. But guilt at the collective memory of his companions also enables Pyper to identify with his comrades, as 'we' against the ultimate enemy, a hangman God: 'Did you intend that we should keep seeing ghosts?' But the 'you' remains accusatory, oppositional. And if, as Pyper surmises like Beckett's Winnie, this God is at least capable of small mercies, then his parting gift may be to remove himself and supply the vacant place with Pyper's dead companions: 'You have bestowed your parting gift. Welcome. You look angry, David' (99). As the 'you' becomes more intimate and personal, the ghosts appear and Pyper starts talking with them, asking a series of questions leading to the ultimate: 'Answer me why we did it. Why we let ourselves be led to extermination?' (100). They do not reply; his questions are only met with silence. Pyper is forced to supply his own answer, that they did it for each other out of a deep self-hatred. With the old man incarcerated in a living hell, he turns in his despair to this collective memory of the past, a memory finally made complete by the appearance of his younger self.

This device of the younger and older Pyper, of two actors required to play the same character, is one already discussed throughout this study. It has been likened to Hugh Leonard's use of the younger and older Charleys in *Da*;[49] and McGuinness has put on record his admiration of Leonard's play. But the older Pyper does not remain continuously present; he fades upon the appearance of his younger self and does not re-enter until the play's closing moments. His presence works to keep the past and the present mutually alive and interdependent, as in Leonard, and to represent a deeply divided self, as in Friel's *Philadelphia*. The framing device helps us to read what follows less as a naturalistic presentation of a group of men going to the Front than as a dramatic re-imagining of an individual and collective trauma. Pyper's theatrical act brings his dead companions back to a ghostly on-stage embodiment; it both acknowledges and counters the audience's supreme awareness that all the characters in the play save one are doomed to die.

The play's second scene, 'Initiation', is in many respects the most conventional, a series of arrivals of the new recruits into the barracks. But an echo-chamber effect has been created by the fragments from the scene which the older Pyper has already intoned like a prelude or overture, so that lines the other characters are hearing or speaking for the first time have already been heard in the audience's head. The effect linguistically is one of ghostly recurrence, of lines circulating amongst speakers rather than being the personal property of the utterer, of something more closely resembling a chorus. This textual recirculation, and the use of a great deal of repetition of such key images as blood, work as the verbal equivalent to the drumming in Pyper's head, an important acoustic resonance.

There is a more formal organisation, too, to the arrival of the men from Ulster. Six of the eight come in pairs, speaking the language and defending the honour of a particular area of their beloved province. Moore and Millen are from Coleraine and obviously lifelong friends, with the first trusting to the guidance of the second. In their most benign aspect, they are a baker and a miller; but they also tell stories of the time when, as members of Sir Edward Carson's Volunteers, they 'battered him [a Catholic youth] down the streets of Coleraine' (123). Anderson and McIlwaine enter late, two raucous workers from the Belfast shipyards, and immediately make a scene. The preacher Roulston and the youngster from Derry, Crawford, do not enter as a duo, but the formal symmetry suggests a pairing will emerge, as it begins to do when the older man defends the younger from the Belfast pair's unprovoked attack. Pyper is already 'on' when the scene begins and David Craig of Enniskillen arrives. The special pairing that will occur here has already been suggested by the brief refrain from 'Enniskillen, Fare Thee Well' in the opening monologue and by the fact that Craig is the only one of the seven that Pyper addresses in personal rather than in formal terms through the use of his first name. The refrain of 'David' throughout the play makes of David Craig the absent presence Pyper has most need of, the one with whom a continuing dialogue is necessary.

As each man arrives he has to run the gauntlet of Pyper's sardonic address. Adopting the role of death's head jester, Pyper simultaneously disavows a personal or collective identity and offers a series of challenges to the very bases of identity itself: by appropriating the name of another man in the room when introducing himself to a newcomer; by returning to Craig a shirt he claims is not his; by giving orders and taking on the persona of a commanding officer. Pyper is particularly adept at this last role, as Craig notices; it raises the question of why someone of Pyper's class and background, which he has only managed partially to conceal, is down among the enlisted men rather than taking his place in the officer class.

The challenge that Pyper most consistently offers is to the notion of a fixed sexual identity. In the opening action of the scene, he cuts his finger with a penknife while peeling an apple and holds the finger up to the entering Craig, as a warning and a sign of the blood that will be shed, but also as a sexual come-on, as a reminder of the beautiful as well as the terrible uses to which flesh can be put. In asking Craig to 'kiss it better' (103) and in the continued presence of the apple, there is a replay of the primary seduction scene, now between two men. Where women *are* figured in the play's second scene is in the barrackroom banter of the other men; Moore and Millen, for instance, enter speaking of sexual conquest in overtly phallic terms; and the soldiers' worst term of abuse is to call one of their comrades an old woman. This polarisation of sexual characteristics whereby women are denigrated with 'feminine' qualities so that the men may assert their masculinity is broken down by the line with which each of the other men is greeted by Pyper upon arrival: 'I have remarkably fine skin, don't I? For a man, remarkably fine' (109). The two stories Pyper tells both have a sexual point: one culminates with his punching Anderson in the groin; in the other, he gains their attention by telling them of a French whore and leading them to confront the literal absurdity of their superstitions. The striptease performed by Pyper's French Catholic whore and former nun on her wedding night revealed that she had three legs, a piece of Protestant mythology which one of the listeners defends amid the general incredulity. Pyper later reveals to Craig that, in the truer version of the story, he was the whore:

CRAIG: Men or women?
PYPER: What's the difference?

(129)

The sexual difference according to Freud is a penis, which Pyper has already confounded by attaching it to a woman and a Catholic. *Sons of Ulster's* questioning of the norms and stereotypes of sexual identity is a crucially recurrent feature in its treatment of traditional political allegiances and oppositions.[50]

This theme of male sexuality is most fully addressed in the play's third section, 'Pairing'. There is a sharp intake of breath when it opens and the audience realises that the third scene is not set, as the rising curve of the second might have suggested, on the battlefield at the Somme itself. Rather, in the space between Parts 2 and 3, a 'long five months' has intervened (140). What we are shown are the men returning from the Front, coming home; but what they have experienced on the battlefield they do not leave behind them. McGuinness dramatises their experience by having them relive and work out together what they have experienced there, in many cases a symbolic if not an actual death. The second surprise of Part 3 is the dramatic daring of the technique used, which has to be seen and heard on the stage to be fully appreciated. Borrowed from the very different context of an Alan Ayckbourn play,[51] McGuinness uses split-staging to represent simultaneously four different duos and four different locations, the various places to which the sons of Ulster have retreated as 'home'.

David Craig brings Pyper to Boa Island, Co. Fermanagh, a special place associated with his earlier life, and claims to have brought him 'home'. This turns out to be true in more than one sense. For Craig has saved Pyper's life on the battlefield, has restored to him something his friend was anxious to lose, and so bears a special responsibility for Pyper's new life. That sense of renewal is celebrated by the rituals they improvise on the island. Craig expresses this in explicitly baptismal terms: 'When I walked into the Erne this morning, I just wanted to wash the muck of the world off myself. I thought it was on every part of me for life. But it's not. I'm clean again. I'm back' (140). But the return and the renewal is far more complicated for Pyper. In all four of the pairings, there is one partner who is reluctant to return, whose experience at the Somme has precipitated a crisis of identity. In Pyper's case, there is his marked ambivalence in going out to serve the gods of Ulster, to take part in Sir Edward Carson's dance on a far vaster scale; and there appears to have been a basis of truth in his story of the French whore.

Her death by suicide, and his guilty involvement, raise not only the issue of Pyper's sexuality, but his creativity, whether he can genuinely make something or whether he is perpetuating a form of personal and cultural barrenness. Kenneth Pyper has been a sculptor; and what Boa Island in Fermanagh is most renowned for is a series of stone carvings. When Pyper views them, he sees the petrified icons of the Ulster gods he has unwittingly replicated. Only when the sexual implications of the statues are activated, in a discussion with Craig as to whether they represent men or women or an hermaphroditic combination of both, does their fixity begin to dissolve. What was apparent to me from Michael Attenborough's production of the play at the Hampstead Theatre Club in London in 1986 was that Pyper and Craig made love on the island, signalled by the transition from stone to flesh and the surfacing of the water imagery. The resolution remains ambiguous for Pyper, not just in sexual but in political terms, since he also sees the love-making as taking part in Carson's dance, embracing a role and a destiny in the life of his tribe. But the sexual union with Craig has taken the dead matter, his self-loathing, and reshaped it into something immaterial and finer than his earlier sculptures.

All four pairings stage a similar struggle between the Ulster-men and their traditions. Moore and Millen are at the rope-bridge at Carrick-a-Rede in County Antrim, with Moore a burnt-out case and refusing to go back.

His gaining the courage to return is given a wonderful enactment in the crossing of the bridge on his own. McGuinness at this point inserts an explicit reference to some canonical lines from Beckett, as he does in almost all his plays, when he has Moore give his own version of 'I can't go on' – 'I don't want to go on. I can't' (143) – and rapidly follows it with 'I keep hearing the dead', which echoes the refrain between Didi and Gogo about 'All the dead voices'.[52] Those lines in Beckett's *Godot* have been taken as one of the play's few direct references to World War II and what has entered the environment we inhabit as a result. Here, the reference is explicitly one to the slaughter at the Somme, but the Beckettian terms are maintained by the interiorisation of those dead voices in Moore's head. In crossing the rope bridge, he has to confront the abyss, dangerous enough in the physical surroundings of rocks, cliffs and sea, but vertiginously more so in what the Somme has disclosed to this terrified man. He manages to cross with a symbolic handclasp from Millen and the conjuring of all his comrades; but he also draws on his artistic experience as a dyer to give a colour and a shape to the void. The fixed co-ordinates of 'here' and 'there' break down as he goes over to the regions of the dead, where 'I can see death as sure as touch your hand' (160). At the close, Millen is the one who is left behind, stranded, and begs his friend to return.

In what may have seemed an all-Protestant play, the religious divide opens up unexpectedly in the exchanges between Roulston and Crawford. Their pairing is set in the Protestant church where Roulston worshipped as a boy and where he has subsequently preached. Crawford reveals that he is at least half-Catholic, since his mother was a 'Fenian. She never converted' (161), and he insists that Roulston hear his confession. But that confession is devoid of Catholic overtones; what Crawford is declaring is his belief in man as an animal who helps his own kind. By actively drawing on the sectarian terms of Catholic-Protestant antagonism, goading Roulston as a 'Proddy' and forcing him to fight, Crawford challenges the older man's self-righteousness. For as with Craig and his island, Roulston's church is a sacred place, dear to the memory of the fathers in whose name he has served, but one which has lost its original meaning for him. It needs, and finds, a new ceremony in the physical scrap between the two men, where Roulston stops preaching at people, and Catholic and Protestant lie down together, made one flesh through fighting. As Crawford puts it at the core of the scene:

> Now you can march out of it [the church] with me, a soldier, a man, a brute beast. You're not Christ. You're a man. One man among many. (162)

The fourth pairing is in ways the most powerful, that in which the Belfast men Anderson and McIlwaine go on their own to the Orange field at Edenderry for a belated two-man celebration of the Twelfth of July. It involves a good deal of drinking, making of speeches against the traditional foe, and on several occasions a savage beating of the huge Lambeg drum by McIlwaine with his bare hands. What Anderson and McIlwaine elaborate is a parallel between the celebration of the glorious Twelfth and their fighting at the Front, chiefly by translating the Germans into Irish Catholics and the River Somme into the Boyne. The demons they have encountered there are transferred to the local enemy in an attempted exorcism: 'I do not speak of the Hun, dire enemy though he may be, when I speak of the enemy now. I speak of the Fenian' (167). But self-doubt causes the confident rhetoric to collapse. The disillusioning predictions made by Pyper are among his lines that the two men now echo and repeat. For the other event that is brought into parallel with their experience at the Somme is the sinking of the *Titanic*. The pride of the Belfast shipyards in which they work, the *Titanic*'s fate invokes a sense of doom and collective disaster for the Protestant people of Ulster. McIlwaine attempts to drown out the doubts by calling for the noise of the shipyards but they only bring the *Titanic* to mind, while the beating of the Lambeg drum leads them back inexorably to the battlefield. At the end of the scene, all eight men rise from their separate locales, their individual crises having been carefully interwoven throughout, and come together on the stage. As Angela Wilcox has put it, this third scene effectively and movingly conveys the 'fierce engagement of the Ulstermen with their native place and their Protestant traditions [...] though the traditions themselves are challenged'.[53]

In the fourth and final part, 'Bonding', the men are returned to the Somme while the audience encounters it for the first time. But no audience comes to such a scene without the collective cultural representation of British soldiers crouching down in the trenches and waiting to go over the top to annihilation. McGuinness gives us this very conventional scene all at once, as a familiar scenario, with Pyper no longer the iconoclast but another member of the troop. Throughout what follows, he is absorbed into the group. But the group identity which the individuals in 'Bonding' wish to own is not so easily determined or stable; for they need at this apocalyptic moment to register a crucial sense of dual allegiance, not only to the British but to Ulster.

Taking the men's predetermined position in the British Army as its point of departure, the final scene presents what is now a theatrical ensemble devising a number of communal activities, a series of rituals through which they bond. The first and most comprehensive is storytelling, the chief means through which humour enters the play. McIlwaine delivers a revisionist cod version of the Easter Rising, relating the story of Pearse as the 'boy who took over a post office because he was short of a few stamps' (175). The narrative proceeds on its zany way, with the imprisoned Pearse pleading to be let off on the grounds that he has a widowed mother and that same mother finally grabbing a rifle from a British officer, not to liberate her son but to shoot him and teach him not to go about robbing post offices. The moral of the story is that 'Fenians can't fight'. What gives the spoof its bite is that it keeps the narrative structure of the heroicising of Pearse on which the legend of 1916 was constructed, right down to the centrality of the mother-son relationship in the nationalist myth, even as it proceeds to mock and invert those structures. And its very anachronistic quality makes the historical point that, in the structuring of Irish history, 1916 belongs to those who occupied the Dublin GPO rather than those who fought at the Somme. There is a double reference to Sean O'Casey here, to his own dramatic questioning of 1916 heroics in *The Plough and the Stars* and to his controversial decision to dramatise World War I as part of the Irish experience in *The Silver Tassie*. In a nice ironic turn, the Abbey came to stage *Sons of Ulster* when others had turned it down, and its director Patrick Mason produced *Tassie* some years later on the Abbey mainstage, underscoring the connection between the plays by casting some of the same actors.

Storytelling has been a central feature of this study, in particular the process by which that storytelling becomes drama as it draws on other elements of ritual such as music, chant and mime. Here, the Ulster soldiers begin singing, when Roulston offers a prayer and settles on a hymn as the most appropriate shared medium. There is a brief stab at a football match, a nice reminder of one of the few shared enthusiasms in Northern Ireland. But the most elaborate play-within-this-play is the group's decision to re-enact the annual Battle of the Boyne at Scarva. When Anderson asks 'How the hell can two men do the Battle of the Boyne?' (181), the theatrical challenge mirrors that of *Sons of Ulster* itself in attempting to stage the Battle of the Somme. The cast is swollen to four, with two of the men assigned the role of horses to King James and King William. There is some irony in the casting. Crawford, whom we know to have Catholic blood in him, is chosen to play William of Orange, despite his protests; Pyper's blonde hair earns him the role of William's trusty steed. Moore and Millen play out a version of their earlier trauma and their real-life relations when Moore is placed on Millen's shoulders. Neither is keen to play the historical loser, but as they are reminded: 'And remember, King James, we know the result, you know the result, keep to the result' (182). There is still some drama in the mock-battle, since it mustn't prove too easy a victory; but events take an unexpected turn when Pyper trips and King William falls. The question is raised, but not answered, as to whether he did it deliberately. King Billy's downfall unnerves the men, since they take it as an ill omen for the forthcoming battle. But in not following the prescribed ritual, that characteristically iconoclastic move by Pyper creates instead a broken, fragmented ritual, and here as in so many Friel plays there is a scavenging through the cultural remnants for what can no longer be completely endorsed. The final theatrical gesture occurs when the men don their individual Orange sashes as they arm themselves, with Anderson offering one to Pyper. This precipitates an exchange of sashes similar to that of Beckett and the bowler hats, an appropriating and playing with a traditional symbol to a more personal end.

The battle of Scarva has served to remind them of home. But so, in a more immediate way, does the smell of the river. The act of remembering within the play is not alone of Pyper remembering the other men, all of whom are present, but an imaginative recalling of all the rivers from their home province, the Bann, the Foyle, the Lagan, flowing together into one possession, an imaginary rather than an actual homeland. Craig disputes with Pyper about what he is doing, since he and all the others know why they joined up; but Pyper, as the one who is both on the inside and on the outside, creates a sense of generosity and space in which the audience can participate. This is like the play itself, in which a seemingly closed off and defensive culture is now being imagined on the stage. The leader Pyper rejoins his younger self to speak in unison their one-word synonym for what they love: 'Ulster' (197).

McGuinness proved astonishingly prolific in the years after *Sons of Ulster.* His life of Caravaggio, *Innocence*, was produced at Dublin's Gate Theatre in 1986. In 1988, *Carthaginians*, a play in which its Derry characters struggle with the legacy of Bloody Sunday on their lives and community, was staged at the Peacock. A Catholic counterpart to the Protestant emphasis of the earlier play, *Carthaginians* was originally to be produced by Field Day in 1987 along with Parker's *Pentecost*. But McGuinness became unhappy with the demands that two full-length plays were putting on the company's resources and withdrew his. *Mary and Lizzie*, which rereads Marx and Engels through the perspective of the two Irish women who work for them, was commissioned and produced by the Royal Shakespeare Company in 1989. His weakest play, *The Bread Man*, followed at the Gate in 1990; it lacked a compelling stage metaphor and the pressure of a specific historic incident to harness McGuinness's prodigality with words and images. *Someone Who'll Watch Over Me* dramatised the fate of the hostages in Beirut through the Beckettian situation of three men waiting in a cell for their murder or release. It gave McGuinness his most successful play to date. In the first production of *Someone Who'll Watch Over Me* at Hampstead's Theatre Club in July 1992, particularly memorable performances were given by Alec McCowen as the Englishman Michael and, in a further twist of McGuinness's near involvement with Field Day, the company's co-founder Stephen Rea played the Irishman Michael.

Someone Who'll Watch Over Me is in part based on the experiences of the Irishman Brian Keenan, who was taken hostage in Beirut by Shi'ite militia and held hostage for four and a half years. His prose account, *An Evil Cradling* (1992), centres on his friendship with English journalist John McCarthy and how they helped each other to survive. McGuinness keeps to an Irishman and an Englishman but adds an American for good measure, and he alters many of the details.[54] The Irishman is a journalist. The Englishman taught Old and Middle English, which McGuinness himself trained in and which Thatcher's England no longer has a place for. McGuinness's Edward, like Brian Keenan, is from Northern Ireland but is from a Catholic rather than a Protestant background; this sharpens the contrast between the Irishman and the Englishman. But the greatest transformation McGuinness has wrought is to move the play away from a realistic depiction of men being held hostage in a Beirut cell, in the direction of a more stylised but also a more spare and stripped drama, one that is markedly Beckettian.

Someone Who'll Watch Over Me is structured as a series of scenes, nine in all. The setting is specified by no more than the simple description 'A cell' and its emptiness is broken only by the chains which keep them tied to the walls but give them sufficient '*freedom of movement for both men to exercise*'.[55] They have also been supplied with copies of the Bible and the Koran. The first scene establishes the dynamic and rhythm, the dramatic shape, of two men in a room and, although the second introduces a third character, we know the number is not going to go much beyond that, if at all (even without looking at the cast list). This allows for dramatic condensation and establishes a two-way conversation as the basis of the way the play's language will operate. But there can be considerable variation in the deployment of the three characters. The audience never knows from one scene to the next who will be there at the raising of the lights and the start of the next one. Hamm may tell Clov that 'Outside of here, it's death!'[56] but since Clov never finally leaves, the truth of the proposition is never fully tested and Hamm may simply be saying that to prevent Clov from going. But the line applies even more to any departure from the cell in *Someone Who'll Watch Over Me*. Adam the American disappears between Scenes Five and Six and does not return. He himself feels that he is marked to die and interprets the behaviour of one of his captors in the light of that conviction: 'one raised his hand and pointed his finger at me' (124). In the immediate aftermath of Adam's disappearance, it falls to Michael to convince Edward that the American who has shared the space with him for months is in fact dead. In so doing, he hopes to persuade Edward to call off the hunger strike he has been on since Adam was taken away. The play's final scene shows Edward leaving the cell, about to be set free, and hence leaving Michael on his own as it concludes.

No other characters appear. At no point are their captors represented onstage. The three men frequently remind each other, however, that they are being listened to, overheard, and that the last thing they want to do is to let their listeners hear them crack up. When they try to persuade the newly arrived Michael, nearly hysterical with the trauma he has just undergone, of the importance of not weeping, they do so by enjoining him to laugh and showing him the lead, with the on-again/off-again laughter of Beckett's double male leads, laughter not as a spontaneous reaction but as a routine calculated to aid survival:

> ADAM: [...] Don't weep. That's what they want. So don't cry. Laugh. Do you hear Me? Laugh.
> MICHAEL: I can't.
> ADAM: Laugh, damn you.
> MICHAEL: No.
> EDWARD: *starts to laugh loudly. He stops.* ADAM *laughs loudly,* EDWARD *joining in. They stop.*
> ADAM: Go, Michael. Laugh.
> *Silence.*
> Laugh.
> MICHAEL *laughs.*
> More.
> MICHAEL *laughs more loudly.* ADAM and EDWARD *join in his laughter. They stop.* ADAM *signals* MICHAEL *to continue laughing. He does so.*
> Good guy. That's what you got to do. They've heard you laughing. [...] They're in a peaceful mood tonight.
> EDWARD: They're not even laughing back.
>
> (104–5)

The men respond with antic humour and generate a great deal of genuine laughter in deliberately defying what Edward describes in the play's opening lines as 'the boredom, the boredom, the bloody boredom' (90). As in *Godot* the theatrical space is a vacuum waiting to be filled, the theatrical action that of waiting and the theatrical challenge to stave off the boredom that is readily acknowledged. In his Preface Brian Keenan speaks of the prisoners using 'insane humour' to stave off despair[57] and later briefly instances rewriting 'old movies that I hated' and 'mixing one with the other'.[58] McGuinness develops this into a major dramaturgic strategy. Adam begins by converting Michael's capture into a conventionally suspenseful Hitchcockian scenario. Edward pushes it into the surreal by imagining the Singing Nun (Adam suggests she be played by Madonna) trying to convert the whole of Lebanon but being shot for her pains in a slow-motion Sam Peckinpah bloodbath ending. Michael counters by bringing on Mahatma Gandhi in a Richard Attenborough movie, which Edward tops by throwing in a disabled artist with a mother 'who finally wins an Oscar' (111). Adam's protest – 'Oh Jesus, not an Irish movie, please' – not only identifies Daniel Day-Lewis's Oscar-winning performance as Irish writer Christy Brown in *My Left Foot* as the source of the parody but shows that what is at play here is intimately bound up with national identity.

When the three characters propose they represent their shared situation as a movie, Edward casts it in the following terms: 'There were three bollockses in a cell in Lebanon. An Englishman, an Irishman and an American' (112). The description is deliberately intended to evoke the many hoary ethnic jokes centring on the intertwined comic narrative of an Englishman, an Irishman and a Scotsman and varying the hierarchy of foolishness depending on the ethnic character of the joke's teller. *Someone Who'll Watch Over Me* deliberately avoids representing the Arab captors onstage. Instead, the dramatic focus is tight on the three captured English speakers. Their differences emerge under pressure, differences which are dramatised by the characters themselves through the medium of national character. The British formality of the first words out of Michael's mouth – 'I'm terribly sorry, but where am I?' (101) – are met by Edward's adoption of an exaggerated Irish idiom; 'So it's yourself, is it?' This aggressively Anglo-Irish exchange is followed by Edward's gratuitous references to Eton and Harrow and the offensive interpellation 'Brit boy'. Michael in turn is inclined to treat Edward as an anthropological Irish specimen, especially in the latter's colourful use of English: 'Hiberno-English can be quite a lovely dialect' (129). Where the Irishman and the Englishman attack each other with the eight hundred years of oppression, the American has to challenge the image of himself as the strong American with the recognition of his own vulnerability.

The challenging of national stereotypes in the play leads directly to a challenging of gender stereotypes. By being put under such pressure, the three characters all feel their manhood is being tested. Adam responds with a strict regimen of physical fitness, Edward by playing the tough-talking 'hard man' who extols soccer and rugby. Michael most challenges the standard construct of masculinity. As a teacher of English and lover of poetry, as a son who remains close to his mother, he recognises that the macho Edward will accuse him of being a 'pansy little Englishman. I don't mind. I've had it before. I can tell you, there were people who were surprised I got married' (122). The moving account of his wife's death in a car accident silences doubts on that score. But the challenging of overdetermined national identities by the fluidity of gender identities is a recurrent element of the play and a hallmark of McGuinness's drama. The gender play develops with Michael's active bodily performance of 'the 1977 Wimbledon Ladies' Final (with) Virginia Wade of Great Britain against Betty Stove of Holland' (146). While Edward objects to Michael '*tossing back his head four times*' as he goes to serve, even though Michael reassures him that 'Virginia always tossed her head at a tense moment' (147–8), he is drawn into crossing national as well as gender boundaries when Michael/Virginia turns and addresses him as the Queen of England:

> EDWARD: I'm now the Queen?
> MICHAEL: Yes.
> EDWARD: Hello.
> MICHAEL: Hello [...]
> EDWARD: And what do you do?
> MICHAEL: What do you mean, what do I do? You've just seen me win Wimbledon.
>
> (149)

In opening up to each other, both men speak of fatherhood, Edward of wanting to father another child, Michael of one of the few exchanges he remembers with his own father. Traumatised by the experience of fighting in World War II, Michael's father bequeathed his son a lesson in coping with the trauma of conflict through his story of the Spartans, those brave soldiers who before they went into battle 'combed each other's hair. The enemy laughed at them for being effeminate. But the Spartans won the battle' (158). When Michael and Edward prepare to take their leave of each other in the final scene, none of their words is as moving as the mutual gesture by which they comb each other's hair before separating.

When the Irishman and the Englishman are at the height of their name-calling the American interrupts them to point out: 'English Arab? Irish Arab? Right, guys? Jesus, these guys don't need to tear us apart. We can tear each other apart' (124). In *Observe the Sons of Ulster* and in *Someone Who'll Watch Over Me*, as in the many fine plays Frank McGuinness has written since,[59] the real enemy is not those who oppose the plays' characters at the Somme or in Beirut but the enemy within, the inherited prejudices with which his people face each other as a prelude to having to confront themselves. The salutary end of his writing is liberation, a catharsis at once dramatic, cultural and personal.

Northern Ireland's changing political landscape in the 1990s was registered in Irish theatre. In 1994, first the IRA and then the loyalist paramilitaries declared a cease-fire; the Abbey, in the first year of Patrick Mason's tenure as artistic director, mounted a mainstage revival of McGuinness's *Sons of Ulster* to mark the breakthrough. Both Field Day and Charabanc Theatre Companies, which had done so much for Irish theatre in the 1980s, came to an end; but they influenced the younger companies and new playwrights who came to the fore in the North. The actor Tim Loane has written of the impact the Field Day production of Kilroy's *Double Cross* had on him when he was a student at Queen's University in Belfast in the 1980s. In 1988, Loane and fellow actor Lalor Roddy co-founded Belfast's Tinderbox Theatre Company with the declared intention of developing and producing new work that would interrogate and explore life in Northern Ireland. On 14 November 1998, there was a symbolic passing of the torch when Tinderbox joined with Field Day (in its last ever production) to co-present a production of Parker's *Northern Star* in a Presbyterian church; the production was directed by Stephen Rea.

In 1993, Tinderbox premiered the first stage play by the most important Northern Irish playwright to emerge during the decade, *Independent Voice* by Gary Mitchell. Mitchell was born in Rathcoole, north Belfast, in 1965 and began (as so many Northern playwrights, including Friel, have done) by writing plays for BBC Radio. In 1994, he was the first playwright from Northern Ireland to win the Stewart Parker Award, set up in the late playwright's memory to honour the best new first play staged in the preceding 12 months. Mitchell came to wider attention with his two plays of the late 1990s, *In a Little World of Our Own* (1997) and *As the Beast Sleeps* (1998), staged first at the Peacock in Dublin with subsequent productions at Belfast's Lyric and London's Tricycle and Donmar Warehouse. Their heavily male casts feature members of the loyalist paramilitary Ulster Defence Association who register a sense of betrayal and confusion that the cause to which they have committed themselves for so many years must now be pursued by other means.

In a Little World of Our Own is as claustrophobic as the title suggests. It features an all-male family of three grown-up brothers with distinct resemblances to the family in Harold Pinter's *The Homecoming* and Tom Murphy's *Whistle in the Dark*. Gordon, Ray and Richard, although well into their twenties and thirties, all still live in the Belfast family home. There is no father present or referred to, while their aged Ma is a frequently invoked figure whose permanently bedridden situation prevents her from putting in an appearance in the permanent living room set of the play. Gordon is the oldest son and the only one with a regular income from a job. He is also struggling to break out of the paralysed family situation through his engagement to Deborah, as Murphy's Michael did through his marriage to Betty. That there is more than economics involved becomes clear when Gordon questions his and Deborah's frequently stated resolve to buy and move into a house of their own. His fiancées retort is immediate: 'If you think for one minute that I'm going to get married and live here in this house with your Mum upstairs and your two brothers doing whatever it is they do, forget it. No chance. No way.'[60] Ray's activities are irregular and illegal, signalled both by the pin he wears and the loyalty it betokens to the Ulster Defense Association and by the Manichean credo he articulates in the play's opening scene: 'The world is a violent place. We know that better than anybody. Whether it's dealing with the IRA or dealing with petty theft or glue sniffing. Whatever' (3–4). The changing political climate for the paramilitaries has seen a shift in emphasis away from a primary engagement with their paramilitary Other; instead, they now run an unauthorised police service to take care of the drug pushers who prey on the local community. The other change, as Ray laments, is that non-violent methods are taking over in the altered political landscape and that people like him are becoming an anachronism. The economic landscape is also shifting, as the discussion at the start of Act Three establishes. When Gordon asks how Ray is going to survive economically after his brother marries and moves out, Ray's reply no longer holds true in 1990s Belfast:

RAY: Whenever you go, we'll drop below the income support minimum [...] and that means we'll qualify for everything that's going.

WALTER: I don't think they do that caper any more. Ray. [...] It's all loans and all now.

(35)

The character who complicates and deepens *Little World* is the third and youngest brother, Richard. Like Rose in *Dancing at Lughnasa*, he is learning-disabled and the risks that simplicity incurs make the family vulnerable in both a political and emotional sense. Both of his brothers are protective of him, as is dramatised when they break into a fierce argument over which house Richard will live in after Gordon and Deborah marry. The private and public realms intersect in the crush young Richard has developed on a 15-year-old girl, who also happens to be the local UDA boss Monroe's daughter. When she is raped and left for dead, Monroe orders his hit men, Ray and his fellow employee Walter, to deal with the culprit. The drama hinges around the identity of the guilty man. Ray's prime suspect is based on the gossip that young Susie Monroe has been going out with a 'taig', a Catholic. But the prime suspect in the community's eyes is his brother Richard, who was seeing (and seen with) the girl on the fateful night. The gormless and guileless Richard stumbles through a speech in which he has clearly been coached by his older brother before blurting out: 'I can't remember what you told me to say.' But a third party has been present, Ray in the usual role of his brother's minder, and he is the one the fatally injured young woman identifies as the rapist. For all that he knows himself to be the guilty man, Ray cannot own up to or even admit his crime and instead consistently scapegoats the young Catholic man as the rapist and hence the one deserving punishment: 'You tell Monroe about the other kids saying she left [the dance] with the taig. [...] And tell him that if he wants something done about it, I'm the man to do it. In fact I might just do it anyway. For Richard, you know' (34). The retributive logic whereby the hit man has now merged with the man he is sent to eliminate is graphically evident when, in the final scene, Ray stumbles into the living room and *'has been shot in the stomach'* (55).

A reading of that final scene is informed by two discussions earlier between Gordon and Deborah. It has been hard to read the latter character, since her primary purpose so far has been to come over and pray with the three men's mother. And her bourgeois ways can clearly make little headway with the tribal violence of the all-male family, again like Betty with Michael and his gang of brothers in *Whistle in the Dark*. But at the start of Act IV Deborah engages her reluctant fiancée in a discussion about God, She may be opposing his rational defences with the mantra 'Let God sort it out. There's nothing you can do' (53) but Deborah is also the one to see through Gordon's repeated suggestion that he send for the police to the piercing recognition that 'you wouldn't be able to live with yourself, Gordon' (53). The Old Testament can be sensed in the 'eye-for-an-eye' logic that operates throughout the play. But it is specifically invoked in several references to Abraham's sacrifice of his son Isaac. Gordon in Act One has admitted his difficulty in talking to God: 'I'm not too good at praying. When I pray God doesn't speak to me or if he does, I don't know what he's saying. [...] What if God did answer me? What if he told me to do something that I didn't want to do, or I couldn't do, what then?' (19–20). Gordon specifically mentions the Biblical precedent of Abraham and how he would have responded to God's command that he kill his own son: 'I would have just said no' (20). This is almost precisely the situation Gordon finds himself in when he is persuaded that, even though Ray is the guilty one, popular prejudice will still hold the innocent Richard to blame and that it is he who will be killed. It is suggested that if Gordon inflicted a wound on Richard then honour would be satisfied and the youngest brother's life would be saved. Gordon takes the gun which is placed in front of him and explains to Deborah what he has been asked to do: 'God asked Abraham. [...] He's asked me to shoot my own brother' (54). This discussion is interrupted by the return of the mortally wounded Ray. Gordon is not off the hook, however, because now it is his other brother he is being asked to kill, if not by the voice of God, then by the voice of that same brother: 'Now I'm asking you as your brother. Do me before I start sinning all over again' (60). When Gordon tries but fails, Richard picks up the gun and completes the act by shooting his dying brother. Ray has made it clear that he is 'not going to prison and I'm not going to spend the rest of my life in a fucking wheelchair' (60). He has come back home to say a last farewell to his mother and to make his peace by confessing the truth of what he has done to Richard, calling Deborah to witness that what he is saying is the truth. His final words, and the play's, – 'You're the man. You're the fucking man' (61) – suggest that Richard has come of age while Gordon has regressed, that a tribal ritual with deep roots in the Old Testament has played itself out without any intervention from a supposedly benevolent God.[61]

Mitchell's next play, *As the Beast Sleeps* (1998), is more ambitious and complex. This expansion is represented through the variety of settings. Although the first of the play's ten scenes is set in the living room which dominated *In a Little World of Our Own*, the second shifts to a punishment room (and a set of different characters), the third to a club and the fourth to an office. The boundary lines between these different settings, already represented as fragile in this opening quartet of scenes, give way and collapse into one another in the play's second half. Scene Five returns to the living room and is followed by another scene in the punishment room, thereby establishing that a pattern is operating and that it is based on a principle of recurrence. The club is never directly visited again; the seventh scene instead moves directly to the office, where a debate is initiated on the robbery that took place in the club the previous evening, the accurate resolution of which will determine the rest of the play. Scenes Eight and Nine oscillate between the living-room and the punishment room, with the concluding scene returning to the living room, where all of the play's concerns come home to roost.

The activity which occupies the plays' four living-room scenes is the wallpapering and painting of the domestic interior. The set-up resembles a Sean O'Casey play, with the husband Kyle and his friend Freddie operating the male double-act while the wife Sandra is away. A similar scenario was used by O'Casey in *The End of the Beginning*, with the two men who are left in charge of the house introducing chaos when they attempt to impose order.[62] Here, the worst that happens when Sandra returns is that both appear to be malingering and the job not far advanced, despite Freddie's protestations to the contrary. Kyle has resolutely remained passive throughout while demanding that someone else, first his mate and then his wife, make him a cup of tea. But as it unfolds Mitchell's play reveals itself to be more in line with O'Casey's full-length tragedies, where relations and activities in the domestic sphere become a microcosm of and turn out to have an intimate correlation with the larger political world. As always in Mitchell, economic discussion comes to the fore and helps to lay bare the underlying realities. When Freddie offers to provide a designer football club quilt for their young son's bed, at a considerably reduced sum from the official one, Kyle points out that it is stolen. It also appears that the family is suffering financially, with Sandra borrowing from her mother and Kyle unemployed. It emerges that he and Freddie traded for a long time in stolen goods, with Kyle now trying to distance himself from his former occupation. The 'wee jobs' which they both performed have been suspended indefinitely in the altered political climate. There is talk of the club, which they regard as their own natural preserve but from which the other two members of their gang have been barred (and with Freddie under a cloud for his behaviour). Even before the scene shifts there, the issue of territoriality which is at the centre of the drama has been broached.

But first the action shifts to the punishment room. The setting is merely a bare room with an old table in the middle and four chairs against the wall. Its original purpose only emerges when one of the two men present,

Larry, waxes nostalgic about the acts of intimidation carried out in the room, how they relied on appealing to the victim's imagination more than on direct acts of violence. The other man, Alec, with an active public role in the new constitutional life of Northern Ireland politics, ripostes that 'we've taken great steps to move people away from rooms like this'.[63] Two key issues are raised in the course of the scene, the first in the following exchange between Larry and Alec:

> LARRY: [...] We've another big fat donation to the cause.
> ALEC: It's the same cause, Larry. Just a different way of fighting for it.
> (18)

Have the changed methods in Northern Ireland resulted in a complete dilution and abandonment of the original cause which motivated a political ideology? Or can a huge social change be engineered without at least some of those involved in it feeling betrayed? The other issue is the political and religious affiliations of those who were tortured in the room. The presumption would be that they were Catholic but, as Larry points out, 'there was only ever three Catholics in this room' (19). The vast majority brought there for punishment were Protestant, members of their own side guilty of acts of disloyalty for which they would be punished and serve as an example to others. The process of social transformation is by no means complete. When Larry indicates that he too would like to act in the political arena, Alec's response is that Larry's strong arm tactics are not appropriate at one level in the new dispensation – 'you can't bring people into this room and make them vote for you; politics doesn't work like that' (22) – and that they are still required at another to maintain discipline and order. Larry's fear is that he will be left behind and as he follows this up by making a threat he paradoxically reinforces how the methods of the punishment room are still operative.

The third scene, the only one set in the club, acts out the issues of territoriality discussed in the first scene through the body language of the protagonists. Freddie is both verbally and physically aggressive, challenging and testing the limits of what behaviour is permitted in a jurisdiction he formerly regarded as his own. The unusual element, in terms of the scene's and the play's gender politics, is Sandra's presence in a club that might have seemed an all-male environment. She is drinking beer and playing pool with Freddie, a game which resonates with sexual connotations. Kyle is going through the motions of ordering drinks while counselling Freddie to curb his aggression. The active female–passive male dynamic established here between Sandra and Kyle, and her sustained defence of Freddie's behaviour, have repercussions later in the play. When the owner Jack finally appears and insists on the new rules, Freddie heads for him with the billiard cue and the terse contradiction: 'I'm not playing pool' (32).

The final new setting, in the fourth scene, is the office where everything at first appears routine: Jack is sitting at his desk working on his computer and then proceeding to interview a candidate for the position of Officer in Charge of Security. But the candidate has entered from the club, which is on the other side of the officer door, and is clearly recognisable as the barman Norman from the scene the night before, now with his arm in a sling. The interview scene generates a good deal of comedy through the verbal slippage and the strain between the familiarity of the old dispensation and the attempt to introduce the new. Jack tells Norman he is the only candidate for the position but indicates that he can no longer draw the dole, since everything now has to go through the books. This comic prelude leads to Kyle's arrival with Larry and the negotiations by which not only Freddie but all four of those who helped to sustain the club in the old days are readmitted. But the price is Larry's request that Kyle do a job for them. These discussions are still taking place in the office but the language of jobs to be done and the body language of physical intimidation increasingly give the lie to its normalisation. The key debate revolves around the issue of us versus them and its redefinition. As Larry outlines it: 'There's been problems keeping some people on board. [...] People without the sense and the patience that you have. Renegades' (42). When Kyle identifies this as the reintroduction of punishment squads by other means and condemns it, Larry points out the logic by which Kyle needs constantly to prove that he is one of 'us' and not one of 'them': 'See, someone might think that if you take that attitude, if you refuse to do this [...] you are a renegade' (48).

The play does not return to the club. But the issue of what happened there the previous night becomes the subject of debate. Was the club robbed? Were two individuals involved? And was Freddie one of them? Certainly, 35 thousand pounds has disappeared, and with that fact comes the possibility that the whole thing has been engineered by the owner. But the onus is squarely on finding a culprit and imposing a punishment; the return of the money is not sufficient. The play's inexorable movement leads back to the punishment room, now no longer an abandoned site but the place where Freddie is being punished by his best friend Kyle, a man who abhors their use. As in the modus operandi described by Larry in Scene Two, Kyle is standing behind rather than in front of the victim. But Larry, Jack and Norman are in turn seated behind him. Larry's earlier romanticisation of the process is undercut by the ever-escalating violence Kyle has to bring to bear on Freddie, as the latter refuses to reveal the identity of the other person involved in the robbery. Finally, unable to face any more, Kyle lets Norman finish the job, waits till the others leave the room and then picks up and carries off his friend's bloodied and insensate body. Nowhere in Mitchell's drama is the powerlessness of the individuals involved, and the extent to which they are counterparts mirroring each other's situation, more strongly realised than here. When Freddie taunts Kyle with their shared history and friendship, he retorts: 'you have it all fucked up, Freddie. This isn't happening to you because of me. This is happening to me because of you' (89). The most Beckettian line of *In a Little World of Our Own* describes the process being enacted and the men's response: 'We have to let this thing take its course' (32).

In the final scene, we are back in the couple's living room. Husband and wife are scarcely speaking – all they manage is to quibble over the issue of whether Sandra and Freddie have finished the job on which we have seen them cooperating in all of the living-room scenes: the wall-papering and painting. But at this final stage all of the settings bleed into one and the dialogue applies equally to the job Freddie and another carried out in the club. The presumption throughout is that his accomplice has been male. But then Kyle makes the connection and asks his wife: 'Did you and Freddie do the club?' (97). The increasingly aggressive questioning to which he subjects Sandra leads her to challenge his tactics by invoking another of the play's settings: 'What are you going to do, take me to one of your wee punishment rooms and beat the shite out of me until I tell you everything you want to know?' (98). But there is no need to shift locations, as Kyle demonstrates when he '*grabs* SANDRA *and forces her against the wall*'. Realising what he is doing and what he has become, he breaks off and his wife walks away. The final image is of a man alone in a room, sitting in a chair, laughing louder and louder, his own prisoner.

Both *In a Little World of Our Own* and *As the Beast Sleeps* seem designed to discomfit any cosy notions that the advent of the peace process in Northern Ireland would transform hardened traditional attitudes and patterns of behaviour overnight. They powerfully dramatise a self-incarcerating logic in which the violence so long doled out to the other side is now turned in on itself and brother betrays brother, wife betrays husband. The plays recognise that a change has had to come but also show the pressure on those for whom an inch is a very long way indeed.

In a case of life imitating art, in 2005 their author Gary Mitchell and his extended families 'were forced to flee their homes [in Rathcoole] after a campaign of death threats and bomb attacks by loyalist paramilitaries'.[64] According to Mitchell, a particular provocation was the screening of a film of *As the Beast Sleeps* on the BBC. As the Northern Irish political process moves uneasily forward, there is a temptation to recommend amnesia in terms of its conflicted history. But all the evidence suggests that the tension between the urge to progress and the pull of that history will continue to generate drama.

6

The 1990s and Beyond: McPherson, Barry, McDonagh, Carr

Although Friel, Murphy and Kilroy continued to write challenging and important works through the 1990s, as earlier chapters have shown, this was also the decade in which a new younger generation of Irish playwrights came to the fore. Some of them had started in the late 1980s – Sebastian Barry dramatising the lives of various family members whose lives had been elided or erased from the national record and Marina Carr graduating from a feminist theatre of the absurd to re-enacting Greek tragedies in the Irish Midlands. The decade also saw the emergence of two exciting new talents whose plays attracted awards and audiences in London and New York – Conor McPherson with his damaged males and their compelling stories and the black comedy of Martin McDonagh.

One concern shared by these four very different playwrights is that their people are damaged or hurt in a profound way, whether this condition is presented with the lyrical tenderness of Sebastian Barry or the bleak laughter of Martin McDonagh. Further, the source of this internal wounding is not directly given in the play, nor is it dealt with in terms of a conventional conflict. Rather, their characters are given the space in which to tell their story – and the act of storytelling is central to all four dramatists – and hence to articulate their sense of pain, injury, betrayal or loneliness, even if the plays themselves offer no conventional resolution or panacea; the drama, however varied, is in the articulation. A considerable irony in all of this is that these plays were staged at a time of increasing affluence in Ireland – the arrival of the 'Celtic Tiger' phenomonon when decades of emigration prompted by economic necessity were not only halted but reversed and Ireland became a country which attracted immigrants from around the world for the first time. And yet the lives portrayed in these plays are of people living in conditions of economic hardship and with little actual or imagined opportunity, scenarios in which even the smallest of hopes do not long survive.

For Vic Merriman in his influential article, 'Decolonisation Postponed: The Theatre of Tiger Trash' (published in 1999), there is no contradiction. Focussing on the plays of Marina Carr and Martin McDonagh, he notes how, 'at a time of unprecedented affluence, [they] elaborate a world of the poorly educated, coarse and unrefined. The focus is tight, the display of violence in the people themselves grotesque and unrelenting.'[1] These representations are offered up as spectacle to be consumed by a society which can relax and laugh precisely because there is such a clear disjunction between what they are watching and how they view themselves. As a result, 'the comfortable echelon of a nakedly divided society is confirmed in its complacency, as it simultaneously enjoys and erases the fact that "our" laughter is at the expense of "them"'.[2] But there is a counter-argument to the effect that these plays continue to raise troubling issues at a time when the culture would prefer amnesia in relation to its historical legacy of poverty and failure in the South, of violence and sectarianism in the North. As Patrick Lonergan puts it, 'how can we reconcile the memory of that troubled past with the desire to enjoy an apparently successful present? And if we let go of our histories, will we still remember who we are?'[3]

In their arguments, Victor Merriman and Patrick Lonergan both assume a predominantly, indeed an exclusively, Irish audience. And yet this overlooks the fact that during the 1990s the Irish play became, in Nicholas Grene's words, 'a distinctly marketable phenomenon [...] a commodity of international currency'.[4] There were notable London productions of a range of Irish plays (after the ground gained there by Field Day productions) and Broadway productions and Tony awards for plays by McDonagh and McPherson. What might be termed the internationalisation of Irish drama was given its greatest momentum by the phenomenal success worldwide of Friel's *Dancing at Lughnasa* in 1990. The play opened to reasonable houses and mixed notices at the Abbey in Dublin, but then began to build and build as Irish audiences responded to the moving plight of these five Donegal women from the 1930s for whom De Valera's Constitution brought little in the way of freedom or progress. That momentum proved unstoppable, and *Lughnasa* went on to be acclaimed at London's National Theatre and in New York, where it was awarded a Tony for Best New Play in 1991.

Both *Dancing at Lughnasa* and Friel's earlier *Faith Healer* achieved their first success in Dublin. But the contrasting trajectories of the two plays in production does much to illuminate the difference between Irish theatre in the 1980s and the 1990s. Where the first period saw a burgeoning cultural self-confidence, with Irish production companies the first to stage and Irish audiences the first to acclaim the important new plays of the decade, the 1990s saw the axis of decision making and cultural endorsement shift significantly to London (in particular) and to New York. In 1993, Noel Pearson produced a TV documentary on the rehearsal and staging of Friel's *Wonderful Tennessee* whose title *From Ballybeg to Broadway* suggested the latter as the inevitable destination and ultimate point of validation for Irish plays. This presumption proved undented by the Friel play's truncated run there and gathered momentum over the decade. Where *Faith Healer* failed abroad in its initial 1979 run on Broadway and then succeeded in Ireland, *Lughnasa*'s great success in London and New York helped to create a fashionable trend for Irish plays in those two capitals which still persists at the time of writing. The axis of power and influence shifted accordingly and the net effect was this: where throughout the 1980s new Irish plays established their reputations and meaning in their country of origin, increasingly Irish plays in the 1990s were first staged and acclaimed in London before coming to Ireland to have that process replicated. I would attribute this loss of cultural nerve and self-confidence at least in part to the phenomenon of the 'Celtic Tiger' with the Irish play as one more exotic import in which to invest for a wealthy clientele looking to show their modernity. But contrary to what Merriman argues, I would identify the decision not to premiere these edgy, inventive, ground-breaking plays, whose originality was what most recommended them, as the mark of a society regressing in its development of a post-colonial consciousness.

The situation I have outlined certainly applied in the case of Conor McPherson. Born in Dublin in 1971, he attended University College Dublin from 1988 to 1993, where he took a BA in English and Philosophy (with a double First) and an MA in Philosophy. In addition to his formal study of drama in his English studies, McPherson was extremely active in the university's dramatic society (Dramsoc), not only writing plays but directing them. He also acted (memorably) in plays by Pinter and Friel. After graduation, McPherson and some college friends founded the Fly by Night theatre company and staged his plays in a variety of fringe venues. All of his efforts to interest the major companies, especially the Abbey, in his original work proved unavailing.[5] When *This Lime Tree Bower* was first staged in Dublin's Crypt Arts Centre during the 1995 Dublin Theatre Festival, in a production directed by McPherson himself, it was seen by a London agent whose excitement at what he witnessed led eventually to an invitation to present the same production at the

Bush in July 1996. This in turn led to a commission from London's Royal Court theatre to write an original work for them, which produced the premiere of *The Weir* in July 1997. The following February the production transferred from the Royal Court Upstairs to the much larger Downstairs auditorium and had an extraordinary two-year run. That production surfaced at the Gate Theatre in Dublin in the summer of 1998. This bears out what Stephen Dedalus says in Joyce's *A Portrait of the Artist as a Young Man*: the shortest route to Tara is still via Holyhead.

Conor McPherson regards himself as a storyteller. This self-perception has frequently resulted in plays which are constructed as a series of monologues, where the primary engagement is between the individual actor and the audience he directly confronts. *This Lime Tree Bower*, the play which established him, was written as a series of interlocking monologues and clearly owed a debt to Friel's *Faith Healer*. McPherson believes it is no accident that the monologue form was favoured in the 1990s and the new millennium not only in his case but in acclaimed work by other younger Irish playwrights like Mark O'Rowe's *Howie the Rookie* (1999) and Eugene O'Brien's *Eden* (2001), which McPherson directed at the Abbey. He argues that these monologue plays were written during a period of uncertainty and trauma in the light of political and clerical scandals: 'Irish drama went "inside" because our stories were fragile, because everything was changing.'[6] These monologue plays were frequently accused of being untheatrical, more akin to displaced prose narrative, because they did not avail of the usual resources of a cast of characters engaged in dialogue with each other in a time-specific setting. But as I have argued throughout this study, particularly in relation to *Faith Healer*, the use of the monologue makes the plays more theatrical, not less, allowing for a more unmediated and intimate relationship between actor/character and audience. As McPherson himself puts it: 'These plays are set "in a theatre". Why mess about? The character is *on stage*, perfectly aware that he is talking to a group of people.'[7]

McPherson's plays are a reminder that theatre in Ireland arguably had its origins as much in the communal art of oral storytelling as practiced in the pub or the hearth as in a fourth wall drama performed on a proscenium stage in a metropolitan centre. A play like Friel's *Dancing at Lughnasa* foregrounds the two theatrical modes with the more conventional representation of the five sisters in the kitchen dramaturgically complicated in its development and relation to the audience by the presence and active intervention of the storyteller-narrator, Michael, and his monologues. Friel fused the monologic storytelling with the drama in his most radical play, *Faith Healer*. But there is also a frequent recourse to monologue in the plays of Beckett. His theatre features formal storytellers like Hamm and Winnie who directly face their audience with little more than their story to tell. These playwrights – Beckett, Friel and Murphy in particular – constitute a considerable resource for a younger contemporary Irish playwright like Conor McPherson.

The Royal Court made it a condition of the commission of *The Weir* that McPherson's not be a monologue play. But although he gathered three men and a woman into a familiar realistic locale (a pub in the Irish countryside) and had them engage in bantering dialogue, McPherson's distinctive handling of monologue remains central to the play's dramatic achievement. Each of the four characters in turn tells a story which engages with issues of life and death and the persistence of life after death in particular. The foregrounding and interweaving of their individual narratives centre on the return of the dead to haunt the living and focus on fear of the irrational in human behaviour. *The Weir* recognises the extent to which traditional Irish storytelling has drawn on the cluster of beliefs surrounding the fairy folk, the 'others', those who enjoy a continued existence after death. It is a theme which fascinates McPherson. In his 1997 play, *St Nicholas*, the seedy drama critic who tells us his story moves from the world of the theatre to a twilight existence with a group of vampires. And McPherson based 2006's *The Seafarer* on the folklore narrative of a game of cards with a stranger who turns out to have cloven hooves instead of feet.[8] In *The Weir*, he plugs in to Irish folklore and its tales of fairy forts to undertake a complex exploration of truth and fiction in storytelling. Since the newcomer is a woman and an urbanite, the three men gathered in Brendan's pub are at some level trying to impress, if not scare, her, especially in this desolate, windswept area, with its dark nights and isolation. But as Angela Bourke has argued, such traditional stories, whether people purport to believe in them or not, are a narrative means by which to 'remind listeners (and readers) of everything in life that is outside human control'.[9]

The beginning of *The Weir* is deceptively low-key. Brendan, a man in his thirties who runs the pub, greets two of the regulars in turn: Jack, a fifty-something bachelor who runs a garage, and Jim, a forty-something gentle man who lives with his mother. They are joined by Finbar, the local man made good, who now runs the big hotel, and is squiring a recently arrived young woman Valerie around the neighbourhood – showing her 'the natives', as Jack sardonically remarks.[10] McPherson has Jack begin the storytelling process with what might best be described as 'pure story' because the least personal and most mythic and traditional in terms of its narrative ingredients. His is an account he once heard (and hence is transmitting) by an old woman Mrs Neylon of how, when she was a girl, she and her mother had been terrified while alone in the house one night by a repeated knocking at front and back, 'very low down the door' (37). The fictive rationale supplied by the people at the time (and the story is very precisely dated to 1910 or 1911) is that the Neylon house had been built on a fairy road, a traditional pathway whose access was generally left clear and unblocked. McPherson has written of how his interest in folklore was stimulated by stories told him as a small child by his grandfather, who lived in County Leitrim (where *The Weir* is set): 'Beside his little house was an overgrown wood full of hawthorn trees. This was a fairy fort. No one would cut into it or chop it down. A roadway had been proposed to run through there one time but the plans had been changed. To this day that fairy fort stands peacefully at the banks of the Shannon which I always think of as such a dark, cold river.'[11] But that traditional taboo has been broken or transgressed in the case of Maura Neylon's family home and that sidelining of the older system of belief has resulted in this return of the irrational.

The story's last few details are brief, telling and enigmatic. 'A priest came and blessed the doors and the windows. And [...] Maura never heard the knocking again except on one time in the fifties when the weir was going up. There was a bit of knocking then she said. And fierce load of dead birds all in the hedge and all this, but that was it. That's the story' (37). The residual elements of pagan belief are banished by the ministrations of Catholicism; but that belief system has in turn been eroded, even in the most remote and rural of areas. The passing reference to the weir is one of the few to the play's enigmatic title. The only sustained reference is when one of the faded photographs on the wall is discussed: '1951. The weir, the river, the weir em is to regulate the water for generating power for the area. [...] This was when the ESB [Electricity Supply Board] opened it. Big thing around here' (32). Not least because it introduced the electrical light, which banished the absolute darkness in which the remnants of the older beliefs could linger. The detail about the dead birds, which is the first mention of death in the play's five narratives, received a wonderful interpretation in Garry Hynes' 2008 revival of the play at the Gate, as Patrick Lonergan has noted: 'Valerie (Genevieve O'Reilly) is arched forward, listening to the ghostly story. Jack (Sean McGinley) is moving closer to her, slowing down his delivery, staring directly into her eyes. But he goes too far, telling her about dead birds being found in bushes – an image that is much too close to reality for Valerie to stomach. O'Reilly sits back suddenly.

McGinley stands up straight. Both retain eye contact with each other, and there is silence, before McGinley says simply: "That's the story."'[12]

That the primary intention of the storytelling is to frighten Valerie is suggested by the disclosure that the house she has just moved into on her own is the same haunted one the Neylons lived in. Finbar, whose married status does not seem to be preventing him from making a play for Valerie, has been the one to suggest that Jack tell the fairy fort story. He is accused by the others of deliberately trying to frighten her when the identity of her house is made known to them. We are never given definitive verification of whether Finbar has done this deliberately or unwittingly; but his protests of innocence and forgetfulness are undermined by the fact that he is the one who sold Maura Neylon's house to Valerie and that he is old enough to know the story associated with it. But it is Valerie who has specifically asked to be told the story when Jack initially demurs and it is at her specific request that each of the two subsequent stories is told before she volunteers her own. Finbar, who continually decries the credulous ignorance fuelling such stories, is hoist with his own petard when the others force him to tell a story, one in which he himself is subject to a supernatural haunting which caused him to move house and quit smoking. Like the apocryphal story of the old Irish woman who was once asked if she believed in the fairies and replied 'I do not, but they're there anyway', Finbar denies that the ghost is there on the stairs but describes how he was unable to move all the same: 'I wouldn't move in case something saw me. [...] But when it was bright then, I was grand, you know? Obviously there was nothing there and everything' (43–4).

The experience that Finbar undergoes is one that he has not directly experienced but had heard about from neighbours. His inability to move while sitting on the stairs precisely mimics or parallels what befalls Mrs Walsh and her teenage daughter Niamh, after the latter believes she has unwittingly conjured a spirit while using an Ouija board: 'the young one, Niamh was going hysterical saying there was something on the stairs' (42). Finbar as the nearest neighbour is summoned to help and, as soon as he enters the Walsh household, the daughter begins 'shouting at me to close the living-room door. Because I was out in the hall where the phone was, and she could see the woman looking at her over the bannister. Like she was that bad, now' (43). A phone call later on reveals that a close family friend who minded Niamh as a child has just fallen down the stairs and died. When Finbar retreats to the apparent safety of his own house, he finds that he is unable to turn around or go to bed because 'I thought there was something on the stairs (*Low laugh.*)' (43).

In actual terms, he has simply gone to a neighbour's house to aid a distressed young woman. But that charged atmosphere has been filled with the daughter's account of something uncanny on the stairs and the phone call supplying the detail of the dead friend. This narrative is decades later than the previous one and has been assimilated into a transmissible narrative and passed on by Finbar himself, with the suggestion of something like imaginative contagion being contracted.

Jim's story is introduced initially as a realistic memory of a story associated with a friend of his who died, and of how 'twenty or more years ago' (48) he and his friend were asked by the priest of the next parish to dig a grave. As the narrative proceeds to relate how the pair of them dug the grave on a relentlessly rainy day, mysterious details accumulate. Why hasn't the priest asked two young men from his own parish to dig the grave? Why are there so few people at the church for the removal, especially since the dead man was only middle-aged? Left on his own to finish the digging, Jim encounters a man who directs him to dig in another area, in the grave of a young girl. From a later description, he finds out that the dead man resembles the man who approached him and that the dead man was known to be a paedophile: 'The fella who'd died had had a bit of a reputation for em ... being a pervert. [...] And he wanted to go down in the grave with the ... little girl. Even after they were gone. It didn't bear ... thinking about' (51). Jim's story dates from the early 1970s and is being recounted in a pub in the early to mid 1990s. During the latter period, the newspapers were filled with court cases in which older men (many of them Catholic priests or brothers placed in charge of young people) were brought to trial charged with sexual abuse of children.[13] Often, they were serial offenders with a large number of victims and in many cases the incidents dated from several decades earlier. The men responsible were left in positions of trust over many years and nothing was done officially. These cases of child abuse were ongoing through the Celtic Tiger years, one of the series of scandals referred to by McPherson in explaining how the monologue form enabled Irish narratives to go 'inside' and explore the society's secrets in ways other than the reportage of the media and documentary drama. McPherson also explored the subject of child abuse in a one-act play, *Come On Over*, in 2001, where the paedophile is a former priest.

Jim's story is skin-crawling in two senses. The folk value being explored is that a ghost will continue to enact his or her crime even after death. In social terms, the fairy story is a means, as Angela Bourke demonstrates in *The Burning of Bridget Cleary*, of finding a narrative form to cope with social aberration in a traditional community: 'Fairies belong to the margins, and so can serve as reference points and metaphors for all that is marginal in human life. Their underground existence allows them to stand for the unconscious, for the secret or the unspeakable, and their constant eaves-dropping explains the need sometimes to speak in riddles, or to avoid discussion of certain topics.'[14] The horrified reaction of the other men to Jim's story (which Valerie receives in silence before asking for the loo) raises the question of how responsible or 'knowing' he is for the narrative he tells. Jim is the most emotionally damaged of the three males in the pub, clearly suffering some kind of psychological disability that renders him somewhat 'simple', one might even say child-like, in the manner if not to the same extent as Friel's Sarah or Rose or Mitchell's Richard. Jack behaves protectively towards him, giving him odd jobs in his garage (about which he reminds him in a way that would otherwise be unnecessary). Jim's straitened existence is suggested by the way he hugs closest to the bar, counting out his pennies for his drink and moving with a good deal less bodily freedom than the other characters. The chief talking-point is the condition of his perennially ageing mother, in whom both Jim's life and emotions seem invested. The part was played by Mark Lambert in the Hynes production as a complete innocent, somebody who was unaware of the implications of the story he was telling. Kieran Ahern was a deal more knowing and unsettling as Jim in Ian Rickson's 1998 Royal Court premiere. As Ahern's Jim told the story, it was at least as if he had been tainted by the experience and was passing that along to his listeners. If Jim is seen as directing the narrative, then he is the one who produces the stranger at the graveside, and makes the visual identification with the dead man and the verbal imputation of paedophilia. The stories being told increasingly converge on identification between the person at the centre of the narrative and the person telling the story; and so it is difficult to avoid some association of this kind between Jim and the story he tells.

The three men vow the stories have gone too far and, when Brendan brings Valerie back from the house (the pub's 'Ladies' being out of order), say there are to be no more fairy stories. But Valerie has her own story to tell and insists on doing so, one about the recent death by drowning of her young daughter and a communication from beyond the grave. The story is the most urban and 'realistic' of the four; and yet it builds and draws on elements from all of the stories so far told. Her daughter Niamh even has the same name as the young haunted woman in Finbar's narrative. On the surface, Valerie, her husband and daughter are leading the 1990s lifestyle of the 'Celtic Tiger', both parents holding down academic positions at Dublin City University. The disparity between urban and rural lifestyles has always been marked in Ireland, but became particularly so during this decade when a rapidly expanding Dublin added satellite housing developments that encroached on several counties. Hynes' 2008 production set the play several years earlier in the 1990s, to make the economic hard times it depicts more credible (and perhaps with an eye to the post-Tiger economic downturn which her audiences would have been entering into).

But for all of the modern lifestyle that Valerie and her husband are able to offer a 'bright, outgoing, happy girl' (57), the night time tells a different, unchanged and primal story: 'She had a problem sleeping at night. She was afraid of the dark. She never wanted you to leave the room.' And in that darkness Niamh is subject to nightmare visions: 'there were people at the window, there were people in the attic, there was someone coming up the stairs. There were children knocking, in the wall. And there was always a man standing across the road who she'd see.' Although Niamh's nightmares share features in common with those of the earlier stories, arguably they are the traditional contents of any ghost narrative. But the details are more precise and uncanny than that, evoking each of the three stories Valerie and the audience have been told in the course of the evening. This is particularly the case with the return of the man stalking a young girl from Jim's narrative, with the association now established of child abuse. It is as if Jim has personally entered her narrative, as if all three of the men who have subjected her to their stories have now been engorged by hers. The setting of the swimming pool in which her daughter is drowned is also a reminder that apart from clerics and teachers another job category from which male paedophiles were charged was that of swimming coach, again because of their being put in sole charge of large numbers of children, less fully clothed than normal. The association is activated by the pause in the sentence describing what greets Valerie when she finds her daughter: 'An ambulance man was giving her the... kiss of life. She was in her bathing suit' (58).

The deliberate and suggestive use of the pause brings Harold Pinter to mind. McPherson read and acted in Pinter's plays while a student at UCD. That association only increases the possibility that Valerie is doing more than just recounting a personal story to some sympathetic listeners. The reason she gives is that her personal tragedy contains a supernatural development which she would normally be reluctant to disclose: 'That just hearing you talk about it tonight. It's important to me. That I'm not... bananas' (57). For after the funeral, and after weeks of sitting around the house in a paralysis of depression, Valerie answers the phone and hears her daughter's voice. Niamh says she is at her grandmother's house and reiterates the details from her earlier visions and the play's first three stories, with the added development that 'the man [who] was standing across the road' is now looking up and 'going to cross the road' (60). Valerie hastens to her husband's mother's house, even though she knows she will not find her daughter there. But what persists is the haunting question: what 'if she's out there? She still... she still needs me.' Perhaps Valerie is solely motivated by recounting a tale of personal grief. But perhaps she is responding to the masculine intimidation of the three stories to which she has been subjected and the sexual competition between Jack and Finbar which her presence in the all-male domain of the pub has ignited. Perhaps she is seeking to turn the tables on them by constructing her own story which deliberately draws on theirs but ups the emotional sweepstakes by centring it on the death of her own daughter, a 'real' incident whose maternal veracity no man would have the nerve to challenge. It is in this respect (and interpretation) that she resembles Ruth in Pinter's *The Homecoming*. In the first half of the play, the all-male household to which her husband Teddy brings Ruth to meet the family instead subjects her to a bewildering mixture of politeness and rudeness, compliments and misogynist abuse. Lenny, the most sophisticated and intimidating brother, tells two stories of women he has had to punish before Ruth responds by ordering him to drink a glass of water and calling him 'Leonard', a family name. Throughout the rest of the play, she continues to seize control by turning the men's own ploys against them, until the patriarch Max is on all fours and she is fondling a young man's head in her lap. Valerie and McPherson's play in the end seem more benign than that, doing no more than exposing and sending off Finbar and leaving Jack, Brendan and Valerie to conclude the play. But one of Jack's last lines, where he alludes to the young woman in excremental terms by remarking that Finbar will be 'sniffing around Valerie every night [...] like a fly on a big pile of shit' (73), suggests that the polite conversational veneer towards Valerie only half conceals something more fundamental and Pinteresque.

There is one final story to be told. As with *Faith Healer*, the storytelling comes round once more, having traversed each of the play's characters, to conclude with the man who began it. Jack assures Valerie that 'it isn't a ghostly story, anyway' (69) but rather a personal one responding to the question of why he never married. It turns out there has been a woman in his life, someone he courted for years but did not ultimately follow to Dublin except for a few carnal weekends until she finally did the sensible thing by marrying a policeman. The story is indeed a realistic one and crowded with socially exact and historicised detail: of the woman choosing to leave the rural backwater where her employment prospects would be even more limited than those of the men, for instance. But the narrative is no less haunted than any of those we have heard. For when Jack is confronted by her challenge to accompany her to Dublin and build a new life there, he is no less paralysed than the figures in the ghost stories, struck by 'an irrational fear, I suppose, that kept me here' (67). Jack is his own dead man in the story, emotionally atrophied at this point and self-condemned to a life that is marginalised in every sense. The pub provides the one possibility of human contact, however minimal, but it still doesn't prevent the fact that he is haunted by her absence and 'there's not one morning I don't wake up with her name in the room' (69).

Like all of the stories told in *The Weir*, Jack's story enacts the trauma of displacement, of the psychic disturbance caused by the inroads of social progress on traditional ways of life. His closing story is a 'true' personal account, to balance the 'fairy' story he told at the outset, but in ways his is the most uncanny and haunting of all. The possibilities, the future story promised at the end of *The Weir*, were differently weighted in the two productions. In Ian Rickson's, Jim Nortan as Jack was actively steering Valerie and Brendan towards each other. In Garry Hynes's, Sean McGinley's Jack (closer to the mid-fifties age the text indicates for the character) more actively challenged Denis Conway's sexual grandstanding as Finbar, going face to face with him over Valerie at one point And the question she asks about his personal life prompts a story that at least suggests Jack may have some life in him yet, that he still has 'his moments' (71), as he puts it to Brendan. This interpretative range of possibility is true not only of the (open) ending but of the play and all of its wonderful details, both in and between the more formal stories, the extent to which the pub-like environment invites us to join in and participate in the play's unfolding.

From *Boss Grady's Boys* in 1988 through *Our Lady of Sligo* in 1998, Sebastian Barry (born in Dublin in 1956) wrote a series of history plays about a series of marginalised figures whose lives did not fit into the accepted grand narrative of Irish history. Roy Foster has described them as 'people left over in the margins and interstices, through religious exclusion, or a change of regime, or a redefinition of loyalty'.[15] They include the two old sheep farmers from *Boss Grady's Boys* (1988); the dying Quaker community on Sherkin Island in *Prayers of Sherkin* (1990); the emigrant Irishman turned American cowboy in *White Woman Street* (1992); a dancer on the English music-hall stage in *The Only True History of Lizzie Finn* (1995); an alcoholic woman in *Our Lady of Sligo* (1998). Further, all of the individuals whose lives they dramatise are members of

Barry's own family about whom little was said because they had in some significant way transgressed the taboos of Catholic Nationalist Ireland and so were consigned to oblivion. The few details Barry had in each instance were just enough to stimulate him into imagining them into a full dramatic existence without tying him down to too many particulars. His first two plays were staged at the Peacock theatre but after that in the 1990s his plays more often premiered in London at theatres like the Bush and the Royal Court and with companies like Max Stafford-Clark's Out of Joint company, which opened his plays in London before sending them on a tour which would include a run at the Abbey or Gate. This was the provenance of the play by Barry I wish to discuss, *The Steward of Christendom*. If the prominence given to a religious sect like the Quakers had limited Irish audiences for Caroline FitzGerald's luminous production of *Prayers of Sherkin* at the Peacock in 1990, then an even smaller audience might have been predicted for the Dublin run of *The Steward of Christendom*, given that its title character was a Southern (rather than the more usual Northern) Unionist. Thomas Dunne was Sebastian Barry's great-grandfather and the last Catholic Superintendent of the Dublin Metropolitan Police who helped curb the crowds involved in the Easter Rising of 1916 and whose son served (and died) in the English army fighting at the Somme. The production of the play did not originate in Dublin, however, but in London with Out of Joint; and by the time it began the first of two lengthy crowded runs at the Gate *The Steward of Christendom* came garlanded with critical superlatives, not least for the central performance of Donal McCann in his greatest role since Frank Hardy in *Faith Healer.*

The Thomas Dunne we see in Barry's play is no longer the proud head of Dublin's B Division of the DMP during the 1913 Lock Out, Easter 1916 and the Civil War. For *The Steward of Christendom* is set in 1930s Ireland, the decade in which the former Republican party Fianna Fáil embraced constitutional politics and came to power with Éamon De Valera (who had fought in 1916) as Taoiseach (prime minister). In the present of the play, Thomas Dunne is incarcerated in a mental home, a forlorn figure in a pair of dirty longjohns. He has been fighting a battle with his personal demons. The play also strongly suggests that he is part of a vanished world and there is increasingly nowhere in the newly formed Irish Free State where Southern Unionists may find a home or refuge:

> A man that loves his King might still have gone to live in Crosshaven or Cobh, and called himself loyal and true. But soon there'll be nowhere in Ireland where such hearts may rest.[16]

Although he never ruled this kingdom, but kept it in trust for king and queen and God, Thomas Dunne bears resemblance to a Hibernian King Lear. He does so not least because he has three daughters who feature in the play, Maud, Annie and Dolly. Although Dolly the youngest is Dunne's favourite, neither he nor the playwright suffer any Manichaean division of the women into 'good' and 'bad' daughters, even where Maud, the moodiest and most difficult, is concerned. Nor do they all three inherit the kingdom when their father ceremonially hands over Dublin Castle to Michael Collins and his Irish army of rebels. We come in on this King Lear relatively late in his tragic trajectory, at the point where he has been shut out in the storm and has run mad. The opening physical movement of *The Steward of Christendom*, with the cowering and babbling Dunne forced to strip naked by the asylum's attendants, evokes the Shakespearean scene in which Lear strips naked to present himself as 'the thing itself. Unaccommodated man is no more but such a poor, bare, forked animal.'[17] The requirement for the actor playing Thomas Dunne to be stripped full frontally naked before a live audience is only one of the part's considerable demands. (Another is that he remain continuously onstage for the entire two and a half hours' duration of the play.) Son of acclaimed Abbey actress Joan O'Hara, Barry writes with a keen theatrical awareness of the actors who will play his imagined ancestors (including his wife Alison Deegan, who played Fanny Hawke in *Prayers of Sherkin* and *Lizzie Finn*). He is well placed to register that a great deal of an actor's life is invested in stripping, putting off (or on) one set of clothes or another. Barry creates a profound link, one that is central to all of his plays, between the theatrical necessity of wearing costume and the different historical roles played by his ancestors. Much of the onstage activity of Act One of *Steward* involves Mrs O'Dea the seamstress trying to take the physical measurements of the recalcitrant Dunne and make up an outfit for him in regulation black. Dunne's stubborn insistence that it contain gold thread bears on his official uniform in the days of the DMP, as he explains to her:

> I have a hankering now for a suit with a touch of gold. There was never enough gold in that uniform. If I had made commissioner, I might have had gold, but that wasn't a task for a Catholic, you understand, in the way of things, in those days.
>
> (245)

The costume made up for Thomas Dunne is a version, a travesty one might almost say, of the uniform he wore as chief superintendent, playfully augmented with gold in the theatrical space of the asylum to compensate for his lack of promotion in the world of historical necessity.

But *The Steward of Christendom* does not stay in a fixed present of the 1930s. To do so would be to give that decade too much of a determining role in the characters' dramatic lives. Rather scenes flash back repeatedly to earlier periods in Dunne's life and career, vivid memories of a historicised time and place in Ireland as much as an individual's personal autobiography. At the beginning of Act Two, Dunne is being arrayed in his smart dress uniform by his daughters, topped off with a white rose for his buttonhole. The historic occasion is the transfer of power at Dublin Castle in 1922, the last occasion on which the outgoing superintendent will wear his uniform. But the play's conscious theatricalising makes it difficult to view the past from a fixed, detached position in the present. What we witness onstage in the play are two acts of theatrical investiture – one in which Dunne is fitted for a suit in an asylum, the other in which he is consciously dressing up for a public role, but one which will end by relegating him to permanent anonymity. As Fintan O'Toole has noted, Sebastian Barry's plays lay great emphasis on the imagery of clothing.[18] I would add that they do so in ways that do not pin or confine the individual character to the social role (and fate) prescribed by such an outfit. His people not only take on and off the costumes they wear in public; they make alterations to them as their whim, size and personal circumstances dictate.

And yet Barry's plays reveal the extent to which in Ireland clothes are read historically as fixed signs proclaiming permanent allegiances, even the core identity, of a man or a woman. This is best illustrated in *Steward* by the discussion between Dunne and his tormentor, the nationalist Smith, over his son Wilie's death in World War I. Frank McGuinness's *Observe the Sons of Ulster Marching Towards the Somme* presents the latter as another blood sacrifice to set beside the Easter Rising. But what his play leaves untouched is the fact that not only Northern Irish Protestants but Southern Irish Catholics were to be found at the Front, following the constitutional nationalist John Redmond's political counsel, and in greater numbers than occupied the GPO. Thomas Dunne has lost his young son in the trenches and has only two mementoes: the uniform and his son's letters home. In Act One Smith taunts Dunne in terms of the historic role the latter has occupied, branding him a 'Castle Catholic'. The outfit Dunne has so much admired is verbally transformed by Smith into a badge of shame and source of recrimination. In Act Two, Smith has undergone his own change of costume by trading in his suit of customary black, worn in his public capacity at the asylum, for an outlandish and historically unlikely costume when he reappears '*dressed like a cowboy complete with six-shooters*' (290). He does so, it turns out, not because he has suddenly decided to emigrate to the United States and thinks that this will pass for acceptable dress there but because he is on his way to a fancy-dress party. Smith's change of costume, and the shift it signifies from historical to theatrical reality, allows for an expansion if not a change of heart. Smith inquires solicitously into Dunne's relations with his son and asks to hear him read the letter from the trenches. As Dunne says when he shyly opens and reads it to him, 'It's an historical document' (292).

In a play which has as much to do with fathers and sons as with fathers and daughters, the national dimension of such a relationship is represented through the figure of Michael Collins. He is something of a hate figure for the Dunne family, memorably anathematised by Annie when she predicts that the 'like of Collins and his murdering men won't hold this place together' (278). Thomas Dunne's own instincts are to repudiate the new political order that is going to take over the ordering of the country in 1922. But the shock of his direct personal encounter with Collins during the exchange at Dublin Castle intimates a possible relationship between Dunne's Ireland and what is replacing it, some measure of continuity between Irish people of strikingly divergent political affiliations:

> I could scarce get over the sight of him. He was a black-haired handsome man, but with the big face and body of a boxer. He would have made a tremendous policeman in other days. [...] I felt rough near him, that old morning, rough, secretly. There never was enough gold in that uniform, never. I thought too as I looked at him of my father, as if Collins could have been my son and could have been my father. (285–6)

But those possibilities are denied in the emerging Irish state by the assassination of Michael Collins and the subsequent reign of 'King DeValera' (262), as Mrs O'Dea refers to him. That possibility, of an enlargement and transfer of traditional allegiances, emerges and vanishes just as briefly within Thomas Dunne: 'I felt a shadow of that loyalty pass across my heart. But I closed my heart instantly against it' (286).

The Steward of Christendom's final image is of a no less impossible reconciliation between biological father and son. Although Willie Dunne would have been in his late teens when he was killed, he appears to his father on several occasions in the play, always at the age of thirteen or so, his voice not yet broken. In the closing moments, the dead son reappears, a child dressed in the costume of the World War I, and climbs into bed beside his father. Thomas Dunne tells Willie of a childhood incident in which he, Dunne, ran away from home with a dog that had killed sheep and was to be put down. When the two strays were rounded up and brought home, he thought they were both 'for slaughter' (301). Instead, he is swept up into his father's arms and embraced: 'And I would call that the mercy of fathers, when the love that lies in them deeply [...] is betrayed by an emergency, and the child sees at last that he is loved' (301). This may well be the only positive representation of a father-son relationship in the canon of contemporary Irish drama. Sebastian Barry's plays are remarkable for their intimate interrelation of private and public event, for the love they bring to bear on relationships which have traditionally been fuelled by ignorance and misunderstanding.

The figure who most raises all of the issues regarding the 1990s development of the Irish play in London is Martin McDonagh (born 1970). He does so in part because his own background throws into question any easy assumptions about national origins. When he made his theatrical debut with *The Beauty Queen of Leenane* in February 1996, it was in a Druid Theatre production directed by Garry Hynes at Galway's Town Hall Theatre. With a playwright bearing a West of Ireland name (the young man who taught Synge Irish on the Aran Islands a century earlier was also named Martin McDonagh),[19] and a title which declared the play's being set there, an audience would assume that its author came from the same locale. But any filmed interview with McDonagh (and there are very few) revealed a punk sensibility and an unmistakably London accent. Though the son of a Galway father and a Sligo mother, McDonagh was born and raised in London when his father moved there seeking employment. McDonagh accordingly is one of the first, and certainly the most high profile, playwrights of the Irish diaspora in England. In the past, such emigrants were generally lost to Ireland. Even as transportation improved, and return visits became possible, they remained invisible in terms of a cultural contribution, at least a contribution that would be recognised as such. Perhaps it was the Irish soccer team of the late 1980s, where second- and third-generation exiles with Irish grandmothers and English accents were encouraged to play for Ireland, that redrew the boundaries of what was acceptably 'Irish'. And a singer-songwriter like Shane McGowan and his band the Pogues, who infused Irish traditional music with a punk sensibility and so recovered some of its anarchic energies, emerged at the head of a London-Irish artistic community. But these were from the popular arenas of football and rock music. Theatre is traditionally more highbrow, especially when dubbed 'national', and the introduction of McDonagh into that context and scene has predictably made waves.

McDonagh has encouraged this tendency by presenting himself in a populist way. In interviews, he has habitually disdained the medium of theatre in favour of movies: 'I always thought theatre was the least interesting of the art forms. I'd much rather sit at home and watch a good TV play or series than go to the theatre. I only used to go to see plays with film stars in them. I think the first play I ever saw, and still probably the best, was David Mamet's *American Buffalo* with Al Pacino when I was about 15.'[20] And even when McDonagh does condescend to mention a playwright, as here, the instances (David Mamet, Harold Pinter) are drawn from American and British theatre. Yet the evidence of McDonagh's plays suggests a deep knowledge of theatre, and of Irish theatre, at that. The title of the third play in McDonagh's Leenane trilogy, *The Lonesome West* (1997), directly quotes Pegeen Mike's father from Synge's *The Playboy of the Western World*: 'Oh, there's sainted glory this day in the lonesome west.'[21] When a character in the second play, *A Skull in Connemara* (1997), has a spade driven into his skull and returns to say that he is not dead yet, we are clearly in the vicinity of Synge's father-son parricidal conflict. The title of this McDonagh play is itself a quote from the closing lines of Lucky's lengthy monologue in Beckett's *Waiting for Godot*: 'the skull the skull the skull the skull in Connemara'.[22] But part of McDonagh's plays' unsettling, post-modernist effect is that they create a hybrid by merging these elements from the classical canon of Irish drama with popular elements from American television and cinema. The mother and daughter in *The Beauty Queen of Leenane* may recall Mommo and her granddaughters in Tom Murphy's *Bailegangaire*, an association promoted by the Druid connection and the shared direction of Garry Hynes; but in the grotesque pouring of the burning cooking oil on one by the other we are also close to the camp melodrama of Bette Davis and Joan Crawford in the Hollywood grand guignol of Robert Aldrich's *Whatever Happened to Baby Jane* (1962). And if the endlessly feuding brothers in *The Lonesome West* are close kin to their equivalents in Lady Gregory's *The Workhouse Ward* (1906), the prissiness of the one and the slobbishness of the other are straight out of Neil Simon's *The Odd Couple*, in its widely disseminated TV incarnation rather than its prior stage or cinematic representations.

For McDonagh, the difficulty he had in becoming a playwright lay in determining how his characters spoke. In a TV documentary broadcast by Radio Telefis Eireann early in 1998, he was asked why he chose to set his plays in the West of Ireland and replied that he had tried setting plays in London and the US but without success. It was when he recalled the setting and conversations from his summer visits as a child to the

West of Ireland that he found his dramatic idiom: 'close to home, but distant', as he put it. The TV documentary is keen to stress McDonagh's authenticity, his claims to a direct experience of life in the West, and is also working to stress the parallel with J. M. Synge and his similar cultural placement at the turn of the previous century. But the parallel with Synge can work both ways, as much against McDonagh's work as in favour of it, either by those who champion Synge's authenticity at the expense of McDonagh's pastiche, or those who condemn both. The staging of *The Beauty Queen of Leenane* by Garry Hynes and Druid in Galway, before it toured the country and transferred to London and New York, led to objections of misrepresentation that eerily echoed those which had been directed at Synge 90 years earlier. The argument ran that what was being dished up here, primarily for non-Irish consumption (since both *The Beauty Queen* and the trilogy as a whole were co-productions with London's Royal Court Theatre) was Stage Irish stereotype and that the plays relied for their theatrical effect on characters of limited intelligence and psychopathic tendencies. The involvement of respected theatre professionals like director Hynes and of actors like Marie Mullan, Anna Manahan, Mick Lally and Maeliosa Stafford served only to give a gloss of authenticity to plays which were not Irish. If Irish audiences responded by turning out in great numbers and responding with laughter to what they saw, then that only served to show how thoroughly colonised they were.[23]

In relation to McDonagh's plays, Hynes gave the 'authenticity' argument short shrift: 'There's this issue about Martin and authenticity – the response that his is not Irish life now and it's not Connemara life. Of course it isn't. It's an artifice. It's not authentic. It's not meant to be. It's a complete creation, and in that sense it's fascinating.'[24] Much of the issue of Synge's authenticity revolved around the way his characters spoke. The TV documentary on McDonagh stressed the authenticity by stating that he remembered and wrote down what he had heard. The account McDonagh gave Fintan O'Toole marks a significant shift of emphasis, where he describes his development of an Irish dramatic idiom as a calculated ploy to disguise the influence of David Mamet and Harold Pinter: 'I wanted to develop some kind of dialogue style as strange and heightened as those two, but twisted in some way so the influence wasn't as obvious. And then I sort of remembered the way my uncles spoke back in Galway, the structure of their sentences. I didn't think of it as structure, just as a kind of rhythm in the speech. And that seemed an interesting way to go, to try to do something with that language that wouldn't be English or American.'[25] If there is a playwright missing from this account who would mediate between the Irishness of Synge on the one hand and the Englishness and Americanness of Pinter and Mamet on the other, it is of course Samuel Beckett. When his official biographer asked Beckett what figure from world theatre had most influenced his own playwriting, the octogenarian author murmured only one word by way of reply: 'Synge'.[26] Pinter and Mamet have both acknowledged a profound debt to Beckett, in particular for the way they deploy a heightened, repetitious dramatic speech and a strategic use of silence. Beckett is named by McDonagh in this acknowledgement of influence and intertextuality, not by name, but by the explicit quotation of the *Skull in Connemara*.

McDonagh says that it was 'as soon as I started writing the first scene'[27] of *Skull* that he found his idiom and rhythm. The opening lines, with the arrival of the 70-year-old Maryjohnny Rafferty in the fifty-something Mick Dowd's cottage to cadge her usual store of free poteen, make this clear:

> MARY: Mick.
> MICK: Maryjohnny.
> MARY: Cold.
> MICK: I suppose it's cold.
> MARY: Cold, aye. It's turning.
> MICK: Is it turning?
> MARY: It's turning now, Mick. The summer is going.
> MICK: It isn't going yet, or is it now?
> MARY: The summer is going, Mick. [...]
> MICK: What summer we had.
> MARY: What summer we had. We had no summer.
> MICK: Sit down for yourself, there, Mary.
> MARY: (*sitting*) Rain, rain, rain, rain, rain we had. And now the cold. And now the dark closing in. The leaves'll be turning in a couple of weeks and that'll be the end of it.[28]

At one level this is an accurate reflection of the Irish weather, or at least of how the Irish talk endlessly about the weather as a medium of dialogue and of the alleged persistence of rain as a confirmation of their gloomiest fears. McDonagh, like Synge and Beckett before him, avoids the summer as a seasonal setting for his drama, preferring late autumn and the onset of winter, with things on the turn and the advent of decay. But at another level this is pure verbal music, an exchange of language at a heightened level. The main activity here is linguistic, with an element in one statement being picked up and responded to by the other speaker. Every variation is touched on in a general context of repetition and minimalism, from questioning through contradiction to agreement (though that will only be temporary before the whole thing starts off again). That heightened degree of verbal exchange will remain the dominant key of McDonagh's drama, for all that subsequent developments may seem to alter it into a succession of violent incidents or an endless stream of verbal 'fecks'. The proliferation of both is something of an illusion. For all of the violent incidents in *The Lonesome West*, for instance, when the brothers face each other with a knife and a shotgun which may or may not be loaded, the play's dramatic energies are primarily concentrated on the comically inventive, verbally and psychologically unrelenting dialogue between the pair. It is the younger male characters who tend to come out with the 'fecks'. They are usually reprimanded for their language by one of the older people and so strive to refrain from using 'feck' again, the comic payoff coming when the older character is usually goaded into unleashing an even greater stream of them.

But it is not just McDonagh's language which has this heightened, repetitive rhythm. Scenes are framed in ways that draw attention to their formal repetition, creating the sense of the characters enacting a circular pattern of behaviour. The opening of *A Skull in Connemara* with the arrival of Maryjohnny Rafferty at Mick Dowd's house also opens the play's fourth and final scene, and the opening exchange of names is once more followed by her declaration that the weather is 'cold'. That play, however, is the weakest of the trilogy; one has the sense of someone tuning their instrument. But *The Beauty Queen of Leenane* and *The Lonesome West* possess a greater degree of dramatic concentration with something actually at stake between mother and daughter and brother and brother. An old woman rocking in a chair and two males locked together in endless verbal interchange dominates each.

Beauty Queen opens with the 70-year-old Mag, memorably played by veteran Irish actor Anna Manahan, '*sitting in the rocking-chair, staring off into space*'[29] as her 40-year-old daughter Maureen enters with the shopping. The scene will end in a row between the pair which causes Maureen to exit furiously and leave her mother alone onstage in the rocking chair to '*stare grumpily out into space*' (7). This dramatic image bears a striking resemblance to two Beckett plays: in terms of two characters, it resembles the combination of master-slave and parent-child relationship between Hamm and Clov in *Endgame* and in the case of mother Mag the old grey-haired woman rocking in her chair and talking to herself in *Rockaby*. What the visual image of the '*stoutish*' old woman sitting and rocking implacably in her chair in the McDonagh play suggests is that, like Hamm in his centrally placed armchair, Mag cannot physically get out of it. But this visual impression is contradicted by the opening exchange where it emerges that the old woman has made her own Complan in her daughter's absence; to which Maureen sardonically responds: 'So you *can* get it yourself so' (1). It is not that Mag cannot get out of the rocking chair; it is that she does not want to. And so on those occasions when she is forced to leave her chair she does so reluctantly and with no good in mind, Anna Manahan deployed her weight just as implacably when she made one of Mag's few decisive movements as when she sat in the chair. The endless petty domestic chores that Maureen is made to perform on her mother's behalf and at the latter's insistence resemble those in the Beckett play, especially when it comes to bodily functions. Hamm's reply that he is urinating onstage, to Clov's relief, is verbally echoed in McDonagh's equal concern with the bodily functions of his old woman. Visually McDonagh is much more graphic and shows Mag coming on at the start of Scene Four '*carrying a potty of urine, which she pours out down the sink*' (25). This is something Maureen has long suspected but which Mag has consistently denied doing, despite the appalling odour.

A reason the mother has given for wanting Maureen to make her the Complan is the fear that she might scald her hands by spilling the boiling water through infirmity. To confirm the possibility, she holds up the '*shrivelled and red*' left hand called for in the stage directions (1), as *Endgame* calls for Hamm to have a '*very red face*' covered by '*a large blood-stained handkerchief*'.[30] Later on, Mag will protest that her only daughter is torturing her, a claim which we interpret as bad-mindedness until we actually see her do it. Despite Hamm's repeated acts of cruelty, Clov in the main responds submissively, though he does rhetorically question himself for doing so. There is one outburst of violence, however, when he lunges at Hamm but forgets that he is standing on the ladder and has to stop himself falling off instead. After Hamm has ordered him to move further away, Clov remarks: 'If I could kill him I'd die happy.'[31] The inhibitions which keep Maureen from doing the same to Mag break down when she discovers that her budding relationship with a returned émigré has been gratuitously nipped in the bud by her mother. At the start of Scene Eight, the set is dark save for the glowing coals of the fire, which illuminates '*the dark shapes of* **Mag**, *sitting in her rocking-chair, which rocks back and forth of its own volition, and* **Maureen** [...] *who idles very slowly round the room, poker in hand*'. Mag is silent throughout Maureen's monologue in which the young woman describes how she effected a last-minute reconciliation with her man. As Maureen finishes speaking, *'the rocking-chair has stopped its motions.* **Mag** *starts to slowly lean forward at the waist until she finally topples over and falls heavily to the floor, dead'* (49). The foregrounding of the rocking-chair moving of its own volition and then ceasing to show that Mag is dead bears closely on the life-and-death synchrony of the moving chair in Beckett's *Rockaby* which is 'slight [and] slow' and 'controlled mechanically without assistance from W', the woman.[32] *Rockaby* formally concludes with W saying 'fuck life' and the chair *'coming to rest of rock'* accompanied by a *'slow fade out'* of light which has never been more than 'subdued' (433). In McDonagh's play, the coming to rest of the rocker and the old woman falling out of the chair are accompanied by a visual trope whose Gothic explicitness marks the difference between the two playwrights by revealing that Maureen has killed the old woman: *'A red chunk of skull hangs from a string of skin at the side of her head'* (51). W in *Rockaby* has spent most of the play, and her life, sitting at the window looking for company. Failing in that quest, she is forced to turn inwards, both 'down the steep stair' (441) and psychologically to go deep within, into her mother's rocking chair and become 'her own other'. This resolution is played out in a more conventional way in the McDonagh, when the last scene reveals the rocking-chair now occupied not by Mag but by her daughter Maureen, even if the play is a little too concerned to point it out: 'The exact fecking image of your mother, you are, sitting there' (60).

Thomas Kilroy has written of Anglo-Irish playwrights from Farquhar to Beckett in terms of a 'characteristic distancing effect, a cool remove of the playwright from his subject matter', and a concomitant concern with form.[33] Hence the restlessness, whether of Synge's vagrants or of Beckett's immobiles.[34] There is a movement in these plays towards 'abstraction and the perfection of the idea, the radical reshaping of human action for particular effects'.[35] But that distancing which Kilroy sees as inhering in the socially hyphenated category of the Anglo-Irish and as culminating in Beckett may have started up again and become relevant with playwrights of the Irish diaspora. As Martin McDonagh puts it, in relation to Ireland and England: 'I always felt somewhere kind of in-between, same way I do now. I felt half and half and neither, which is good.'[36] At their best, his plays operate so successfully in the theatre because their greatest engagement is between the playwright and the audience, because of the superb calibration of their formal dramatic effects. Maureen's burning of her mother's hands, to some, is a device out of melodrama. But it is shorn of almost all surrounding plot complications that would have accompanied its nineteenth-century manifestation and is explored in almost purely formal Beckettian terms at both the verbal and dramaturgic levels. And the outbreaks of melodramatic violence invariably achieve nothing and leave the basic situation unaltered. Maureen has not freed herself from Mag's psychological domination by killing her. In *The Lonesome West*, when Coleman and Valene cease from quarrelling, we know it is only temporary and that they will soon resume goading each other. Beckett and McDonagh are both dark comedians and as such have scant interest in any great psychological particularity of character. People in stage comedies usually remain fundamentally unchanged from start to finish; indeed, much of the comedy derives from a character's determined efforts to change along with the certainty that they will be unable to do so and that fundamental patterns of behaviour will reassert themselves. In the two playwrights' work, the sense of the play's duration is of a temporary entering (and leaving) a ceaseless continuum in which a lifetime of such activity between the characters is deftly suggested. The only possible change or outcome is death, hence the blackness.

In relation to what may best be termed 'the McDonagh phenomonon', the aesthetic merits of his plays may be (and are being) endlessly debated. They work with a calculated cunning and an utter lack of sentimentality towards the Irish theatrical canon as much as anything else. My own feeling is that the smash-and-grab theatrics of McDonagh have, through their impact in Ireland and throughout the world, changed the landscape of the Irish play irrevocably. They have done so in particular because they have dispensed with nostalgia and any evocation of a consoling Irish past. There is no clear time frame for his plays: they draw on images from traditional Irish drama of decades past – the country cottage, the crucifix on the wall, the graveyard, the West of Ireland – but their younger people endlessly discuss characters from popular American TV series of the 1970s; while other references suggest a more contemporary provenance. Their cultural references are no more secure and prove equally free-floating. McDonagh's plays suggest that what now constitutes 'Irish theatre' either on the Dublin or London stages is a range of signifiers which may be played with at random rather than anything which bears a direct relation to social reality or historical continuity and can be construed as a 'national' theatre. In this sense, Martin McDonagh has had a discernible impact and effect on many Irish playwrights, no doubt envious of the trail his plays have blazed around the world.

Outside of the Northern Irish context, almost all of the playwrights considered in this study have been male. With the change that I have discussed in relation to Friel, Murphy and Kilroy to strong female characters, the question must be asked: is this not a move by Irish male playwrights to appropriate and colonise the concerns of woman in a feat of expert ventriloquism, a pre-emptive creative strike against the increasing number of women writing plays and having them staged? David Grant has written that the 'list of recognized living Irish playwrights, [even when] far from complete, [reflects] the common tendency to exclude women' and 'fails to take account of a truly enormous group of talented younger or less established playwrights'.[37] There has been a considerable contribution by women to contemporary drama in the North, as the last chapter has shown; but the theatrical scene in the Republic of Ireland has remained much more male-dominated. In 1989, Glasshouse Productions was founded in Dublin as a feminist collective and during their five years of existence produced new plays by Clare Dowling, Trudy Hayes and novelist and short story writer Emma Donoghue (her lesbian dramas, *I Know My Own Heart* and *Ladies and Gentlemen*). Their two stage anthologies of work done by earlier writers was ironically entitled *There Are No Irish Women Playwrights*.[38] Despite such movements as this, the only contemporary Irish woman playwright to break through the glass ceiling during the period under review was Marina Carr, a fact confirmed by her sole presence among the array of male playwrights lined up during the Abbey Theatre Centenary celebrations in 2004 for productions of their works. I would like to close this book by examining the work of Marina Carr not just as representative of the radical experimentation of that younger generation and as an important playwright who is also a woman but above all because her formidable body of work has already established Carr as 'one of the most powerful, haunting voices on the contemporary Irish stage'.[39]

In theatrical terms, Marina Carr would acknowledge a debt to Samuel Beckett more readily than many of the other playwrights I have discussed. Having graduated from University College Dublin with a BA in 1988, she enrolled in the MA in Anglo-Irish Literature and Drama to do a major thesis on Samuel Beckett; but her playwriting took over and it was never completed.[40] In so doing, she managed a more creative play with Beckett's theatre than academic procedures would have allowed. In her earliest work she applied gender politics to the theatre of the absurd. *Low in the Dark* (1989) questions whether it is the most natural thing in the world to have a baby and does so through a variety of absurdist techniques, including the presence of a pregnant male. The mother and daughter are named Bender and Binder, sounding like characters from a late Beckett play, the Bam, Bem, Bim and Bom of *What Where* (1983). Bender and Binder resemble female versions of Vladimir and Estragon. But there is an increased awareness of gender issues, as Carr's version of their hat-swapping indicates in its switching between male and female voices and personae: 'Listen, I have my work. [*Takes off the hat.*] What about me? [*hat back on.*] Don't I spend all the time I can with you? [*Hat off.*] It's not enough, I miss you.'[41]

Among all of the role and gender swapping of the two women and the two men in *Low in the Dark* there is a fifth character, Curtains by name and Curtains by 'nature', since Carr is careful to specify that 'she is covered from head to toe in heavy, brocaded curtains' and 'not an inch of her face or body is seen throughout the play' (5). Through Curtains, Carr raises the issue of on-stage representation of the female body and displays a keen sense of how that body may be read by the male gaze. Beckett had done something similar in *Happy Days* by burying Winnie, first up to her waist, where an audience could still see her '*low bodice [and] big bosom*',[42] then up to her neck in sand. He had gone even further in *Not I* where only the actor's mouth is visible as she tells her story over and over. (This covering up of Mouth's body went missing in Neil Jordan's film of *Not I* for the *Beckett on Film* project, where the camera was erotically fixated on Julianne Moore's mouth and red lips.) Carr's Curtains (and hence the actor playing her) can be any age and runs no risk of being reduced to her physical characteristics. But in a significant swerve or revision Curtains does not undergo the physical confinement to which Beckett subjects his women characters, increasingly paralysed as to mobility and strapped into various contraptions. Curtains still has and indeed enjoys the greatest autonomy of all, experiencing an orgasm which we register through her breathing and movement rather than through any overt visual sign (55).

Curtains is also, significantly, the play's storyteller, telling a tale (the tale?) of a man and woman roaming the earth. Beckett's Winnie had described such a couple who came upon her buried in the sand and argue as to her meaning. The man directs an elaborate series of aggressive questions at Winnie which his woman companion then turns on him by asking 'and you, she says, what's the idea of you, she says, what are you supposed to mean?'[43] Curtains' couple chance 'upon a woman singing in the ditch. "Sing us your song," the man said. The woman sang' (45). Any MA student of Anglo-Irish Literature and Drama at UCD would be familiar with Beckett's remark about Irish writers singing in the last ditch, which was quoted at the end of Chapter 1. In *Low in the Dark*, during an extended round of story-telling involving all of the characters, there is a further, more explicitly gendered reference to ditches:

BONE: He hinted at desperation sung in ditches.
BINDER: She hinted at desperation not sung at all.

(59)

In this early work, Marina Carr acknowledges the possibility for women that their condition of 'desperation' may never be expressed or dramatically represented at all. The plays she has written since do not flinch from that ethical obligation nor from the extremes which must be confronted to do so. Carr's plays are centred on women's experience and, while adopting no Manichean feminist line with regard to gendered stereotypes (some of the women can be just as destructive of the heroine's desire as the men), provide a challenge and alternative to the male-dominated world of Irish playwriting. They claim the right for her and by extension other women playwrights to be among the singers of contemporary Irish drama, not at the expense of but in the company of precursors like Beckett and Friel.

Where *Low in the Dark* and *Ullaloo* (1991) displayed a radical experimentalism in the vein of a Beckettian absurdist drama, with *The Mai* (1994) and *Portia Coughlan* (1996) it has been widely recognised that Marina Carr found her own voice and created a distinctive dramatic world. These two plays, both staged at the Peacock, have also met with a high degree of critical and popular acclaim. They established the Midlands of Ireland and its strong dialect as her dramatic terrain. But presumptions that she had unequivocally rejected absurdism to embrace realism turned out to be premature and (increasingly) inaccurate. Rather than exorcising Beckett's influence, Marina Carr may be said rather to have absorbed it. Beckett's influence remains subtly present throughout her work, for example in the way Act Three of *Portia Coughlan* repeats Act One after the chronological disruption of the middle Act, and resurfaces forcibly in 2006's *Woman and Scarecrow*.

In *The Mai*, its most traditional element is the troubled relationship between a married couple. The great equality that now exists between men and women is signalled by the relative equality of their ages, where in Irish society earlier in the twentieth century and the plays written during that period the husband tended to be much older than the wife; an outstanding example would be Synge's *Shadow of the Glen*. Here, the Mai is 40, Robert in his early forties. They have been married for 17 years and have four children, though the only one we get to see in the play is the eldest, their 16-year-old daughter Millie. The first act of the play signals the Mai's joyous expectation of her errant husband's return home after five years' absence. Despite the Mai's delight at his return, Robert's presence turns out to be as provisional and qualified as before he left. He declares at one point that he needs more time on his own, that he would prefer his own company or that of other people to the at-home demands of the woman he has married.

What is less conventionally dramatic than the man's grandiose entrances and exits every few months or years is the woman's pattern of internal withdrawal while she remains physically confined within the house. Robert concentrates on playing his cello as he pursues his romantic dream of being a great musician. The psychological transference was expressed in Brian Brady's production in a surreal moment when Olwen Fouere as the Mai was substituted for the cello. The idea for this visual substitution is contained in a scene where the Mai plucks at herself as if she *were* a cello. To offset the title character's isolation, Carr has filled the dramatic space with an ensemble support system of female energy: two sisters, two aunts, a daughter and a 100-year-old grandmother. When Grandma Fraochlan makes her memorable entrance, she does so bearing a big colourful oar which she is determined to intrude and which is the flamboyant sign of her unorthodox and romantic life with her husband, the nine-fingered fisherman. She bears in her first name an archetypal female status and in her second the name of the island on which she reared the Mai when the natural mother, her daughter Ellen, died young. Grandma Fraochlan enlarges the scope of the play's environment with her exotic presence and a fund of stories drawn from her 'ancient and fantastical memory',[44] her connection to the world of myth and legend. She offers a matrilineal line of support and continuity rather than a substitute patriarchy, acting as a living conduit to the dead. But Grandma Fraochlan can only do so much; and her surviving daughters argue that she may not have been the best of mothers, since her husband had all her love.

The play's crucial figure is the Mai's daughter, Millie, who bears the brunt of the tensions between her mother and father. But as Millie talks, we realise that she is not merely responding to what is occurring onstage but narrating to the audience an entire drama that has occurred many years in the past. To this extent, she resembles Michael in Friel's *Dancing at Lughnasa*. When Millie reveals before the play is half over that the Mai took her own life, the effect resembles that moment when Friel's Michael discloses the squalid death of two of the women who are so vibrantly alive before us. Nowadays, when Millie meets her father, 'we shout and roar till we we're exhausted or in tears or both, and then crawl away to lick our wounds already gathering venom for the next bout' (128). This speech not only establishes Millie as the play's storyteller but shows that, far from being aligned with the words and deeds of her father, the Mai's husband, she is tussling with him verbally for control of the Mai's narrative, her legacy of the future. Carr's play remains open to that future in ways that Friel's does not, not only through Millie's survival and determination to tell her story but in the references to Joseph, her five-year-old son.

The final image is of the Mai, standing alone at the window of the house. This is the dream house she has constructed from her teacher's salary, from the cleaning jobs she has taken, and from getting the builder to let her have it for a song. It has been built according to her specifications, to house herself and her four children and to prepare a place for the long-desired return of her husband. Its central feature in terms of the staging and Kathy Strachan's design is a huge window centre stage which gives out on to Owl Lake, the pattern of the lake reflected around the stage. When characters appear up the steps, it would seem more natural for them to step through than to go round to the door. The Mai is most often to be found standing in that window, as much looking in as looking out, not fully contained by the house she has built. We are never directly given the Mai's death; it would be unnecessary and untrue to the way in which her story is told. Her search is like that of the old woman in Beckett's *Rockaby*. Like her, the Mai is looking 'for another/ at her window/ another like herself /a little like'.[45]

At first glance, *Portia Coughlan* bears certain striking resemblances to *The Mai* in its casting requirements and stage situations. But the sound and the feel of the play are very different and mark yet another advance for Carr as a dramatist. Once again, there is a troubled young woman at its centre and a drama that draws in an extraordinary supporting cast of family and friends. Once again, the woman is trapped in an unsatisfactory marriage, though the reasons here are harder to find, since Raphael (unlike Robert) is in most respects a model husband to Portia. In *The Mai* Millie was a crucial presence, but the other sons and daughters were never seen. Here, all three children remain offstage, relegated to the sidelines and their minder. Portia's friend Stacia. The compelling onstage child is 15-year-old Gabriel Scully, Portia's twin brother, who was drowned at that age in the Belmont River and continues to haunt her. There is once again a wonderfully diverse line-up of female relatives, including a mother of 50 and a grandmother of 80 whose power of audacious speech goes beyond even Grandma Fraochlan's. Although the Mai and Portia Coughlan are to a degree unreachable and living in a world of their own, the former did receive and acknowledge a consistent degree of support from the women in her life. But Portia's mother and grandmother are either indifferent or hostile to her. Instead they are locked into an internecine feud over family secrets and over which family line caused the queerness of the twins, Portia and Gabriel. The sole unlikely source of support for Portia from among her female relatives is from her aunt,

Maggie May, whose first entrance is described in the following terms: '*Enter Maggie May Doorley, an old prostitute, black mini skirt, black tights, sexy blouse, loads of costume jewellery, high heels, fag in her mouth.*'[46] Even more unlikely, perhaps, is the old man who trails in her wake, fussy, nervous, skinny, endlessly asking for cups of tea with which he can wash down his supply of digestive biscuits. This is Senchil Doorley, Maggie May's husband; and their affectionate, loving relationship is as credible as it is initially surprising. As Maggie May remarks, in one of the play's best lines: 'Senchil wasn't born, he was knitted on a wet Sunday afternoon' (240). Their relationship extends the range of Marina Carr's characterisation and shows that marriage *per se* is not what bedevils Portia Coughlan.

With their physical mutilations and their invective-fuelled encounters the characters in *Portia Coughlan* are in many ways grotesque: Raphael walks with a limp; Stacia has lost an eye and is referred to as the Cyclops of Coolinarney. And yet they remain rooted in a detailed and recognisable social world, in which children have to be collected, meals prepared, in which old school rivalries and attractions persist into adulthood, in which there is a definite class structure separating the owner of the local factory from the young man working in the local pub as wooers of Portia. This is a world in which Portia Coughlan is present more in body than in spirit, a world from which she threatens to float free and to which most of the other characters seek to recall her, mostly by insisting on her gendered duties as daughter, wife and mother. Like the plays of Synge, though rooted in the real, Carr and her characters hanker after the mythic. Portia says at one point that she married Raphael because he was named after an angel. So, too, was her dead brother, Gabriel. And it is in the direct staging of the dead brother's presence, established from the start, that the play most breaks with a conventional or narrow realism. As the light comes up on Portia Coughlan at home with a drink in her hand, the '*other light comes up simultaneously on Gabriel Scully, her dead twin. He stands at the bank of the Belmont River singing. They mirror one another's movement in an odd way, unconsciously*' (239).

Gabriel is never far away throughout the play and is evoked in a variety of ways: by his physical presence onstage while a more realistic scene is in progress, troubling Portia's concentration; as an acoustic presence, through his high-pitched beautiful singing, a music which sometimes drowns or tunes out the dialogue in which Portia is involved; and through Gabriel's association with the Belmont river, the site which draws Portia repeatedly throughout the play and to which she is ineluctably destined. There is a strong similarity between this ghostly boy in the Marina Carr play and the ghost of young Willie Dunne in Barry's *The Steward of Christendom*, a coincidence underlined by the plays' being staged simultaneously at Dublin's Peacock and Gate Theatres in 1996. The simultaneity underscores the extent to which onstage ghosts have been a feature of plays in the 1990s. In one sense, they are a dramatic means of dramatising the persistence of the past in the present, a particularly if not exclusively Irish obsession. But they also figure in these plays as troubling reminders that economic prosperity leaves a good many atavisms unslaked, that the decline in the power of the Catholic church leaves the way open for a return of the irrational which the surface of Irish life no longer acknowledges.

Portia Coughlan is caught midway between the worlds of the living and the dead, with the combined weight of nine living family and friends on the one side and the uncanny and singular presence of her dead alter ego on the other. What complicates this equation, and the play's time sequence, is the short second act. It is clear from the start that the appeal of the Belmont river is going to prove fatally irresistible to Portia; and that what we are witnessing are the last hours or (at most) days of her life. Indeed, as I indicated earlier, the third act is a Beckettian repetition of the first, taking us through the encounters of two successive days with variations. But the play disrupts chronology in its startling second act, which shows the dead body of Portia Coughlan being winched up from the Belmont river in front of the silent assembled cast of characters. In so disrupting linear time, as well as by bringing the ghost of Gabriel onstage, Carr makes apparent that she is writing a drama of ritual rather than of realism. She achieves this by dramatising death, not considered as an end in itself or as a single self-contained incident but as a process extending over (and disrupting) time and involving a wide group of family and friends as mourners.

My critical account of death as process is deliberately drawing on terms from a valuable study by Fiona McIntosh, which makes a series of sustained parallels between the tragedies of Classical Greece and the plays of Yeats, Synge and O'Casey.[47] Although McIntosh does not deal with contemporary Irish plays, a work like *Portia Coughlan* benefits from her analysis in showing how, while the precise moment of the heroine's suicide is elided, its implications inform her every living gesture in the world of the play. For in McIntosh's account of such a drama the dying character meets his or her death not once but many times. Often, as the intensity of their suffering escalates, they are seen to occupy a liminal zone in which it is difficult to determine whether they are living or dead: 'Ah'm dead, Maggie May, dead an 'whah ya seen this long time gone be a ghost who cha'nt fin' her restin' place, is all' (293). In the 'big speeches' which the title characters of Greek tragedy deliver, their auditors are as likely to be the ghosts they are going to encounter as the living they are leaving behind. Here, the ghost of Gabriel is increasingly addressed directly by Portia in an exchange which allows for a greater range of levels and registers of language (as it did in the Barry). *Portia Coughlan* is a modern-day tragedy where the death of the individual character spirals out to implicate the fate of a family line and the spiritual well-being of the entire community.

What is finally so striking about the play is its language, what Frank McGuinness rightly calls its 'physical attack on the conventions of syntax, spelling and sounds of Standard English'.[48] In first attending the play, the audience had to tune in to what immediately strikes the listener as the heavy accent of the Irish Midlands and a dialect which bends the vowels and either strikes out the final 't' in words or turns them into an 'h'. The following is a representative exchange:

PORTIA: Busy ah tha factory?
RAPHAEL: Aye.
PORTIA: Ud's me birtha taday.
RAPHAEL: Thah a fac?
PORTIA: Imagine, ah'm thirty...Jay, half me life's over.
RAPHAEL: Me heart goes ouh ta ya.
PORTIA: Have wan wud me an me birtha.
RAPHAEL: Ah this hour, ya mus' be ouha yar mine.

(240)

The language also contains recognisable features of Hiberno-English pronunciation and usage: 'me' for 'my', 'seen' and 'done' for 'saw' and 'did'. But it is one thing to hear it spoken onstage, another to read the script from the printed page, where some of it borders on the incomprehensible. Perhaps this is the reason why the subsequent printings of the play by Faber and Faber and the Gallery Press removed much of the dialect, with the third line above now standardised as 'It's me birthday today.'[49] But the dialect remains a strong feature of most of Carr's plays to date. No more than Synge stopped at transcribing Hiberno-English in his plays, Marina Carr is doing more than merely replicating Midlands speech. Frank McGuinness, who first printed the dialect version of *Portia Coughlan*, describes her language as a 'haunting'[50] and Carr herself describes the Midlands as 'a metaphor for the crossroads between the worlds'[51]. She is fashioning a dramatic speech all her own, bending and shaping the forms of Standard English in ways that allow for the disruption of the linguistic present by the Irish past and evolving a flexible linguistic medium to address the living and the dead, the real and the mythic.

In 1998, *By the Bog of Cats* was the first of Marina Carr's plays to break through from the smaller Peacock Theatre to the Abbey main stage, in a powerful epic production by Patrick Mason. The play was her most confident and ambitious to date, deepening and extending the concerns of her earlier plays while perfecting a dialogue of pitched intensity and mordant wit between the characters. Its dramatic action centres on a marriage between Carthage Kilbride, Oedipally dominated by a ferocious mother, and Caroline Cassidy, the weak daughter of a strong farmer. The omens for their marriage are not good. But they are seriously complicated by the presence of the play's central character, Hester Swane, the woman with whom Carthage has shared his life for over ten years and who is not willing simply to disappear along with their daughter, Josie. The personal dimensions of this conflict are deepened by the categorisation of Hester by those who oppose her, notably Carthage's mother, as a tinker:

> I warned him [her son] about that wan, Hester Swane, that she'd get her claws in, and she did, the tinker. That's what yees are, tinkers [she is addressing her granddaughter].[52]

In writing a programme note several years ago for an Abbey production of Synge's *The Playboy of the Western World,* Marina Carr drew on her teacher Gus Martin's distinction between two categories of character in Synge's drama, the householders like Pegeen Mike's father Michael James Flaherty and the poetry people, notably Christy Mahon and his father – and the profound unease between them.[53] Her own play develops and complicates the distinction. Hester Swane occupies two dwellings in the play, the caravan by the Bog of Cats, in which she was raised by her mother, and the bourgeois house she has shared with Carthage and her daughter. Much is made in the play by the household people of the legal procedures by which Hester has signed over her rights in the property. Like a latter-day Antigone, Hester responds by insisting with a stubborn personal authority on her right to dwell by the Bog of Cats, whether in a housed or an unhoused condition, as the place in which she was born and reared. In that conflict lie the seeds of the tragedy.

But first there is the high comedy of Act Two, the wedding from Hell. It brings on the disconsolate bride, Caroline. While wandering around in her wedding dress on what should be the happiest day of her life, Caroline is increasingly aware that her new husband is still obsessed with the woman he claims to have left for her. But there is more than one bride at this wedding; the mother-in-law Mrs Kilbride has insisted on decking herself out in white. And at a key point in the action, the spurned Hester Swane makes her entrance, as the stage directions have it, '*in her wedding dress, veil, shoes, the works*' (311). The showdown between the forces of respectability and those castigated as tinkers erupts in the following exchange where Mrs Kilbride reveals why she is so anxious to keep her social distance:

> MRS KILBRIDE: A waste of time givin' chances to a tinker. All tinkers understands is the open road and where the next bottle of whiskey is coming from.
> MONICA: Well, you should know and your own grandfather wan!
> MRS KILBRIDE: My grandfather was a wanderin' tinsmith –
> MONICA: And what's that but a tinker with notions!
>
> (314–15)

The social order is not confirmed or ratified by the wedding of Carthage and Caroline, but rather exposed and ripped apart. There is still the play's third act to come where the tragedy will play itself out to a Medea-like climax.

In his programme note for *By the Bog of Cats…*, Frank McGuinness argued for the presence of the Greeks in the deep structure of Marina Carr's imagination: 'I wonder what [she] believes? I think it might be the Greek gods – Zeus and Hera, Pallas Athene. […] I can't say for certain, but I am certain in this play she writes in Greek.'[54] From the beginning of the play, Hester Swane has been under a sentence of death, to be fulfilled in the succeeding 24 hours. She is a figure fated to die but also the instrument and vehicle of that fate. This play presses further with the point made by Fiona McIntosh in *Dying Acts* that many of the central characters of Greek and Irish tragedy occupy a liminal zone, where they are as likely to communicate with the dead as with the living. In the first scene, she encounters 'a ghost fancier' (265) who claims he has come to take her to the regions of the dead. When he realises that he has come at dawn rather than at dusk, he formally apologises, like Emily Dickinson's gentleman caller, and says he will see her later. The first character we meet in Act Two is also a ghost, of Hester's dead brother, leading the Cat Woman – another of the poetry people – to remark; 'Ah Christ, not another ghost' (299). The one dead or absent figure who never returns is Hester's mother, Josie Swane. Carr's stage plays host to the living and the dead. In the play's tragic climax Hester and Josie, mother and daughter, cross over from one realm to the other and do so through Hester's agency. When the daughter declares that she does not want to be adopted by her father and stepmother (as is threatened) but to stay with her birth mother Hester – 'I want to be where you'll be' (338) – Hester realises that there is only way for her daughter to escape her own fate, a life spent waiting for a mother who never returns:

> Alright, alright! Shhh! (*Picks her up.*) It's alright, I'll take ya with me, I won't have ya as I was, waitin' a lifetime for somewan to return, because they don't, Josie, they don't.
>
> (339)

And Hester cuts Josie's throat. The echo of Medea, of child-slaying by a mother, is reworked by the fact that mother soon follows daughter. When the ghost fancier reappears, he and Hester '*go into a death dance with the fishing knife, which ends plunged into Hester's heart. She falls to the ground*' (341). As she dies, she whispers her mother's name over and over. Carthage denounces her a moment earlier, when confronted with his dead daughter as 'a savage' (340). But Hester's daughter-slaying registers more as an act of compassion than savagery and when she completes the act by killing herself the two dead bodies bespeak a comprehensive indictment of the patriarchal world presided over by the weak Carthage and the brutal Xavier Cassidy, Caroline's father. *By the Bog of Cats...* never wavers in the degree of discipline and control Carr brings to the handling of her Graeco-Hibernian tragedy and the emotional through-line she maintains in elaborating the dramatic process by which her heroine gets her death.

The relationship between the legacy of Greek tragedy and the representation of a contemporary Ireland is a good deal more problematic in *Ariel* (2002), again presented on the Abbey main stage. In its account of an ambitious politician who sacrifices his 16-year-old daughter to gain power, there is a strong debt to Euripides' *Iphigenia at Aulis*. Carr's play covers a time-span of over a decade. It begins with Fermoy Fitzgerald on the night before his election to a Dáil seat in a tightly fought contest; this is the same night as his daughter Ariel's sixteenth birthday when her father takes her on a fatal car ride in her brand new birthday present. Act Two resumes ten years later when, in a television interview, Fermoy discusses his meteoric rise to political success, his record of achievement in three government ministries, and his ambition to be Taoiseach (prime minister). This is also the night on which he will be killed by his Clytemnestra, Fermoy's wife Frances, when she discovers what he has done to their daughter. Act Three extends the time span posthumously with the return of Fermoy's ghost, still searching for a reunion and atonement with his dead daughter. The mix of the contemporary and the mythic is *Ariel*'s most daring stylistic feature, and also its most unstable. This is a recognisably contemporary Ireland, more so than in many of Carr's other plays. The Fitzgeralds are a Celtic Tiger family, one that has prospered from the success of its cement factory and which can now afford to send its children to university and to buy them brand new cars for their sixteenth birthdays. But it is also a world where incest and murder are a part of the family legacy, where the ghost of a dead child can talk to her father from the bottom of a lake, and where a contemporary politician can speak of his one-to-one relationship with God. The play could be said to be a probing of the contemporary political unconscious. The overt emphasis in public debate is on economic development while the sub-structure of religious belief which still underpins the Irish state and moulds its discourse remains unaddressed or even acknowledged.

There is a huge area of *Ariel* to be explored in terms of its belief systems, which I can do no more than touch on. If Fermoy claims (as he does) that he is merely following God's plan in sacrificing his daughter, the question remains: what God? The theistic issue is raised by his wife Frances when she accuses Fermoy of the murder in Act Two. But it emerges even more powerfully in Act One when he is still contemplating the deed and discusses it with his brother, the monk Boniface. There is some suggestion that in a Yeatsian sense the 2000-year-old Christian dispensation is at an end and what will succeed is the 'rough beast', when Fermoy explicitly invokes a 'blood-dimmed tide'[55]:

FERMOY: [...] I'll give him [God] whah he wants for ud's hees in the first place anyway.
BONIFACE: And whah is ud he wants?
FERMOY: I tould ya, blood and more blood, blood till we're dry as husks, then pound us down, spread us like salt on the land, begin the experiment over, on different terms next time.
BONIFACE: We've moved beyond the God a Job, Fermoy.[56]

This invocation of Job is the only Old Testament reference in *Ariel*. But in a Biblical context where Abraham is called on to sacrifice his son Isaac, the Job reference provides a context in which human sacrifice to the deity can be considered. The primary reference would appear to be the Greeks mentioned by Frank McGuinness where, even though the play retains the singular reference, what is being talked about is 'the gods', those supernatural figures who take a particular delight in intervening in human affairs and toying with the destinies of mortal individuals. But it is quite a task to translate such mythic concepts into the social scene of contemporary Ireland and I am not sure the play succeeds in doing so. If there is no endorsement of Fermoy's views by the play, then he is a raving megalomaniac, a psychotic creature beyond the reach of the other characters or our belief or sympathy. If there are gods to whom contemporary Irish politicians have sacrificed in order to achieve their worldly success, then I am not convinced that addressing them in borrowed Greek robes, in terms that are not fully translated into the post-Catholic Ireland of the twenty-first century, is the best or most dramatically effective way of presenting them.[57]

2006's *Woman and Scarecrow* marked a triumphant return to form, in part because it seriously addressed the issue of form. The presence of Greek tragedy, which had come so much to the fore in Carr's drama that it threatened to overthrow everything else, is much more muted. The strain placed on the realistic base of the Midlands plays by the increasingly fantastic events of the drama is abolished at a stroke by the presence of both Woman and Scarecrow, the dying character and her alter ego, on the same bare stage. It is the return of Beckett to Carr's theatre[58] as mediated by Friel's *Philadelphia, Here I Come!* Death, not hope, is the thing with feathers which lurks in the wardrobe and starts to emerge at the close of Act One: '*The wardrobe door creaks open.* WOMAN *and* SCARECROW *turn to look. A wing droops from the wardrobe, then a clawed foot hovers, then lights down.*'[59] In Act Two, death has retreated, Woman is bloodied and clutching black feathers in her hand, and Scarecrow is keeping death at bay in the cupboard to gain the heroine a brief stay of execution. That imminence of death keeps the absurd elements of the play anchored and in the service of a drama where something is at stake. As *Woman and Scarecrow* unfolds, two other characters enter: Woman's maternal aunt (Auntie Ah) and her husband (merely referred to as 'Him'). On the other side of that door is the real world: Woman's eight children and domestic duties, her maternal relatives come from the west of Ireland determined on a traditional Irish wake.

But the play is absolutely centred on the dramatic co-presence of Woman and Scarecrow, the female double-act which Carr had lightly sketched in the mother and daughter of *Low in the Dark* but here represents with depth and complexity. As with Gar Private, no one else can see Scarecrow and so, when she is addressed by Woman when other characters are present, it appears as if Woman is speaking to herself. But the effect is different from the Friel. There, the play retained many of the trappings of the traditional Irish play with the theatrical element of Private Gar inserted. Here, the female double-act predominates with the traditional elements (family members, domestic setting) banished offstage, kept on the other side of the door. It is as if the entire play is set in the bedroom, with the dying Woman largely confined to bed. The dramatic mainstay is the dialogue between Woman and Scarecrow, but that recedes when either husband or aunt enter for lengthy exchanges. Auntie Ah and Him represent the claims of family and kin, of the life that one part of Woman very much wanted to lead; but Scarecrow is there to represent and articulate alternatives, even when Woman doesn't want to hear them.

That conflict is present in the very opening exchange, with one character contradicting the other. Woman's opening statement, 'I ran west to die', is immediately called into question by Woman's refractory reply: 'You ran south – and you didn't run, you crawled' (11). If Woman is characterised as the emotional one, then Scarecrow provides the cold, questioning foil, to articulate what Woman is feeling but also to question it. She allows no sentiment to colour the claims of children and husband and is particularly insistent that Woman has made a decision to die:

> WOMAN: I didn't decide to die. How dare you!
> SCARECROW: You can lie to everyone except me.
>
> (17)

Once Woman has come around to agreeing with this proposition, Scarecrow can define her primary role as that of preparing Woman to die. Here, it is death not Godot who is awaited. The death of the heroine will be co-terminous with the end of the play, as it is in a Beckett prose work like *Malone Dies*.

The question arises on more than one occasion: does Scarecrow know something about what comes after? Her presence brings with it shades of the ghost in Shakespeare's *Hamlet* and a probing of the boundaries between life and death. Shakespeare's tragedy is more to the fore in *Woman and Scarecrow* than any Greek precursor and arises quite naturally in the dramatically heightened liminal zone in which the play is staged. In thinking of death as something romantic, 'magnificent', Woman finds the example of Ophelia initially appealing: 'Ophelia now. She had a good death' (17). But the limited, circumscribed roles available to the heroines in Shakespearean tragedy are acknowledged and gone beyond when Woman subsequently appropriates one of Hamlet's own lines: 'Alas, poor Yorick' (33). The invocation of the Gravediggers' scene allows the physical process of death to be confronted: '[Watching] the rat and the worms have their smelly feast'. And when Woman abjures Scarecrow to 'desist in thy crazy orisons' (42), it is Scarecrow who is being cast as Ophelia with Woman retaining the Hamlet role. Carr undertakes a rewriting of Shakespearean tragedy, to admit of a tragic heroine claiming centre stage rather than being pushed to the margins in the company of Ophelia and Lady Macbeth. But in the swapping of roles there is also culturally astute gender play which argues finally for an androgyny at its heart rather than any strictly defined binary (and a neat reversal and update of the historical fact that on the Elizabethan stage the female roles were played by men).

This alter-ego is more overtly metaphysical than Friel's. The wings which Scarecrow wears as part of her costume may link her to the thing in the cupboard but they also hint at angelic provenance. The speech by Scarecrow which most speaks to this adumbrates one of the feminist creation myths that have featured in many of her plays (the legend attaching to Owl Lake in *The Mai*, for instance). The central metaphor is weaving and Scarecrow as the figure of the traditional guardian angel is recast as a female collaborator who chooses to work on the weave of Woman's character:

> I truly believed when I latched on to you before the weaver's throne, I truly believed that you and I would amount to something. I was wrong. Yes, your bitterness was a flaw in the weave. I noticed it, but I never thought it would bring us down. It looked such a small inconsequential thing, no more than a slipped stitch.
>
> (19)

This provides a rich retrospective gloss on Maggie May's line about her husband in *Portia Coughlan*: 'Senchil wasn't born, he was knitted on a wet Sunday afternoon.'[60]

What is foregrounded here more than in the Friel is the intimations of a same-sex relationship between Woman and Scarecrow. Gar Private is particularly hostile to Katy Doogan and prefers to see Gar Public remaining a confirmed 'bachelor'.[61] Scarecrow is actively jealous of the relationship between Woman and her husband and criticises him in the tones and terms of a spurned lover. The metaphysical declaration that 'I was there before you and I'll be there after' gives way to the passionate utterance: 'I have loved you so long' (20). These intimations of same-sex love underscore Scarecrow's relentless anatomisation of the failures of Woman's marriage. If Woman and Him manage a moment of shared intimacy in Act Two from which Scarecrow is necessarily excluded, the play concludes with, as the stage directions specify, Woman dying '*in* SCARECROW's *arms*' (68), and in Selena Cartmell's Peacock production in a distinctly Pieta-like pose.

It is in Scarecrow's psychological role that the play is most Frielian, specifically in prompting Woman to remember. Initially, these memories have to do with the various lovers Woman has taken over the years, men who appeared so fleetingly in her life she is hard-pressed to remember their details but also lovers whose existence is at odds with her own self-image. Scarecrow's motive is the delight she takes in undermining Woman's claims on behalf of her marriage. But the key act of remembering is centred on the dead mother, which brings Friel's play into the foreground. Where the memory on which father and son failed to agree hinged on the blue boat incident, what Woman has clung to is the memory of the red coat she wore when visiting her dying mother in the hospital, a coat they had bought a few days before. Woman questions Auntie Ah and is told she came to visit her mother in the hospital with no red coat, 'only the clothes on your back' (47). In the exchange between Woman and Scarecrow which follows, the latter makes clear why she has all along opposed the fiction of the red coat: because it has kept Woman from focusing not so much on what actually occurred but more precisely on the figure of her mother as she lay dying in the hospital bed. As Scarecrow sensitively prompts Woman's reconstruction of the scene, a reciprocity is achieved between mother and daughter:

> As I stand there, I see myself here. Now. I see my own death day [...] and now she wakes and looks at me. I swim in her eyes, she in mine. (48)

This memory may be no more real than that enjoyed by Gar O'Donnell but it achieves what that memory spectacularly failed to do, a moment of atonement with the parent.

Woman and Scarecrow was premiered at London's Royal Court in June 2006 with Fiona Shaw and Brid Brennan in the title roles. For all of the performances of these two strong actors, the production failed to cohere. It was only in the Dublin Peacock production in October 2007 that the combination of direction, design and performance worked in harmony to convey the play's dramatic and metaphysical power. Central to this were the complementary and equally powerful performances of Olwyn Fouere as Woman and Barbara Brennan as Scarecrow. It was moving to see the older but still beautiful Fouere reenter Carr's imaginative world, having given life to the Mai and Hester Swane in their first theatrical incarnations. Fouere has said of playing her roles in a Carr play that each of them 'was already there in front of me. I just needed to step into her. Her inner world was a furnace, and I had no questions about her. [...] I never felt I had to tell the audience anything about her, because she was already so alive.'[62] From performances directed by Steven Berkoff in Wilde's *Salomé* through devised work with the Operating Theatre company she co-founded to her collaborations with Selina Cartmell on Shakespeare (a total of seven parts in 2008's *Macbeth,* including all three of the witches), Olwyn Fouere has made it abundantly clear how in the contemporary Irish theatre scene an actor can be as much of an auteur as a playwright or director. It is her ongoing collaboration with Marina Carr that I choose to close with, as a potent reminder of how the plays of the contemporary Irish stage on which this study has concentrated would not have achieved their memorable embodiment and articulation without such performances.

Notes

Introduction

1. The other two critical studies of the subject are: Michael Etherton, *Contemporary Irish Dramatists* (London: Macmillan, 1989) and Margaret Llewellyn-Jones, *Contemporary Irish Drama and Cultural Identity* (Bristol: Intellect Books, 2002). As an English Marxist, Etherton's approach is necessarily very different from mine, giving a great deal of space to the work of Margaretta D'Arcy and John Arden, for instance. But our two studies agree and converge on the central achievement of Brian Friel and Tom Murphy in the field of contemporary Irish drama (see Etherton, Introduction, p. xv). Llewellyn-Jones casts her net wide in terms of the number of plays considered, but gives most emphasis to the drama of the 1990s. She deploys a range of critical theories (performance, postcolonial theory, feminism, psychoanalysis) in her analysis.
2. Richard Pine, 'Brian Friel and Contemporary Irish Drama', *Colby Quarterly*, Vol. 28, No. 4 (December 1991), p. 190. The claim underwrites Pine's full-length study of the dramatist, *Brian Friel and Ireland's Drama* (London: Routledge, 1990), revised as *The Diviner: The Art of Brian Friel* (Dublin: University College Dublin Press, 1999).
3. See Fintan O'Toole, *Tom Murphy: The Politics of Magic* (Dublin: New Island Books; London: Nick Hern Books, rev. edn 1994), p. 10.
4. See Anthony Roche, 'John B. Keane: Respectability At Last!', *Theatre Ireland* 18 (April–June 1989), pp. 29–32.
5. Alan Simpson, *Beckett and Behan and a Theatre in Dublin* (London: Routledge and Kegan Paul, 1962), p. xi.
6. The same is even more the case for Behan's second play, since it exists in two versions and two languages, *An Giall* and *The Hostage*.
7. See John P. Harrington, 'Samuel Beckett and the Countertradition', in *The Cambridge Companion to Twentieth-Century Irish Drama*, edited by Shaun Richards (Cambridge: Cambridge University Press, 2004), pp. 164–76. Crucial to this process was the decision by Michael Colgan, Director of Dublin's Gate Theatre, to stage all 19 of Beckett's plays in 1991.
8. See Christopher Murray, *Twentieth-Century Irish Drama: Mirror up to the Nation* (Manchester: Manchester University Press, 1997).
9. A detailed listing of the critical monographs published to date on Friel is given in Chapter 2.
10. See Tom Maguire, *Making Theatre in Northern Ireland: Through and Beyond the Troubles* (Exeter: University of Exeter Press, 2006).
11. See Imelda Foley, *The Girls in the Big Picture: Gender in Contemporary Ulster Theatre* (Belfast: The Blackstaff Press, 2003).
12. For a detailed account of Field Day's early years, see Marilynn J. Richtarik, *Acting Between the Lines: The Field Day Theatre Company and Irish Cultural Politics 1980–1984* (Oxford: Clarendon Press, 1994). For discussion of all of the company's

work, see Carmen Szabo, *Clearing the Ground: The Field Day Theatre Company and the Construction of Irish Identities* (Newcastle: Scholars Publishing, 2007).
13. See Eamonn Jordan, *The Feast of Famine: The Plays of Frank McGuinness* (Bern: Peter Lang, 1997) and Helen Heusner Lojek, *Contexts for Frank McGuinness's Drama* (Washington, DC: The Catholic University of America Press, 2004).
14. Christopher Morash, *A History of Irish Theatre 1601–2000* (Cambridge: Cambridge University Press, 2002), p. 271.

1 Beckett and Behan: Waiting for Your Man

1. Deirdre Bair, *Samuel Beckett* (New York and London: Harcourt Brace Jovanovich, 1978), p. 339.
2. The details in what follows are drawn from Deirdre Bair's biography of Beckett and Ulick O'Connor's *Brendan Behan* (London: Black Swan, 1985). Only specific quotations will be cited. James Knowlson's authorised biography, *Damned to Fame: The Life of Samuel Beckett* (London: Bloomsbury, 1996), contains no reference to Behan. In Anthony Cronin's *Samuel Beckett: The Last Modernist* (London: HarperCollins, 1996), there is a brief discussion of meetings between Behan and Beckett which chiefly emphasises the latter's desire to avoid such meetings (pp. 497–8).
3. O'Connor, *Brendan Behan*, p. 81.
4. See Clair Wills, *That Neutral Island: A Cultural History of Ireland during the Second World War* (London: Faber and Faber, 2006), p. 305: 'The theatres flourished, and more plays were presented annually in Dublin than ever before.'
5. Information from Souvenir Catalogue marking Golden Jubilee of Edwards, MacLiammoir and the Gate Theatre, Richard Pine (ed.) (Dublin: 1978). On the Gate's war years, see Christopher Fitz-Simon, *The Boys: A Biography of Micheál MacLiammóir and Hilton Edwards* (Dublin: New Island Books, 2002), pp. 121–42.
6. Hugh Leonard, *Out After Dark* (London: Andre Deutsch, 1989), p. 188.
7. Personal interview, 1989. See my article on Keane in *Theatre Ireland* 18 (April–June 1989), pp. 29–32.
8. Leonard, *Out After Dark*, p. 188.
9. Leonard, *Out After Dark*, p. 191.
10. Alan Simpson, *Beckett and Behan and a Theatre in Dublin* (London: Routledge and Kegan Paul, 1962), p. 2. Information on the Pike Theatre is drawn from Simpson's book and from Carolyn Swift's later complementary account, *Stage by Stage* (Dublin: Poolbeg Press, 1985). The Pike's manifesto is quoted in full in Swift, p. 105.
11. O'Connor, *Brendan Behan*, p. 143.
12. O'Connor, *Brendan Behan*, p. 166.
13. O'Connor, *Brendan Behan*, p. 311.
14. Bair, *Samuel Beckett*, pp. 557–8.
15. Swift, *Stage by Stage*, p. 180.
16. O'Connor, *Brendan Behan*, p. 85.
17. O'Connor, *Brendan Behan*, pp. 87–8.
18. Swift, *Stage by Stage*, p. 138.

19. For the full complicated story see John Brannigan, *Brendan Behan: Cultural Nationalism and the Revisionist Writer* (Dublin: Four Courts Press, 2002), p. 80. One source cited by Brannigan argues that Blythe claimed never to have received the three-act manuscript.
20. Swift, *Stage by Stage*, p. 142.
21. Swift, *Stage by Stage*, p. 194.
22. Samuel Beckett, *The Complete Dramatic Works* (London: Faber and Faber, 1990), p. 23. All future references to *Waiting for Godot* are to this edition and will be incorporated in the text.
23. *Brendan Behan: The Complete Plays*, introduced and edited by Alan Simpson (London: Eyre Methuen, 1978), pp. 43–4. All future references to *The Quare Fellow* are to this edition and will be incorporated in the text.
24. J. M. Synge, 'The Dramatic Movement in Ireland', *The Abbey Theatre: Interviews and Recollections*, edited by E. H. Mikhail (London: Macmillan, 1984), p. 54.
25. J. M. Synge, *Collected Works, Volume IV: Plays 2*, edited by Ann Saddlemyer (London: Oxford University Press, 1968; Gerrards Cross: Colin Smythe, 1982), pp. 107–9.
26. Brian Friel, *Plays: One* (London: Faber and Faber, 1994), p. 446.
27. W.B. Yeats, *Collected Plays* (London: Macmillan, rev. edn 1952), p. 82.
28. The text we have is that of the 1956 Joan Littlewood production. While the textual changes are nothing as great as those between *An Giall* and *The Hostage* the ending of Act One of *The Quare Fellow* has been considerably altered. In the 1954 Pike production, the attempted hanging/suicide was by the other prisoner, not Lifer, and succeeded rather than being narrowly averted. But my essential point about the two hangings remains the same, since the first is not the 'official' one. My thanks to the late Carolyn Swift for showing me the prompt script of the Pike production in 1989. It is now lodged as part of the Pike Theatre Archive in Trinity College, Dublin.
29. Stephen Watt queries – and queers – my reading in order to take seriously the description of Behan as 'the Irish Jean Genet'. He explores the common ground between the prisoner and the bisexual in a reading which is given support by the exploration of Behan's complex sexuality in Michael O'Sullivan, *Brendan Behan: A Life* (Dublin: Blackwater Press, 1997). See Stephen Watt, 'Love and Death: A Reconsideration of Behan and Genet', in *A Century of Irish Drama: Widening the Stage*, edited by Stephen Watt, Eileen Morgan and Shakir Mustafa (Bloomington and Indianapolis: Indiana University Press, 2000), pp. 130–45.
30. Martin Esslin, *The Theatre of the Absurd* (London: Penguin Books, 3rd edition, 1980), p. 20.
31. Simpson, *Beckett and Behan and a Theatre in Dublin*, p. 175.
32. On this controversy, see Nicholas Grene, *The Politics of Irish Drama: Plays in Context from Boucicault to Friel* (Cambridge: Cambridge University Press, 1999), pp. 157–65.
33. Cited in Introduction by Declan Kiberd, *Brendan Behan: Poems and a Play in Irish* (Dublin: The Gallery Press, 1981). Also on Behan see Kiberd, *Inventing Ireland* (London: Jonathan Cape, 1995).
34. Simpson, *Beckett and Behan and a Theatre in Dublin*, pp. 20–1.
35. Andre Tarkovsky, *Sculpting in Time: Reflections on the Cinema*, translated from the Russian by Kitty Hunter-Blair (London: The Bodley Head, 1986), p. 43.

36. Simpson, *Beckett and Behan and a Theatre in Dublin*, p. 100.
37. Bair, *Samuel Beckett*, p. 282.
38. For the full details, see Gerard Whelan with Caroline Swift, *Spiked: Church-State Intrigue and 'The Rose Tatoo'* (Dublin: New Island Books, 2002).

2 Friel's Drama: Leaving and Coming Home

1. Seamus Heaney, *Friel Festival Programme* (Dublin: April–August 1999), p. 23.
2. Chronologically, they are: D. E. S. Maxwell, *Brian Friel* (Lewisburgh, Pa: Bucknell University Press, 1973); Ulf Dantanus, *Brian Friel: A Study* (London and Boston: Faber and Faber, 1988); George O'Brien, *Brian Friel* (Dublin: Gill and Macmillan, 1989; Boston, MA: Twayne Publishers, 1989); Richard Pine, *Brian Friel and Ireland's Drama* (London and New York: Routledge, 1990), subsequently revised and expanded as Richard Pine, *The Diviner: The Art of Brian Friel* (Dublin: University College Dublin Press, 1999); Elmer Andrews, *The Art of Brian Friel: Neither Reality Nor Dreams* (Basingstoke and London: Macmillan, 1995); Martine Pelletier, *Le Theatre de Brian Friel: Histoire et Histoires* (Villeneuve d'Ascq: Presses Universitaires du Septentrion, 1997); F. C. McGrath, *Brian Friel's (Post) Colonial Drama: Language, Illusion and Politics* (Syracuse, NY: Syracuse University Press, 1999); Nesta Jones, *Brian Friel: Faber Critical Guides* (London: Faber and Faber, 2000); Tony Corbett, *Brian Friel: Decoding the Language of the Tribe* (Dublin: The Liffey Press, 2002); Tony Coult, *About Friel: The Playwright and the Work* (London and New York: Faber and Faber, 2003); David Grant, *The Stagecraft of Brian Friel* (London: Greenwich Exchange, 2004); Scott Boltwood, *Brian Friel, Ireland and the North* (Cambridge: Cambridge University Press, 2007).
3. Pine, *Brian Friel and Ireland's Drama*, p. 1; Pine, *The Diviner*, p. 3.
4. Thomas Kilroy, 'Friel's Plays', introduction to Brian Friel, *The Enemy Within* (Dublin: The Gallery Press, 1979), p. 8.
5. Pine, *Brian Friel and Ireland's Drama*, p. 56; Pine, *The Diviner*, p. 84.
6. Brian Friel, 'In Interview with Fintan O'Toole (1982)', in *Brian Friel: Essays, Diaries, Interviews: 1964–1999*, edited by Christopher Murray (London and NewYork: Faber and Faber, 1999), p. 109.
7. O'Brien, *Brian Friel*, p. 6.
8. Friel, *The Enemy Within*, p. 33.
9. Brian Friel, 'Self-Portrait (1972)', in *Brian Friel: Essays, Diaries, Interviews: 1964–1999*, edited by Murray, p. 42.
10. Friel, *Volunteers* (Oldcastle, Co. Meath: The Gallery Press, 1989), p. 22.
11. Friel, *Plays: One* (London and Boston, Faber and Faber, 1996), p. 27. All references are to this edition and will be incorporated within the text.
12. Richard Allen Cave, 'Friel's Dramaturgy: The Visual Dimension', in *The Cambridge Companion to Brian Friel*, edited by Anthony Roche (Cambridge: Cambridge University Press, 2006), p. 133.
13. Seamus Deane, Introduction, Brian Friel, *Plays: One*, p. 14.
14. The loss of Maire's freedom consequent upon her marriage to S. B. is discussed by Scott Boltwood in relation to 1930s Ireland and de Valera's 1937 Constitution in *Brian Friel, Ireland and the North*, p. 58.
15. Samuel Beckett, *The Complete Dramatic Works* (London: Faber and Faber, 1990), p. 400.

16. See especially Chapter 5, 'Plays of Language', Pine, *Brian Friel and Ireland's Drama*, pp. 144–84; Chapter 6, 'Plays of Language and Time', Pine, *The Diviner*, pp. 179–254.
17. Seamus Heaney discusses this 'self-conscious translation of the famous Sean O'Casey scene' in Heaney, 'For Liberation: Brian Friel and the Use of Memory', in *The Achievement of Brian Friel*, edited by Alan J. Peacock (Gerrards Cross: Colin Smythe, 1993), p. 238.
18. Brian Friel, 'Extracts from, a Sporadic Diary (1979): *Translations*', in *Brian Friel: Essays, Diaries, Interviews: 1964–1999*, edited by Murray, p. 75.
19. Letter from Brian Friel to Terence Brown, 28 January 1992.
20. *Faith Healer* finally had its Broadway success in the 2005 production which originated at the Gate Theatre in Dublin, starring Ralph Fiennes as Frank Hardy and directed by Jonathan Kent. The actor Ian McDiarmid won a Tony for his performance as Teddy.
21. Lecture on Brian Friel by Frank McGuinness at the Patrick McGill Summer School, Glenties, County Donegal, August 1991.
22. Brian Friel, *Plays: One*, p. 331. All future references are to this edition and will be incorporated in the text.
23. Beckett, *The Complete Dramatic Works*, p. 98. For the question to which this is the answer, see the first epigraph to this book.
24. For a vivid account of life with the fit-ups, see the memoir, 'Mac', Harold Pinter's account of working as an actor with the Shakespearean actor-manager Anew McMaster in Ireland during the 1950s. See Harold Pinter, *Various Voices: Prose, Poetry, Politics 1948–2005* (London: Faber and Faber, 2005), pp. 27–34.
25. Scott Boltwood argues that my 'treatment of Grace fluctuates between accepting Frank's identification of her as Yorkshire mistress and her self-representation as Irish gentry', *Brian Friel, Ireland and the North*, p. 126. At no point does my argument 'accept' what I have just described in the text as 'Frank's distorting simplification'.
26. F. C. McGrath adds to the post-colonial interpretation advanced here (which he finds 'persuasive') by arguing that the 'monologues of the three characters constitute hybrid enunciations in Bhabha's sense'. McGrath, *Brian Friel's (Post) Colonial Drama*, p. 175.
27. J. M. Synge, *Collected Works Volume IV, Plays: Book II*, edited by Ann Saddlemyer (London: Oxford University Press, 1968; Gerrards Cross: Colin Smythe, 1982), p. 173.
28. This notion has also been addressed by Richard Kearney, 'The Language Plays of Brian Friel', in *Brian Friel: A Casebook*, edited by William Kerwin (New York and London: Garland Publishing, Inc., 1997), p. 88.
29. Declan Kiberd, 'Brian Friel's *Faith Healer*', in *Brian Friel: A Casebook*, edited by William Kerwin, pp. 212–13.
30. Jennifer Johnston, *The Gates* (London: Hamish Hamilton, 1973) where the gates front the entrance to the Major's Big House and large estate, now reduced by the Troubles and the family's diminishing wealth.
31. J. M. Synge, *Collected Works Volume I: Prose*, edited by Alan Price (London: Oxford University Press, 1966; Gerrards Cross: Colin Smythe, 1982), p. 231.
32. Cited in David H. Greene and Edward M. Stephens, *J. M. Synge 1871–1909* (New York: New York University Press, 1959, rev. edn 1989), p. 300.

33. Seamus Heaney, 'Punishment', *North* (London: Faber and Faber, 1975), p. 38.
34. Stewart Parker, *Northern Star, Plays: 2* (London: Methuen, 2000).
35. Brian Friel, *Plays: Two* (London: Faber and Faber, 1999), p. 1. All further references are to this edition and will be incorporated in the text.
36. Friel, *Plays: Two*, p. 354.
37. See Csilla Bertha, 'Six Characters in Search of a Faith: The Mythic and the Mundane in *Wonderful Tennessee*', Special Issue: Brian Friel, edited by Anthony Roche, *Irish University Review*, Vol. 29, No. 1 (Spring/Summer 1999) pp. 119–35.
38. Christopher Murray, 'Introduction', Brian Friel, *Plays: Two*, p. xix.
39. Murray, 'Introduction', Friel, *Plays: Two*, p. xx.
40. Brian Friel, '6 – Translations', 'Seven Notes for a Festival Programme (1999)', in *Brian Friel: Essays, Diaries, Interviews: 1964–1999*, edited by Murray, p. 179.

3 Murphy's Drama: Tragedy and After

1. Interview with the author, Dublin, September 1985. Subsequent remarks by Tom Murphy which are not attributed are from this interview.
2. In a lecture given at the UCD Drama Centre/Irish Theatre Archive Conference on 'Themes, Absences and Hidden Agendas in Irish Theatre and Drama' held at University College Dublin in November 1992.
3. In the programme for Tom Murphy's *The Blue Macushla* at the Abbey in 1980.
4. Cited in Declan Kiberd's script for Seán Ó Mordha's documentary on Samuel Beckett, *Silence to Silence* (Radio Telefis Eireann, 1986).
5. Cited in Fintan O'Toole, *Tom Murphy: The Politics of Magic* (Dublin: New Island Books; London: Nick Hern Books, rev. edn 1994), p. 22.
6. The term comes from an early play by Williams, *Fugitive Kind*, produced in 1936.
7. Tom Murphy, *Plays: 4* (London: Methuen, 1997), p. 180.
8. Article by Noel O'Donoghue, 'On the Outside', in the programme for a production of *On the Outside* and its 1975 sequel/companion piece, *On the Inside*, at the Peacock Theatre in Dublin in 1992.
9. Cited in O'Toole, *The Politics of Magic*, p. 7.
10. See O'Toole, *The Politics of Magic*, pp. 7–19.
11. The term, from director Peter Brook's highly influential *The Open Space*, is applied by editor Christopher Murray to Murphy's drama in his Introduction to the Tom Murphy Special Issue of the *Irish University Review*, Vol. 17, No. 1 (Spring 1987), p. 9.
12. The title Murphy himself gave to three of his plays published in a single volume by Methuen in 1988.
13. William Shakespeare, *Macbeth*, Act 1, Scene 5, lines 46–7, *The Arden Shakespeare Complete Works*, edited by Richard Proudfoot, Ann Thompson and David Scott Kastan (Walton-on-Thames, Surrey: Arden, 1998), p. 771.
14. On *A Whistle in the Dark* as a tragedy, see O'Toole, *The Politics of Magic*, pp. 60–2, 73–4.
15. Tom Murphy, *Plays: 4*, p. 60. All future references are to this edition and will be incorporated in the text.

16. Samuel Beckett, *Disjecta: Miscellaneous Writings and a Dramatic Fragment*, edited with a foreword by Ruby Cohn (London: John Calder, 1983), p. 82.
17. On the play's relationship to Synge's *Playboy*, see O'Toole, *The Politics of Magic*, pp. 69–70.
18. Nicholas Grene, 'Talking, Singing, Storytelling: Tom Murphy's *After Tragedy*', Special Issue on Contemporary Irish Drama, edited by Anthony Roche, *Colby Quarterly*, Vol. XXVII, No. 4 (December 1991), p. 216.
19. Tom Murphy, *Plays: 3* (London; Methuen, 1994), p. 167. All future references are to this edition and will be incorporated in the text.
20. Ivor Browne, 'Thomas Murphy: The Madness of Genius', *Irish University Review*, Vol. 17, No. 1 (Spring 1987), p. 131.
21. Grene, 'Talking, Singing, Storytelling', p. 217.
22. Otto Rank, *The Double: A Psychoanalytical Study*, edited and translated by Harry Tucker Jr. (Chapel Hill: Beacon Press, 1971).
23. *The Bodley Head Bernard Shaw: Collected Plays with their Prefaces*, Volume II, edited by Dan Laurence (London: Max Reinhardt, The Bodley Head Ltd., 1971), p. 919.
24. Grene, in 'Talking, Singing, Storytelling', finds Mona 'awkward third party to the drama' (p. 216), arguing that 'she is sidelined in the play' and concluding that she 'seems to stray by and ends up looking merely left out' (p. 217). He modifies this view in 'Talking It Through: *The Gigli Concert, Bailegangaire*', *Talking About Tom Murphy*, edited by Nicholas Grene (Dublin: The Carysfort Press, 2002), p. 70, citing the 'persuasive' argument by Declan Kiberd 'that Mona is in fact crucial as the feminine principle that both the men repress'. See Declan Kiberd, 'Theatre as Opera: *The Gigli Concert*', *Theatre Stuff: Critical Essays on Contemporary Irish Theatre*, edited by Eamonn Jordan (Dublin: Carysfort Press, 2000), pp. 145–58. Patrick Mason, the play's first director, in an interview with Christopher Murray in the *Irish University Review*, Vol. 17, No. 1 (Spring 1987), p. 106, says 'many people said that [Mona should have been cut]'. Mason, however, goes on to argue that he does not think it works without her: 'Because Mona is a sort of fulcrum point. She is a touchstone to JPW and to the Irish Man. Without Mona there is no centre to the play, there are only polarities. She is important also because she is the only woman and has the female energy to counterbalance these two men'.
25. For a full discussion of Mona's role in *The Gigli Concert*, see José Lanters, 'Gender and Identity in Brian Friel's *Faith Healer* and Thomas Murphy's *The Gigli Concert*', *Irish University Review*, Vol. 22, No. 2 (Autumn/Winter 1992), pp. 279–83.
26. O'Toole has a lengthy discussion of this change and its dramatic implications. See *The Politics of Magic*, pp. 219–22.
27. Christopher Murray, Introduction, *After Tragedy: Three Irish Plays* (London: Methuen, 1988), p. vii.
28. Tom Murphy, *Plays: 2* (London: Methuen, 1993), p. 170. All future references are to this edition and will be incorporated in the text.
29. Murphy directly staged the encounter in a separate play, *A Thief of a Christmas*, presented in Dublin at the Abbey in December 1985, the same month as *Bailegangaire*'s premiere at the Druid Theatre in Galway. The texts of the two plays are printed together in Tom Murphy, *Plays: 2*.

30. Nicholas Grene has also commented on Cathleen ni Houlihan as a dramatic precursor of Mommo and discusses the relationship between *Baileganaire* and *Riders to the Sea* in *The Politics of Irish Drama: Plays in Context from Boucicault to Friel* (Cambridge: Cambridge University Press, 1999), pp. 227–31.
31. 'In stories of this kind he always speaks in the first person, with minute details to show that he was actually present at the scenes that are described', in J. M. Synge, *Collected Works: Prose,* edited by Alan Price (London: Oxford University Press, 1966; Gerrards Cross: Colin Smythe, 1982), p. 72.
32. J. M Synge, *Collected Works, Plays: Book II,* edited by Ann Saddlemyer (London: Oxford University Press, 1968; Gerrards Cross: Colin Smythe, 1982), p. 169.
33. Kristin Morrison, *Canters and Chronicles: The Use of Narrative in the Plays of Samuel Beckett and Harold Pinter* (Chicago: University of Chicago Press, 1983), p. 3.
34. Alec Reid, 'Impact and Parable in Beckett: A First Encounter with *Not I*', 'Beckett at Eighty: A Birthday Tribute', edited by Terence Brown and Nicholas Grene, *Hermathena* CXLI (Winter 1986), p. 16.
35. Morrison, *Canters and Chronicles,* p. 28.
36. Samuel Beckett, *The Complete Dramatic Works* (London: Faber and Faber, 1990), p. 113.
37. This line is in the original script but is missing from the published version. I would like to thank Tom Murphy for supplying me with a copy.
38. As Nicholas Grene elaborates, in 1984 'the death in labour of a 15-year-old schoolgirl at the end of a pregnancy completely concealed from her family, and the notorious "Kerry Babies" case in which not one but two murdered infants were found in a single area of Kerry, highlighted the continuing plight of rural Irish women', *The Politics of Irish Drama,* p. 226.
39. Thomas Kilroy, 'A Generation of Playwrights', *Irish University Review,* Vol. 22, No. 1 (Spring/Summer 1992), p. 140.
40. Tom Murphy, *Plays: 5* (London: Methuen, 2006), p. 107. All future references are to this edition and will be incorporated in the text.
41. Grene, Introduction in Tom Murphy, *Plays: 5,* pp. xi–xii.
42. Tom Murphy, *Alice Trilogy* (London: Methuen, 2005), p. 3. All future references are to this edition and will be incorporated in the text. *Alice Trilogy* is also available in *Plays: 5* but has been revised. I have preferred to remain with the original text, with its more detailed stage directions.
43. Elin Diamond, 'Feminist Readings of Beckett', *Palgrave Advances in Samuel Beckett Studies,* edited by Lois Oppenheimer (Basingstoke: Palgrave Macmillan, 2004), p. 60.
44. Jacques Lacan, *Ecrits: A Selection,* translated by Alan Sheridan (London and New York: Routledge, 1989), p. 46.
45. Lacan, *Écrits,* p. 6.

4 Kilroy's Doubles

1. Anthony Roche, 'Thomas Kilroy: An Interview', Special Issue: Thomas Kilroy, edited by Anthony Roche, *Irish University Review,* Vol. 32, No. 1 (Spring/Summer 2002), p. 156.

2. Thomas Kilroy, 'The Irish Writer: Self and Society 1950–1980', *Literature and the Changing Ireland*, edited by Peter Connolly (Gerrards Cross: Colin Smythe, 1982), p. 177.
3. Thomas Kilroy, 'Groundwork for an Irish Theatre', *Studies* Vol. 48 (1959), p. 193.
4. Kilroy, 'Groundwork for an Irish Theatre', p. 195.
5. Thomas Kilroy, *The Death and Resurrection of Mr Roche* (Oldcastle, County Meath: The Gallery Press, 2002), pp. 42–3. All future references are to this edition and will be incorporated in the text.
6. Kilroy, 'The Irish Writer: Self and Society 1950–1980', p. 181.
7. Thomas Kilroy, *The O'Neill* (Oldcastle, County Meath: The Gallery Press, 1995), p. 12. All future references are to this edition and will be incorporated in the text.
8. See Anne Fogarty, 'The Romance of History: Renegotiating the Past in Thomas Kilroy's *The O'Neill* and Brian Friel's *Making History*', *Irish University Review*, Vol. 32, No. 1, pp. 25–6.
9. Interview with Declan Hughes, June 1988.
10. Hugh Kenner, *The Stoic Comedians: Flaubert, Joyce and Beckett* (Berkeley: University of California Press, 1962), p. 83.
11. Thomas Kilroy, *Tea and Sex and Shakespeare* (Oldcastle, County Meath: The Gallery Press, 1998), p. 21.
12. Thomas Kilroy, *Talbot's Box* (Dublin: The Gallery Press, 1979), p. 11. All future references are to this edition and will be incorporated in the text.
13. 'It was not Death', *A Choice of Emily Dickinson's Verse*, selected by Ted Hughes (London: Faber and Faber, 1968), p. 34.
14. Thomas Kilroy, *Double Cross* (London: Faber and Faber, 1986), p. 6. All future references are to this edition and will be incorporated in the text.
15. Oscar Wilde, *Plays* (Harmondsworth: Penguin, 1954), p. 305.
16. W. J. McCormack, *The Battle of the Books: Two Decades of Irish Cultural Debate* (Mullingar: The Lilliput Press, 1986), p. 66.
17. Thomas Kilroy, *The Madame MacAdam Travelling Theatre* (London: Methuen, 1991), p. 29. All future references are to this edition and will be incorporated in the text.
18. Marjorie Garber, *Vested Interests: Cross-Dressing and Cultural Anxiety* (New York: Routledge, 1992), p. 22.
19. Garber, *Vested Interests*, p. 29.
20. Richard Pine, *The Thief of Reason: Oscar Wilde and Modern Ireland* (Dublin: Gill and Macmillan, 1995), p. 137.
21. Cited in Michael Holroyd, *Bernard Shaw: The One-Volume Definitive Edition* (London: Chatto and Windus, 1997), p. 593.
22. Thomas Kilroy, *The Secret Fall of Constance Wilde* (Oldcastle, County Meath: The Gallery Press, 1997), p. 66. All future references are to this edition and will be incorporated in the text.
23. Nicholas Grene, 'Staging the Self: Person and Persona in Kilroy's Plays', *Irish University Review*, Vol. 32, No. 1, p. 70.
24. Richard Ellmann, *Oscar Wilde* (London: Hamish Hamilton, 1987), p. 262.
25. Anna McMullan, 'Masculinity and Masquerade in Thomas Kilroy's *Double Cross* and *The Secret Fall of Constance Wilde*', *Irish University Review*, Vol. 32, No. 1, p. 136.

26. McMullan, 'Masculinity and Masquerade in Thomas Kilroy's *Double Cross* and *The Secret Fall of Constance Wilde*', p. 136.
27. Thomas Kilroy, *The Shape of Metal* (Oldcastle, County Meath: The Gallery Press, 2003), p. 40. All future references are to this edition and will be incorporated in the text.
28. Anthony Cronin, *Samuel Beckett: The Last Modernist* (London: HarperCollins, 1996), p. 279. Cronin provides a valuable account of the artistic affinities and relationship between Giacometti and Beckett.
29. See Csilla Bertha, 'Thomas Kilroy's *The Shape of Metal*: "Metal ... Transformed into Grace" – Grace into Metal', in *The Brazilian Journal of Irish Studies/ABEI Journal*, Number 9 (June 2007), pp. 85–97.
30. Samuel Beckett, *The Complete Dramatic Works* (London and Boston: Faber and Faber, 1990), p. 105.
31. Adrian Frazier, 'The Body of Life', Programme Note, Thomas Kilroy, *The Shape of Metal*, Abbey Theatre, October 2003.
32. Thomas Kilroy, 'Two Playwrights: Yeats and Beckett', *Myth and Reality in Irish Literature*, edited by Joseph Ronsley (Waterloo, Ontario: Wilfred Laurier University Press, 1977), pp. 183–95.
33. Grene, 'Staging the Self: Person and Persona in Kilroy's Plays', p. 81.

5 Northern Irish Drama: Imagining Alternatives

1. Stewart Parker introduced and edited the late Sam Thompson's ground-breaking play of the late 1950s, *Over the Bridge*, set in the Belfast shipyards. *Over the Bridge* was staged in 1960 at Belfast's Empire Theatre, and aroused considerable Protestant opposition. John Boyd, for many years literary advisor to the Lyric Theatre, had a succession of naturalistic tragedies staged there in the 1970s, the most successful of which was *The Flats* (1971). On Thompson and Boyd, see D. E. S. Maxwell, *A Critical History of Modern Irish Drama 1891–1980* (Cambridge: Cambridge University Press, 1984), pp. 179–80.
2. Stewart Parker, *Dramatis Personae*, John Malone Memorial Lecture (Belfast: Queen's University, 1986), pp. 18–19. See Claudia W. Harris, 'From Pastness to Wholeness: Stewart Parker's Re-Inventing Theatre', *Colby Quarterly*, Vol. XXVII, No. 4 (1991), Special Issue on Contemporary Irish Drama, guest-edited by Anthony Roche, pp. 233–41.
3. For my analysis of Parker's stage *oeuvre*, see Anthony Roche, 'Stewart Parker's Comedy of Terrors', *A Companion to Modern British and Irish Drama 1880–2005*, edited by Mary Luckhurst (Oxford: Blackwell, 2006), pp. 289–98.
4. Stewart Parker, Foreword, *Plays: 2* (London: Methuen, 2000), p. xiii.
5. Parker, *Plays: 2*, p. 200. All future references are to this edition and will be incorporated in the text.
6. As a ghost, Lily initially assumes the form in which she died. In subsequent appearances, which reflect different stages of her life, her dress and appearance are altered accordingly.
7. See the Appendix on Irish Women Playwrights, especially Part II on Contemporary Women Playwrights [i.e. post 1980], which lists 148 titles, in *Women in Irish Drama: A Century of Authorship and Representation*, edited by Melissa Sihra (Houndmills: Palgrave Macmillan, 2007).

8. Christina Reid, *Plays: 1* (London: Methuen, 1997), p. 10. All future references are to this edition and will be incorporated in the text.
9. Julia Kristeva, 'Women's Time', *The Kristeva Reader*, edited by Toril Moi (Oxford: Basil Blackwell, 1986), p. 187. The description is from the introductory note by Toril Moi.
10. Kristeva, 'Women's Time', p. 191.
11. 'Long Line of Storytellers', *Richmond and Twickenham Times*, 4 June 1993.
12. Kristeva, 'Women's Time', p. 202.
13. Kristeva, 'Women's Time', p. 204.
14. Kristeva, 'Women's Time', p. 195.
15. Anne Devlin, *Ourselves Alone* (London and Boston: Faber and Faber, 1990), p. 22. All future references are to this edition and will be incorporated in the text.
16. See Helene Keyssar, *Feminist Theatre: An Introduction to Plays of Contemporary British and American Women* (London: Macmillan, 1985), pp. 1–21.
17. See Enrica Cerquone, 'Women in Rooms: Landscape of the Missing in Anne Devlin's *Ourselves Alone*', *Women in Irish Drama*, edited by Melissa Sihra, p. 167. Cerquone describes me as one of the critics 'who have concentrated on the ostensibly "realist" limits of Devlin's early play' (p. 170). The thrust of my analysis of *Ourselves Alone* is, on the contrary, to question and to go beyond such limits.
18. On Charabanc, see Anna McMullan, 'Irish women playwrights since 1958', *British and Irish Women Dramatists since 1958*, edited by Trevor R Griffiths and Margaret Llewellyn-Jones (Buckingham: Open University Press, 1993), pp. 121–2.
19. Eleanor Methven; quoted in Claudia W. Harris, 'Reinventing Women: Charabanc Theatre Company', *State of Play: Irish Theatre in the Nineties*, edited by Eberhard Bort (Trier: Wissenschaftlicher Verlag, 1996), p. 111.
20. See 'A Night at the Theatre 7: *Translations*, Guildhall, Derry, 23 September 1980' in Christopher Morash, *A History of Irish Theatre 1601–2000* (Cambridge: Cambridge University Press, 2002), pp. 233–41.
21. Seamus Deane, 'Preface', *Ireland's Field Day Theatre Company* (London: Hutchinson, 1985), p. vii.
22. For a detailed account of Field Day's early years, see Marilynn J. Richtarik, *Acting Between the Lines: The Field Day Theatre Company and Irish Cultural Politics 1980–1984* (Oxford: Clarendon Press, 1994). For coverage of the theatre company's full 17 years, see Carmen Szabo, *'Clearing the Ground': The Field Day Theatre Company and the Construction of Irish Identities* (Newcastle: Cambridge Scholars Publishing, 2007).
23. My text throughout is William Shakespeare, *King Henry IV, Part 1*, Act 3 Scene 1, *passim*, *The Arden Shakespeare Complete Works*, edited by Richard Proudfoot, Ann Thompson and David Scott Kastan (Walton-on-Thames, Surrey: Arden, 1998), pp. 377–9.
24. Brian Friel, *Plays: One* (London and Boston: Faber and Faber, 1996), p. 384. All future references are to this edition and will be incorporated in the text.
25. Brian Friel, 'Extracts from a Sporadic Diary (1979): *Translations*,' *Brian Friel: Essays, Diaries, Interviews: 1964–1999*, edited by Christopher Murray (London and New York: Faber and Faber, 1999), p. 74.

26. At another level it retains the link, since 'burn' means 'river' in Scots Gallic. Owen may not have been aware of this, but Friel and his text are. I am grateful to Terence Brown for this point.
27. See Richard Kearney, 'Language Play: Brian Friel and Ireland's Verbal Theatre', *Brian Friel: A Casebook*, edited by William Kerwin (New York and London: Garland Publishing, 1997), p. 97.
28. The play's direct quotations from George Steiner, *After Babel: Aspects of Language and Translation* (London and New York: Oxford University Press, 1975) are recorded by Richard Pine in *The Diviner: The Art of Brian Friel* (Dublin: University College Dublin Press, 1999), pp. 359–62. See also Richard Kearney, 'Language Play', pp. 112–15.
29. George Steiner, *Antigones* (Oxford: Clarendon Press, 1984), p. 138.
30. Conor Cruise O'Brien, 'Views', *The Listener*, 24 October 1968, p. 526.
31. In Conor Cruise O'Brien, *States of Ireland* (London: Hutchinson and Co., 1972; revised edition, 1974), p. 158.
32. Tom Paulin, review of Conor Cruise O'Brien, *Neighbours: The Ewart-Biggs Memorial Lectures, 1979–1979*, reprinted as 'The Making of a Loyalist' in Tom Paulin, *Ireland and the English Crisis* (Newcastle-upon-Tyne: Bloodaxe Books, 1984), pp. 23–8.
33. Paulin, 'The Making of a Loyalist', p. 29.
34. Cruise O'Brien, *States of Ireland*, p. 159.
35. Paulin, 'Introduction', *Ireland and the English Crisis*, p. 16.
36. Tom Paulin, *The Riot Act: A Version of Sophocles' 'Antigone'* (London: Faber and Faber, 1985), p. 16. All future references are to this edition and will be incorporated in the text.
37. Fintan O'Toole, 'Field Day on the Double', *The Sunday Tribune*, 23 September 1984; reprinted in *Critical Moments: Fintan O'Toole on Modern Irish Theatre*, edited by Julia Furay and Redmond O'Hanlon (Dublin: Carysfort Press, 2003), p. 30.
38. Mitchell Harris, 'Friel and Heaney: Field Day and the Voice of the Fifth Province', MLA paper, Washington, DC, December 1984, p. 10.
39. Paulin, *Ireland and the English Crisis*, p. 33.
40. Richard Kearney, *Myth and Motherland*, Field Day Pamphlet 5 (Derry: Field Day, 1984); reprinted in *Ireland's Field Day*, p. 67.
41. Sophocles, *The Three Theban Plays: Antigone, Oedipus the King, Oedipus at Colonnus,* translated by Robert Fagles, introduction and notes by Bernard Knox (London: Allen Lane, 1982; Penguin Classics, 1984), p. 106.
42. W. B. Yeats, *The Poems: A New Edition*, edited by Richard J. Finneran (Dublin: Gill and Macmillan, revised edition 1983), p. 184.
43. Tom Paulin, *Liberty Tree* (London: Faber and Faber, 1983), p. 13.
44. Paulin, *Ireland and the English Crisis*, p. 17.
45. Paulin, *Ireland and the English Crisis*, p. 17.
46. Paulin, 'Under Creon', *Liberty Tree*, p. 13. Henry Joy McCracken (featured in Stewart Parker's *Northern Star*), James Hope and Joseph Biggar were all Belfast Presbyterians involved in the republican struggle for Irish independence. Paulin identifies the first of the three in 'Brief Lives' at the close of *Ireland and the English Crisis*: 'Henry Joy McCracken (1767–98). Northern Irish republican leader executed after the failure of the 1798 rebellion', p. 217. In the Antrim insurrection led by McCracken, 'a party of the insurgents,

"the Spartan band", led by James Hope, a weaver, displayed outstanding determination': R. B. McDowell, *Ireland in the Age of Imperialism and Revolution* (Oxford: Clarendon Press, 1979), p. 637. The *Northern Star* was their newspaper. Joseph Biggar became MP for Cavan in 1874; on page 145 of *Ireland and the English Crisis*, Paulin cites James Joyce's description of Biggar as 'the inventor of parliamentary obstructionism'.

47. 'Speaking for the Dead: Playwright Frank McGuinness talks to Kevin Jackson', *The Independent*, 27 September 1989.
48. Frank McGuinness, *Plays: One* (London: Faber and Faber, 1996), p. 98. All future references are to this edition and will be incorporated in the text.
49. By Patrick Burke in his review of the play, 'Dance Unto Death', *Irish Literary Supplement*, Vol. 6, No. 2 (Fall/Winter 1987), p. 45.
50. Although much has been written on McGuinness' challenging of gender stereotypes since, the ground-breaking essay in this regard is Helen Lojek, 'Myth and Bonding in Frank McGuinness's *Observe the Sons of Ulster Marching Towards the Somme*', *Canadian Journal of Irish Studies*, Vol. 14, No. 1 (1988), pp. 45–53. See also Helen Heusner Lojek, *Contexts for Frank McGuinness's Drama* (Washington, DC: The Catholic University of America Press, 2004).
51. See Angela Wilcox, 'The Temple of the Lord is Ransacked', *Theatre Ireland* 8 (1984), p. 87.
52. Samuel Beckett, *The Complete Dramatic Works* (London: Faber and Faber, 1990), p. 58.
53. Wilcox, 'The Temple of the Lord is Ransacked', p. 88.
54. As Helen Lojek recounts in her discussion of the play's origins: 'McGuinness's play is based loosely on Keenan's story. [...] McGuinness was drawing on Keenan's story as it had appeared in the media, though, not as it is told in *An Evil Cradling*, which was not published until two months after the play premiered.' Helen Heusner Lojek, *Contexts for Frank McGuinness's Drama*, p. 83.
55. Frank McGuinness, *Plays: Two* (London and New York: Faber and Faber, 2002), p. 89. All future references are to this edition and will be incorporated in the text.
56. Beckett, *The Complete Dramatic Works*, p. 126.
57. Brian Keenan, *An Evil Cradling* (London: Hutchinson, 1992), p. xiii.
58. Keenan, *An Evil Cradling*, p. 64.
59. 1994's *The Bird Sanctuary*, premiered at the Abbey, centres on a woman painter and makes a fascinating comparison with Kilroy's later *Shape of Metal*. 1997's *Mutabilitie* at London's National Theatre staged the meeting between cultures during the Elizabethan plantation of Ireland and brought Shakespeare on a little-known visit to join the poet Spenser in encountering the native Irish. 1999's *Dolly West's Kitchen*, directed by Mason at the Abbey, is McGuinness's World War II play, set just over the Border from Derry in Donegal. 2002's *Gates of Gold* was a fictionalised version of Hilton Edwards' and Micheál MacLiammóir's lifelong love affair, with each other and with the theatre, and was staged appropriately at the Gate. 2005's *Speaking With Magpies* was his original contribution to the RSC's Gunpowder Plot season. *There Came A Gipsy Riding* in 2007 dealt powerfully with that most pressing of tragedies in contemporary Irish society, young male suicide. It was staged at the Almeida in a production by Michael Attenborough in January 2007 and has not yet been staged in Ireland.

60. Gary Mitchell, *Tearing the Loom and In a Little World of Our Own* (London: Nick Hern Books, 1998), p. 22. All future references are to this edition and will be incorporated in the text.
61. Tim Miles notes that Richard has been taught 'to commit the biblical sin of killing one's own brother'. See Tim Miles, 'Understanding Loyalty: The English Response to the Work of Gary Mitchell', in *Irish Theatre in England*, edited by Richard Cave and Ben Levitas (Dublin: Carysfort Press, 2007), p. 99.
62. Beckett was a great admirer of this play, writing that Sean O'Casey 'is a master of knockabout in this very serious and honourable sense – that he discerns the principle of disintegration in even the most complacent of solidities, and activates it to their explosion. [...] This impulse of material to escape and be consummate in its own knockabout is admirably expressed [...] in "The End of the Beginning", where the entire set comes to pieces and the chief character, in a final spasm of dislocation, leaves the scene by the chimney.' See Samuel Beckett, 'The Essential and the Incidental', *Disjecta: Miscellaneous Writings and a Dramatic Fragment*, edited by Ruby Cohn (London: John Calder, 1983), pp. 82–3.
63. Gary Mitchell, *As the Beast Sleeps* (London: Nick Hern Books, 2001), p. 18. All future references are to this edition and will be incorporated in the text.
64. Angelique Chrisafis, 'Loyalist paramilitaries drive playwright from his home', *The Guardian*, 21 December 2005.

6 The 1990s and Beyond: McPherson, Barry, McDonagh, Carr

1. Vic Merriman, 'Decolonisation Postponed: The Theatre of Tiger Trash', *Irish University Review*, Vol. 29, No. 2 (Autumn/Winter 1999), p. 312.
2. Merriman, 'Decolonisation Postponed', p. 317.
3. Patrick Lonergan, Introduction, *The Methuen Anthology of Irish Plays* (London: Methuen, 2008), p. vii.
4. Nicholas Grene, *The Politics of Irish Drama: Plays in Context from Boucicault to Friel* (Cambridge: Cambridge University Press, 1999), p. 262.
5. The details in this paragraph are drawn from personal knowledge. McPherson discusses them in the 'Afterword: An interview with Carol Vander' in Conor McPherson, *Plays: Two* (London: Nick Hern Books, 2004), pp. 207–20.
6. Conor McPherson, 'Will the Morning After Stop Us Talking to Ourselves?', *The Irish Times*, 3 May 2008.
7. Conor McPherson, Author's Note, *This Lime-Tree Bower: Three Plays* (Dublin: New Island Books; London: Nick Hern Books, 2006), p. 5.
8. Conor McPherson, 'The Pagan Landscape', Programme Note for the Abbey Theatre production of *The Seafarer*, April 2008.
9. Angela Bourke, *The Burning of Bridget Cleary: A True Story* (London: Pimlico, 1999), p. 29.
10. Conor McPherson, *Plays: Two*, p. 16. All future references are to this edition and will be incorporated in the text.
11. Conor McPherson, 'The Pagan Landscape', Abbey programme note.
12. Patrick Lonergan, Review of Conor McPherson, *The Weir*, directed by Garry Hynes, Gate Theatre Dublin, 10 June 2008, *Irish Theatre Magazine*, Vol. 8, No. 35 (Summer 2008), p. 85.

13. See Brian Lalor (ed.), *The End of Innocence: Child Sexual Abuse in Ireland* (Cork: Oak Tree Press, 2001).
14. Bourke, *The Burning of Bridget Cleary*, p. 28.
15. Roy Foster, '"Something of us will remain": Sebastian Barry and Irish History', *Out of History: Essays on the Writings of Sebastian Barry*, edited by Christina Hunt Mahony (Dublin: Carysfort Press; Washington, DC: The Catholic University of America Press, 2006), p. 183.
16. Sebastian Barry, *Plays: 1* (London: Methuen, 1997), p. 262. All future references to the play are to this edition and will be incorporated in the text.
17. William Shakespeare, *King Lear*, Act 3, Scene 3, lines 106–8, *The Arden Shakespeare Complete Works*, edited by Richard Proudfoot, Ann Thompson and David Scott Kastan (Walton-on-Thames, Surrey: Arden, 1998), p. 651.
18. See Fintan O'Toole's Introduction to an earlier volume comprising three of Barry's plays, *The Only True History of Lizzie Finn; The Steward of Christendom; White Woman Street* (London: Methuen, 1995), p. ix. O'Toole also makes the *King Lear* comparison.
19. See W. J. McCormack, *Fool of the Family: A Life of J. M. Synge* (London: Weidenfeld and Nicolson, 2000), pp. 209–10.
20. Fintan O'Toole, 'Nowhere Man', *The Irish Times*, 26 April 1997.
21. J. M. Synge, *Collcted Works* Volume IV: *Plays* Book II, edited by Ann Saddlemyer (London: Oxford University Press, 1968; rept. Gerrards Cross: Colin Smythe, 1982), p. 65.
22. Samuel Beckett, *The Complete Dramatic Works* (London: Faber and Faber, 1990), p. 43.
23. See Vic Merriman, 'Decolonisation Postponed: The Theatre of Tiger Trash', which I discussed at the outset of this chapter. The Merriman piece has been reprinted in *The Theatre of Martin McDonagh: A World of Savage Stories*, edited by Lilian Chambers and Eamonn Jordan (Dublin: Carysfort Press, 2006), pp. 264–80. This volume combines a great deal of the material already published on McDonagh with newly commissioned articles and is virtually unique among critical studies of writers in representing the case for the prosecution as well as the defence.
24. Cited in Introduction, *Martin McDonagh: A World of Savage Stories*, p. 10.
25. O'Toole, 'Nowhere Man'.
26. James Knowlson, *Damned to Fame: The Life of Samuel Beckett* (London: Bloomsbury, 1996), pp. 56–7.
27. O'Toole, 'Nowhere Man'.
28. Martin McDonagh, *Plays: 1* (London: Methuen, 1999), p. 63. All future references are to this edition and will be incorporated in the text.
29. Martin McDonagh, *Plays: 1*, p. 1. All future references are to this edition and will be incorporated in the text.
30. Beckett, *The Complete Dramatic Works*, p. 93.
31. Beckett, *The Complete Dramatic Works*, p. 105.
32. Beckett, *The Complete Dramatic Works*, p. 434. All future references are to this edition and will be incorporated in the text.
33. Thomas Kilroy, 'The Anglo-Irish Theatrical Imagination', *Bullán: An Irish Studies Journal*, Vol. 3, No. 2 (Winter 1997/Spring 1998), p. 6.
34. Beckett, *The Complete Dramatic Works*, p. 378.
35. Kilroy, 'The Anglo-Irish Theatrical Imagination', p. 9.

36. O'Toole, 'Nowhere Man'.
37. David Grant, Introduction, *The Crack in the Emerald: New Irish Plays* (London: Nick Hern Books, 1990), p. vii.
38. See Caroline Williams, Katy Hayes, Sian Quill and Clare Dowling, 'People in Glasshouse: An Anecdotal History of an Independent Theatre Company', *Druids, Dudes and Beauty Queens: The Changing Face of Irish Theatre*, edited by Dermot Bolger (Dublin: New Island Books, 2001), pp. 132–47.
39. Introduction, *The Theatre of Marina Carr: "before rules was made"*, edited by Cathy Leeney and Anna McMullan (Dublin: Carysfort Press, 2003), p. xv.
40. See Biographical Note on Marina Carr in *A Crack in the Emerald.*
41. Marina Carr, *Plays One* (London: Faber and Faber, 1999), p. 66. All future references are to this edition and will be incorporated in the text.
42. Beckett, *The Complete Dramatic Works*, p. 138.
43. Beckett, *The Complete Dramatic Works*, p. 156.
44. Carr, *Plays One*, p. 121. All future references are to this edition and will be incorporated in the text.
45. Beckett, *The Complete Dramatic Works*, p. 437.
46. Marina Carr, *Portia Coughlan, The Dazzling Dark: New Irish Plays*, edited by Frank McGuinness (London and Boston: Faber and Faber, 1996), pp. 241–2. All future references are to this edition (rather than the version in *Plays One*) and will be incorporated in the text. I have preferred this original version for a number of reasons, not least because it retains the dialect.
47. Fiona Macintosh, *Dying Acts: Death in Ancient Greek and Modern Irish Tragic Drama* (Cork: Cork University Press, 1994).
48. Frank McGuinness, Introduction, *The Dazzling Dark*, p. ix.
49. Carr, *Plays One*, p. 194.
50. McGuinness, Introduction, *The Dazzling Dark*, p. ix.
51. Carr, Afterword, *Portia Coughlan, The Dazzling Dark*, pp. 310–11.
52. Carr, *Plays One*, p. 279. All future references are to this edition and will be incorporated in the text.
53. See Augustine Martin, 'Christy Mahon and the Apotheosis of Loneliness', *Bearing Witness: Essays on Anglo-Irish Literature*, edited by Anthony Roche (Dublin: University College Dublin Press, 1996), pp. 32–43.
54. Frank McGuinness, Programme Note for *By the Bog of Cats*, directed by Patrick Mason, the Abbey Theatre, 1998, *The Theatre of Marina Carr*, pp. 87–8.
55. W. B. Yeats, 'The Second Coming', *The Poems: A New Edition*, edited by Richard J. Finneran (Dublin: Gill and Macmillan, 2nd rev. edn 1983), p. 160.
56. Marina Carr, *Ariel* (Oldcastle, County Meath: The Gallery Press, 2002), p. 19.
57. For a more positive reading of the world of *Ariel*, see Cathy Leeney, 'Marina Carr: Violence and Destruction: Language, Space and Landscape', in *A Companion to Modern British and Irish Drama 1880–2005*, edited by Mary Luckhurst (Oxford: Blackwell, 2006), pp. 509–18.
58. Melissa Sihra argues for Beckett's importance to *Woman and Scarecrow* and for 'a Beckettian senibility' throughout Carr's work. Melissa Sihra notes the links between *Woman and Scarecrow* and the two Gars in *Philadelphia, Here I Come!* See Melissa Sihra, 'The Unbearable Darkness of Being: Marina Carr's

Woman and Scarecrow', *Irish Theatre International*, Vol. 1, No. 1 (April 2008), pp. 22–37.

59. Marina Carr, *Woman and Scarecrow* (Oldcastle, County Meath: The Gallery Press, 2006), p. 40. All future references are to this edition and will be incorporated.
60. Carr, *Portia Coughlan, The Dazzling Dark*, p. 240.
61. Brian Friel, *Plays: One* (London: Faber and Faber, 1996), p. 82.
62. Olwen Fouere, 'On Playing in *The Mai* and *By the Bog of Cats*', *The Theatre of Marina Carr*, p. 160.

Select Bibliography

Featured dramatists

SEBASTIAN BARRY

Plays: 1, introduced by Fintan O'Toole with a Preface by the Author (London: Methuen, 1997). Contains *Boss Grady's Boys, Prayers of Sherkin, White Woman Street, The Only True History of Lizzie Finn, The Steward of Christendom.*

SAMUEL BECKETT

The Complete Dramatic Works (London: Faber and Faber, 1990).

BRENDAN BEHAN

The Complete Plays, with an introduction by Alan Simpson (London: Methuen, 1978).

Poems and a Play in Irish, with an Introduction by Declan Kiberd (Oldcastle, County Meath: The Gallery Press, 1981).

MARINA CARR

Portia Coughlan, The Dazzling Dark: New Irish Plays, selected and introduced by Frank McGuinness (London and Boston: Faber and Faber, 1996).

Plays One, introduced by the author (London: Faber and Faber, 1999). Contains *Low in the Dark, The Mai, Portia Coughlan, By the Bog of Cats ....*

Woman and Scarecrow (Oldcastle, County Meath: The Gallery Press, 2006).

ANNE DEVLIN

Ourselves Alone (London and Boston: Faber and Faber, 1990).

BRIAN FRIEL

The Enemy Within, with an introduction by Thomas Kilroy (Oldcastle, County Meath: The Gallery Press, 1979).

Plays One, with an introduction by Seamus Deane (London: Faber and Faber, 1996). Contains *Philadelphia, Here I Come!*; *The Freedom of the City*; *Living Quarters*; *Aristocrats*; *Faith Healer* and *Translations*.

Plays Two, with an introduction by Christopher Murray (London: Faber and Faber, 1999). Contains *Dancing at Lughnasa, Fathers and Sons, Making History, Wonderful Tennessee, Molly Sweeney.*

THOMAS KILROY

Talbot's Box (Oldcastle, County Meath: The Gallery Press, 1979).

The Seagull by Anton Chekhov, in a new version by Thomas Kilroy (London: Methuen, 1981).

Double Cross (London: Faber and Faber, 1986; Oldcastle, County Meath: The Gallery Press, 1994).

The Madame MacAdam Travelling Theatre (London: Methuen, 1991).

The O'Neill (Oldcastle, County Meath: The Gallery Press, 1995).

The Secret Fall of Constance Wilde (Oldcastle, County Meath: The Gallery Press, 1997).
Tea and Sex and Shakespeare (Oldcastle, County Meath: The Gallery Press, 1998).
The Death and Resurrection of Mr. Roche (Oldcastle, County Meath: The Gallery Press, rev. edn 2002).
The Shape of Metal (Oldcastle, County Meath: The Gallery Press, 2003).

FRANK MCGUINNESS
Plays One, introduced by the author (London: Faber and Faber, 1996). Contains *The Factory Girls, Observe the Sons of Ulster Marching Towards the Somme, Innocence, Carthaginians, Baglady.*
Plays Two, introduced by the author (London: Faber and Faber, 2002). Contains *Mary and Lizzie, Someone Who'll Watch Over Me, Dolly West's Kitchen, The Bird Sanctuary.*

MARTIN MCDONAGH
Plays: 1, introduced by Fintan O'Toole (London: Methuen, 1999). Contains *The Beauty Queen of Leenane, A Skull in Connemara, The Lonesome West.*

CONOR MCPHERSON
Plays: Two, with an afterword by the author (London: Nick Hern Books, 2004). Contains *The Weir, Dublin Carol, Port Authority, Come On Over.*

GARY MITCHELL
Tearing the Loom and In a Little World of Our Own (London: Nick Hern Books, 1998).
As The Beast Sleeps (London: Nick Hern Books, 2001).

TOM MURPHY
After Tragedy: Three Irish Plays (*The Gigli Concert, Bailegangaire* and *Conversations on a Homecoming*), with a preface by Christopher Murray (London: Methuen, 1988).
Plays: 2, with an introduction by Fintan O'Toole (London: Methuen, 1993). Contains *Conversations on a Homecoming, Bailegangaire, A Thief of a Christmas.*
Plays: 3, with an introduction by Fintan O'Toole (London: Methuen, 1994). Contains *The Morning After Optimism, The Sanctuary Lamp, The Gigli Concert.*
Plays: 4, with an introduction by Fintan O'Toole (London: Methuen, 1997). Contains *A Whistle in the Dark, A Crucial Week in the Life of a Grocer's Assistant, On the Outside, On the Inside.*
Alice Trilogy (London: Methuen, 2005).
Plays: 5, with an introduction by Nicholas Grene (London: Methuen, 2006). Contains *Too Late for Logic, The Wake, The House* and *Alice Trilogy.*

STEWART PARKER
Plays: 1, introduced by Lynne Parker (London: Methuen, 2000). Contains *Spokesong, Catchpenny Twist, Nightshade, Pratt's Fall.*
Plays: 2, introduced by Stephen Rea with a foreword by Lynne Parker (London: Methuen, 2000). Contains *Northern Star, Heavenly Bodies, Pentecost.*

TOM PAULIN
The Riot Act: A Version of Sophocles' 'Antigone' (Faber and Faber, London 1985).

CHRISTINA REID

Plays: 1, with an introduction by Maria M. Delgado (London: Methuen, 1997). Contains *Tea in a China Cup, Did You Hear the One about the Irishman?, Joyriders, The Belle of the Belfast City, My Name, Shall I Tell You My Name?, Clowns.*

Other dramatists

Boyd, John, *Collected Plays: One* (Belfast: The Blackstaff Press, rev. edn, 1981).

Eagleton, Terry, *Saint Oscar* (Derry: Field Day, 1989).

Heaney, Seamus, *The Cure at Troy: A Version of Sophocles' 'Philoctetes'* (London: Faber and Faber in association with Field Day, 1990).

Johnston, Denis, *The Old Lady Says 'No'!* (ed.) Christine St Peter (Washington, DC: The Catholic University of America Press,1992).

Keane, John B., *Sive* (Dublin: Progress House, rev. edn, 1986).

Leonard, Hugh, *Three Plays (Da/A Life/Time Was)* (Harmondsworth: Penguin, 1981).

O'Casey, Sean, *Three Plays (Shadow of a Gunman, Juno and the Paycock, The Plough and the Stars)* (London: Faber and Faber, 1998); *The Silver Tassie*, in *Landmarks in Irish Drama* (ed.) Brendan Kennelly (London: Methuen 1988).

Pinter, Harold, *Plays: Three* (London: Faber and Faber, 1991).

Robinson, Lennox, *Selected Plays* (ed.) Christopher Murray (Gerrards Cross: Colin Smythe, 1982).

Shakespeare, William, *Henry IV, Part One*; *King Lear*; *Macbeth*, in *The Arden Shakespeare Complete Works* (eds) Richard Proudfoot, Ann Thompson and David Scott Kastan; Consultant Editor: Harold Jenkins (Walton-on-Thames, Surrey: Arden, 1998).

Shaw, Bernard, *The Bodley Head Bernard Shaw: Collected Plays with Their Prefaces*, Vol. II (ed.) Dan H. Laurence (London: Max Reinhardt, The Bodley Head, 1971).

Sophocles, *The Three Theban Plays* (*Oedipus the King, Antigone, Oedipus at Colonnus*), translated by Robert Fagles, introduction and notes by Bernard Knox (Harmondsworth: Penguin, 1984).

Synge, J. M., *Collected Works Volumes III and IV* (*Plays One and Two*) (ed.) Ann Saddlemyer (Oxford: Oxford University Press, 1968; Gerrards Cross: Colin Smythe, 1982).

Thompson, Sam, *Over the Bridge* (ed.) Stewart Parker (Dublin: Gill and Macmillan, 1970).

Wilde, Oscar, *Plays* (Harmondsworth: Penguin, 1954).

Yeats, W. B., *Collected Plays* (London: Macmillan, 2nd rev. edn, 1952).

Secondary and related materials

Bair, Deirdre, *Samuel Beckett: A Biography* (New York: Harcourt Brace Jovanovich, 1978).

Beckett, Samuel, *Murphy* (London: Picador/Pan, 1973).

———, *Disjecta: Miscellaneous Writings and a Dramatic Fragment* (ed.) Ruby Cohn (London: John Calder, 1983).

Benjamin, Walter, *Illuminations* (ed.) Hannah Arendt, translated by Harry Zohn (London: Fontana, 1973).

Bertha, Csilla, 'Six Characters in Search of a Faith: The Mythic and the Mundane in *Wonderful Tennessee*', Special Issue on Brian Friel (ed.) Anthony Roche, *Irish University Review*, Volume 29 Number 1 (Spring/Summer 1999).

———, 'Thomas Kilroy's *The Shape of Metal*: "Metal ... Transformed into Grace"', *The Brazilian Journal of Irish Studies/ABEI Journal* Number 9 (June 2007).

Boltwood, Scott, *Brian Friel, Ireland and the North* (Cambridge: Cambridge University Press, 2007).

Bourke, Angela, *The Burning of Bridget Cleary: A True Story* (London: Pimlico, 1999).

Brannigan, John, *Brendan Behan: Cultural Nationalism and the Revisionist Writer* (Dublin: Four Courts Press, 2002).

Cave, Richard Allen, 'Friel's Dramaturgy: The Visual Dimension', in *The Cambridge Companion to Brian Friel* (ed.) Anthony Roche (Cambridge: Cambridge University Press, 2006).

Chambers, Lilian, and Eamonn Jordan (eds), *The Theatre of Martin McDonagh: A World of Savage Stories* (Dublin: Carysfort Press, 2006).

Cronin, Anthony, *Samuel Beckett: The Last Modernist* (London: HarperCollins, 1996).

Dantanus, Ulf, *Brian Friel: A Study* (London: Faber and Faber, 1988).

Deane, Seamus (ed.), *The Field Day Anthology of Irish Writing, Volume III* (Derry: Field Day, 1991).

Diamond, Elin, 'Feminist Readings of Beckett', in *Palgrave Advances in Samuel Beckett Studies* (ed.) Lois Oppenheimer (Basingstoke: Palgrave Macmillan, 2004).

Dickinson, Emily, *A Choice of Emily Dickinson's Verse*, selected by Ted Hughes (London: Faber and Faber, 1968).

Ellmann, Richard, *Four Dubliners* (London: Hamish Hamilton, 1987).

———, *Oscar Wilde* (London: Hamish Hamilton, 1987).

Esslin, Martin, *The Theatre of the Absurd* (Harmondsworth: Penguin, 3rd edn, 1980).

Etherton, Michael, *Contemporary Irish Dramatists* (London: Macmillan, 1989).

Fitz-Simon, Christopher, *The Boys: A Biography of Micheál MacLiammóir and Hilton Edwards* (Dublin: New Island, 2002).

Foley, Imelda, *The Girls in the Big Picture: Gender in Contemporary Ulster Theatre* (Belfast: The Blackstaff Press, 2003).

Foster, Roy, '"Something of us will remain": Sebastian Barry and Irish History', in *Out of History: Essays on the Writings of Sebastian Barry* (ed.) Christina Hunt Mahony (Dublin: Carysfort Press; Washington, DC: The Catholic University of America Press, 2006).

Friel, Brian, *Essays, Diaries, Interviews: 1964–1999* (ed.) Christopher Murray (London: Faber and Faber, 1999).

Garber, Marjorie, *Vested Interests: Cross-Dressing and Cultural Anxiety* (New York: Routledge, 1992).

Greene, David H., and Edward M. Stephens, *J. M. Synge 1871–1909* (New York: New York University Press, rev. edn, 1989).

Grene, Nicholas, 'Talking, Singing, Storytelling: Tom Murphy's *After Tragedy*', *Colby Quarterly*, Volume 32 Number 4 (December 1991).

———, *The Politics of Irish Drama: Plays in Context from Boucicault to Friel* (Cambridge: Cambridge University Press, 1999).

———, 'Talking It Through: *The Gigli Concert, Bailegangaire*', in *Talking about Tom Murphy* (ed.) Nicholas Grene (Dublin: Carysfort Press, 2002).

Guthrie, Tyrone, *A Life in the Theatre* (New York: Limelight Editions, 1985).

Harrington, John P., *The Irish Beckett* (Syracuse, NY: Syracuse University Press, 1991).
———, 'Samuel Beckett and the Countertradition', in *The Cambridge Companion to Twentieth-Century Irish Drama* (ed.) Shaun Richards (Cambridge: Cambridge University Press, 2004).
Harris, Claudia, 'From Pastness to Wholeness: Stewart Parker's Re-Inventing Theatre', *Colby Quarterly*, Volume 32 Number 4 (December 1991).
———, 'Reinventing Women: Charabanc Theatre Company', in *State of Play: Irish Theatre in the Nineties* (ed.) Eberhard Bort (Trier: Wissenschaftlicher Verlag, 1996).
Heaney, Seamus, *North* (London: Faber and Faber, 1975).
Holroyd, Michael, *Bernard Shaw: The One-Volume Definitive Edition* (London: Chatto and Windus, 1997).
Johnston, Jennifer, *The Gates* (London: Hamish Hamilton, 1973).
Jordan, Eamonn, *The Feast of Famine: The Plays of Frank McGuinness* (Bern: Peter Lang, 1997).
Kearney, Colbert, *The Writings of Brendan Behan* (Dublin: Gill and Macmillan, 1977).
Kearney, Eileen, 'Current Women's Voices in the Irish Theatre', *Colby Quarterly*, Volume 32 Number 4 (December 1991).
Kearney, Richard, *Myth and Motherland*, Field Day Pamphlet No. 5; see *Ireland's Field Day* (London: Hutchinson, 1985).
——— 'Language Play: Brian Friel and Ireland's Verbal Theatre', in *Brian Friel: A Casebook* (ed.) William Kerwin (New York and London: Garland, 1997).
Keenan, Brian, *An Evil Cradling* (London: Hutchinson, 1992).
Kenner, Hugh, *The Stoic Comedians: Flaubert, Joyce and Beckett* (Berkeley: University of California Press, 1962).
Keyssar, Helene, *Feminist Theatre: An Introduction to Plays of Contemporary British and American Women* (London: Macmillan, 1985).
Kiberd, Declan, 'Brian Friel's *Faith Healer*', in *Brian Friel: A Casebook* (ed.) William Kerwin (New York and London: Garland, 1997).
———, 'Theatre as Opera: *The Gigli Concert*', in *Theatre Stuff: Critical Essays on Contemporary Irish Theatre* (ed.) Eamonn Jordan (Dublin: Carysfort Press, 2000).
Kilroy, Thomas, 'Groundwork for an Irish Theatre', *Studies*, Volume 48 (1959).
———, 'Two Playwrights: Yeats and Beckett', in *Myth and Reality in Irish Literature* (ed.) Joseph Ronsley (Waterloo, Ontario: Wilfred Laurier University Press, 1977).
———, 'The Irish Writer: Self and Society 1950–1980', in *Literature and the Changing Ireland* (ed.) Peter Connolly (Gerrards Cross: Colin Smythe, 1982).
———, 'A Generation of Playwrights', *Irish University Review*, Volume 22 Number 1 (Spring/Summer 1992).
———, 'The Anglo-Irish Theatrical Imagination', *Bullán: An Irish Studies Journal*, Volume 3 Number 2 (Winter 1997/Spring 1998).
Knowlson, James (ed.), *Happy Days: Samuel Beckett's Production Notebook* (London: Faber and Faber, 1985).
———, *Damned to Fame: The Life of Samuel Beckett* (London: Bloomsbury, 1996).
Kristeva, Julia, 'Women's Time', *The Kristeva Reader* (ed.) Toril Moi (Oxford: Basil Blackwell, 1986).
Lacan, Jacques, *Écrits: A Selection* (trans.) Alan Sheridan (London and New York: Routledge, 1989).

Lalor, Kevin (ed.), *The End of Innocence: Child Abuse in Ireland* (Cork: Oak Tree Press, 2001).

Lanters, José, 'Gender and Identity in Brian Friel's *Faith Healer* and Thomas Murphy's *The Gigli Concert*', *Irish University Review*, Volume 22 Number 2 (Autumn/Winter 1992).

Leeney, Cathy and Anna McMullan (eds), *The Theatre of Marina Carr: "before rules was made"* (Dublin: Carysfort Press, 2003).

Leonard, Hugh, *Out After Dark* (London: Andre Deutsch, 1989).

Llewellyn-Jones, Margaret, *Contemporary Irish Drama and Cultural Identity* (Bristol: Intellect Books, 2002).

Lojek, Helen, 'Myth and Bonding in Frank McGuinness's *Observe the Sons of Ulster Marching Towards the Somme*', *Canadian Journal of Irish Studies*, Volume 14 Number 1 (1988).

———, *Contexts for Frank McGuinness's Drama* (Washington, DC: The Catholic University of America Press, 2004).

Lonergan, Patrick (ed.), *The Methuen Anthology of Irish Plays* (London: Methuen, 2008).

Luckhurst, Mary (ed.), *A Companion to Modern British and Irish Drama 1880–2005* (Oxford: Blackwell, 2006).

Macintosh, Fiona, *Dying Acts: Death in Ancient Greek and Modern Irish Tragic Drama* (Cork: Cork University Press, 1994).

Maguire, Tom, *Making Theatre in Northern Ireland: Through and Beyond the Troubles* (Exeter: University of Exeter Press, 2006).

Martin, Augustine, 'Christy Mahon and the Apotheosis of Loneliness', in Augustine Martin, *Bearing Witness: Essays in Anglo-Irish Literature* (ed.) Anthony Roche (Dublin: University College Dublin Press, 1996).

Maxwell, D. E. S., *A Critical History of Modern Irish Drama 1891–1980* (Cambridge: Cambridge University Press, 1984).

McCormack, W. J., *The Battle of the Books: Two Decades of Irish Cultural Debate* (Mullingar: Lilliput Press, 1986).

———, *Fool of the Family: A Life of J. M. Synge* (London: Weidenfeld and Nicolson, 2000).

McGrath, F. C., *Brian Friel's (Post)Colonial Drama: Language, Illusion and Politics* (Syracuse, NY: Syracuse University Press. 1999).

McMullan, Anna, 'Irish women playwrights since 1958', in *British and Irish Women Playwrights since 1958* (eds) Trevor R. Griffiths and Margaret Llewellyn-Jones (Buckingham: Open University Press, 1993).

Mercier, Vivian, *Beckett/Beckett* (New York: Oxford University Press, 1977).

Merriman, Vic, 'Decolonisation Postponed: The Theatre of Tiger Trash,' *Irish University Review*, Volume 29 Number 2 (Autumn/Winter, 1999).

Mikhail, E. H. (ed.), *The Abbey Theatre: Interviews and Recollections* (London: Macmillan, 1988).

Miles, Tim, 'Understanding Loyalty: The English Response to the Work of Gary Mitchell', in *Irish Theatre in England* (eds.) by Richard Cave and Ben Levitas (Dublin: Carysfort Press, 2007).

Morash, Christopher, *A History of Irish Theatre 1601–2000* (Cambridge: Cambridge University Press, 2002).

Morrison, Kristin, *Canters and Chronicles: The Use of Narrative in the Plays of Samuel Beckett and Harold Pinter* (Chicago: University of Chicago Press, 1983).

Murray, Christopher (ed.), Special Issue on Tom Murphy, *Irish University Review*, Volume 17 Number 1 (Spring 1987).
———, Introduction, Tom Murphy, *After Tragedy: Three Irish Plays* (London: Methuen, 1988).
———, *Twentieth-Century Irish Drama: Mirror up to the Nation* (Manchester: Manchester University Press, 1997).
O'Brien, Conor Cruise, *States of Ireland* (London: Hutchinson, rev. edn, 1974).
O'Brien, George, *Brian Friel* (Dublin: Gill and Macmillan, 1989).
O'Connor, Ulick, *Brendan Behan* (London: Black Swan, 1985).
O'Sullivan, Michael, *Brendan Behan: A Life* (Dublin: Blackwater Press, 1997).
O'Toole, Fintan, *Tom Murphy: The Politics of Magic* (Dublin: New Island Books; London: Nick Hern Books, rev. edn 1994).
———, *Critical Moments: Fintan O'Toole on Modern Irish Theatre* (ed.) by Julia Furay and Redmond O'Hanlon (Dublin: Carysfort Press, 2003).
Parker, Stewart, *Dramatis Personae*, John Malone Memorial Lecture (Belfast: Queen's University, 1986).
Paulin, Tom, *Liberty Tree* (London: Faber and Faber, 1983).
———, *Ireland and the English Crisis* (Newcastle-upon-Tyne: Bloodaxe Books, 1984).
Peacock, Alan (ed.), *The Achievement of Brian Friel* (Gerrards Cross: Colin Smythe, 1993).
Pine, Richard, *Brian Friel and Ireland's Drama* (London: Routledge, 1990).
———, 'Brian Friel and Contemporary Irish Drama', *Colby Quarterly*, Volume 28 Number 4 (December 1991).
———, *The Thief of Reason: Oscar Wilde and Modern Ireland* (Dublin: Gill and Macmillan, 1995).
———, *The Diviner: The Art of Brian Friel* (Dublin: University College Dublin Press, 1999).
Pinter, Harold. *Various Voices: Prose, Poetry, Politics 1948–2005* (London: Faber and Faber, 2005).
Rank, Otto, *The Double: A Psychoanalytical Study* (ed. and trans.) Harry Tucker, Jr. (Chapel Hill: Beacon Press, 1971).
Reid, Alec, 'Impact and Parable in Beckett: A First Encounter With *Not I*' 'Beckett at Eighty: A Birthday Tribute' (eds) Terence Brown and Nicholas Grene, *Hermathena*, CXLI (Winter 1986).
Richtarik, Marilynn J., *Acting Between the Lines: The Field Day Theatre Company and Irish Cultural Politics 1980–1984* (Oxford: Clarendon Press, 1994).
Roche, Anthony, 'John B. Keane – Respectability At Last!', *Theatre Ireland* 18 (April–June 1989).
———, (Guest ed.) Special Issue on Contemporary Irish Drama, *Colby Quarterly*, Volume 32 Number 4 (December 1991).
———, (ed.) Special Issue on Brian Friel, *Irish University Review*, Volume 29 Number 1 (Spring/Summer 1999).
———, (ed.) Special Issue on Thomas Kilroy, *Irish University Review*, Volume 32 Number 1 (Spring/Summer 2002).
Sihra, Melissa (ed.), *Women in Irish Drama: A Century of Authorship and Representation* (Houndmills: Palgrave Macmillan, 2007).
———, 'The Unbearable Darkness of Being: Marina Carr's *Woman and Scarecrow*,' *Irish Theatre International*, Volume 1 Number 1 (April 2008).

Simpson, Alan, *Beckett and Behan and a Theatre in Dublin* (London: Routledge and Kegan Paul, 1962).

Steiner, George, *After Babel: Aspects of Language and Translation* (London: Oxford University Press, 1975).

———, *Antigones* (Oxford: Clarendon Press, 1984).

Swift, Carolyn, *Stage by Stage* (Dublin: Poolbeg Press, 1985).

Synge, J. M., *Collected Works: Prose* (ed.) Alan Price (London: Oxford University Press, 1966; Gerrards Cross: Colin Smythe, 1982).

Szabo, Carmen, *Clearing the Ground: The Field Day Theatre Company and the Construction of Irish Identities* (Newcastle: Scholars Publishing, 2007).

Tarkovsky, André, *Sculpting in Time: Reflections on the Cinema*, translated from the Russian by Kitty Hunter-Blair (London: The Bodley Head, 1986).

Watt, Stephen, 'Love and Death: A Reconsideration of Behan and Genet', in *A Century of Irish Drama: Widening the Stage* (eds) Stephen Watt, Eileen Morgan and Shakir Mustafa (Bloomington and Indianapolis: Indiana University Press, 2000).

Whelan, Gerard, with Carolyn Swift, *Spiked: Church-State Intrigue and 'The Rose Tatoo'* (Dublin: New Island Books, 2002).

Wilcox, Angela, 'The Temple of the Lord is Ransacked', *Theatre Ireland* 8 (Winter 1984).

Williams, Caroline, Katy Hayes, Sian Quill and Clare Dowling, 'People in Glasshouse: An Anecdotal History', in *Druids, Dudes and Beauty Queens: The Changing Face of Irish Theatre* (ed.) Dermot Bolger (Dublin: New Island Books, 2001).

Wills, Clair, *That Neutral Island: A Cultural History of Ireland during the Second World War* (London: Faber and Faber, 2006).

Worth, Katharine, *The Irish Drama of Europe from Yeats to Beckett* (London: The Athlone Press, 1978).

Yeats, W. B., *The Poems: A New Edition* (ed.) Richard J. Finneran (Dublin: Gill and Macmillan, rev. edn 1983).

Index

Abbey Theatre, Dublin, 1, 2, 3, 6, 16, 17, 20, 21, 37, 42, 45, 47, 59, 61, 75, 89, 97, 130, 211, 223, 232, 244, 252, 267, 273
Aeschylus, 91
  Agamemnon, 89
Ahern, Kieran, 228
Aldrich, Robert, 237
Almeida Theatre, London, 273
All-Ireland Amateur Drama Festival, Athlone, 2, 87
Arena Stage, Washington, D.C., 174
Arnold, Matthew, 68
Astaire, Fred, 63, 66, 79
Attenborough, Michael, 202, 273
Attenborough, Sir Richard, 209

Bair, Deirdre, 262
Barry, Sebastian, 10, 11, 220–1
  Boss Grady's Boys, 231
  Hinterland, 10
  Our Lady of Sligo, 231
  Prayers of Sherkin, 231, 232, 233
  The Only True History of Lizzie Finn, 231, 233
  The Steward of Christendom, 232–6, 250
  White Woman Street, 231
Beckett, Samuel, 3, 4, 9, 10, 11, 13–41, 48, 49, 62, 67, 76–7, 80–3, 86, 88, 99, 114, 118, 133, 136, 146, 155–6, 157, 159, 179, 203, 208, 218, 224, 239, 242, 243, 244, 246, 256, 274
  All That Fall, 76
  Eleuthéria, 20
  Endgame, 63, 76–7, 115–6, 197, 207–8, 224, 240, 241
  Footfalls
  Happy Days, 76, 77, 224, 245
  Krapp's Last Tape, 58–9, 197
  Malone Dies, 198, 257
  Not I, 77, 124, 126, 127, 128, 245
  The Old Tune (after Pinget), 33
  Play, 29–30, 66–7
  Rockaby, 124, 126, 166, 240, 242, 248
  Waiting For Godot/En Attendant Godot, 3, 4, 15, 28–30, 32–6, 38–41, 63, 80–2, 141, 164, 174, 203, 209, 237, 257
  Watt
  What Where, 244
Behan, Brendan, 3, 4, 8, 13–41, 48, 69, 75, 81, 88, 159
  The Hostage/An Giall, 261
  The Landlady, 19
  The Quare Fellow, 3, 4, 20–8, 30–2, 37–41
Bell, The, 19
Berkoff, Steven, 260
The Bible, 33, 37–9, 71–2, 158, 207, 214, 255, 274
  Book of Ecclesiastes, 39
Biggar, Joseph, 272–3
Blau, Herbert, 30
Blin, Roger, 20
Blythe, Ernest, 2, 3, 15, 16, 19, 25, 45, 87, 263
Bogart, Humphrey, 98
Boltwood, Scott, 264, 265
Boyd, John, 161, 270
  The Flats, 270
Brady, Brian, 247
Brannigan, John, 263
Brecht, Bertolt
  Mother Courage and Her Children, 121
Brennan, Barbara, 260
Brennan, Bríd, 259
Brennan, Jane, 129
Bronte, Emily
  Wuthering Heights, 15
Brown, Christy, 209
Brown, Terence, 272
Burke, Angela, 224
  The Burning of Bridget Cleary, 227–8

Burke, Edmund, 51–2
Reflections on the Revolution in France, 51–2, 55–6
Bush Theatre, London, 232

Carr, Marina, 4, 220–1
Ariel, 10, 220, 221, 254–6
By the Bog of Cats, 252–4
Low in the Dark, 244–6, 256
The Mai, 246–8, 258
On Raftery's Hill, 10
Portia Coughlan, 246, 248–51, 258, 276
Ullaloo, 246
Woman and Scarecrow, 126, 246, 256–9
Carroll, Lewis, 126
Carson, Sir Edward, 200, 202
Cartmell, Selina, 259, 260
Cave, Richard Allen, 48
Cerquone, Enrica, 271
Chandler, Raymond
The Big Sleep, 98
Charabanc Theatre Company, 9, 177–9, 211
Lay Up Your Ends, 178
Oul' Delf and False Teeth, 178
Now You're Talkin', 178
Gold in the Streets, 178
The Girls in the Big Picture, 178
Somewhere Over the Balcony, 178
Chekhov, Anton, 6, 34, 47, 59, 60, 83, 123, 176
The Cherry Orchard, 60, 123
The Seagull, 8, 130
Three Sisters, 7, 47, 60
Uncle Vanya, 6, 47
Churchill, Caryl, 169
Cluchey, Rick, 30
Colgan, Michael, 261
Collins, Michael, 233, 235
Conway, Denis, 231
Conway, Frank, 97
Costello, John A., 14
Crawford, Joan, 237
Crypt Arts Centre, Dublin, 222
Curtis, Simon, 174
Cusack, Cyril, 20

Davis, Bette, 237
Day-Lewis, Daniel, 209
De Valera, Éamon, 13, 79, 221, 232, 235
Deane, Seamus, 52, 179
Deegan, Alison, 233
Dermody, Frank, 37
Devlin, Anne, 159, 168
Ourselves Alone, 9, 174–7
Devlin, Bernadette, 190, 191
Diamond, Elin, 127
Dickinson, Emily, 142, 253
Dietz, Donna, 174
Donmar Warehouse, London, 211
Donoghue, Emma
I Know My Own Heart, 244
Ladies and Gentlemen, 244
Dowling, Clare, 244
Dowling, Joe, 61
Drake, Sylvie, 34
Druid Theatre, Galway, 97, 109, 236, 238, 267
Dublin Theatre Festival, 42, 130, 160, 168, 222

Edwards, Hilton, 15, 16, 36, 273
Egan, Desmond, 4
Ellmann, Richard, 153
Empire Theatre, Belfast, 270
Esslin, Martin, 30
Etherton, Michael, 261
Euripides
Iphigenia at Aulis, 254
Medea, 89, 254

Faber and Faber, 251
Fallon, Gabriel, 18
Farquhar, George, 1, 75, 132, 145
Field Day Anthology of Irish Writing, 180
Field Day Theatre Company, Derry, 5, 6, 9, 42, 75, 76, 130, 131, 144–50, 159, 160, 179–197, 206, 207, 211, 221
Fields, Gracie, 68
Fiennes, Ralph, 265
FitzGerald, Caroline, 232
Fly By Night Theatre Company, Dublin, 222
Foster, Roy, 231
Fouere, Olwen, 247, 259–60
Frazier, Adrian, 156

Freud, Sigmund, 103, 114
Friel, Brian, 1, 2, 5–7, 9, 11, 42–46, 84, 99, 158, 206, 220, 222, 243
  Afterplay, 6, 43, 47
  Aristocrats, 34, 47, 59
  The Blind Mice, 45
  The Communication Cord, 61, 80
  Crystal and Fox, 68
  Dancing at Lughnasa, 5, 6, 42–3, 59, 61, 72, 75, 77–82, 157, 169, 213, 221–2, 223, 228, 247
  The Enemy Within, 2, 45, 46
  Faith Healer, 5, 6, 10, 32, 43, 46, 59, 61–75, 114, 221, 222, 223, 230, 232, 265
  The Francophile/A Doubtful Paradise, 45
  The Freedom of the City
  Give Me Your Answer, Do!, 6
  The Home Place, 7
  The Gold in the Sea, 43
  The Loves of Cass McGuire
  Making History, 6, 76, 137, 179
  Molly Sweeney, 6, 10
  The Mundy Scheme, 61
  Performances, 7
  Philadelphia, Here I Come!, 1, 2, 5, 6, 33, 42–3, 45, 46, 47–59, 61, 69, 77, 84, 139, 164, 169, 199, 256–9
  The Saucer of Larks, 43
  Three Sisters (after Chekhov), 7
  Translations, 5, 6, 9, 25, 42, 46, 59, 61, 75, 79, 80, 179–89, 193, 198, 272
  Uncle Vanya (after Chekhov), 6
  Volunteers, 47
  Wonderful Tennessee, 5, 10, 61, 79–83, 222
Fugard, Athol
  The Island, 194

Gaiety Theatre, Dublin, 15, 20–1
The Gallery Press, 251
Gandhi, Mahatma, 209
Garber, Marjorie, 149–50
Gas Company Theatre, Dublin, 21
Gate Theatre, Dublin, 15, 20, 21, 206, 223, 232, 250, 265, 273
Giacometti, Alberto, 155
Gielgud, Sir John, 47
Glasshouse Productions, Dublin, 244
Goldsmith, Oliver, 1
  She Stoops To Conquer, 33
Gore-Booth, Constance, 193
Grant, David, 244
Greene, Graham, 6, 43
Gregory, Augusta Lady, 71, 111
  Cathleen ni Houlihan, 25, 111
  The Workhouse Ward, 237
Grene, Nicholas, 96, 123, 151, 157, 221, 267, 268
Group Theatre, Belfast, 45
Guthrie, Sir Tyrone, 46–7

Halpin, Mary
  Semi-Private, 168
Hammett, Dashiell
  The Maltese Falcon, 98
Hammond, David, 179
Hampstead Theatre Club, London, 42, 202, 207
Hardy, Thomas, 117
Harris, Mitchell, 192
Hayes, Trudy, 244
Heaney, Seamus, 9, 42, 75, 179, 265
Hickey, Tom, 139
Hope, James, 272
Hubbard, L. Ron, 100
Hughes, Declan, 139
Hume, John, 191
Hunt, Hugh, 3, 87
Hurd, Douglas, 192
Huston, John, 98
Hyde, Douglas
  Casadh an tSúgáin/The Twisting of the Rope, 19
Hynes, Garry, 88, 96, 97, 109, 225–6, 228, 229, 231, 236, 237, 238

Ibsen, Henrik, 167
  A Doll's House, 174
  Ghosts, 131
Iremonger, Valentin, 16

Jebb, Richard, 193
John Player Theatre, Dublin, 160
Johnston, Jennifer
  The Gates, 70, 265

Jones, Marie, 178–9
  The Hamster Wheel, 178
  A Night in November, 179
  Stones in His Pockets, 179
  Women on the Verge of HRT, 179
Jordan, Neil, 245
  Angel, 161
Joyce, James, 146, 273
  Finnegans Wake, 70
  A Portrait of the Artist as a Young Man, 2, 133, 223
  Ulysses, 26, 89

Kani, John
  The Island, 194
Kavanagh, John, 61
Keane, John B., 2, 5, 15
  Sive, 2
Kearney, Richard, 195
Keenan, Brian, 207, 209
Kennedy, Jimmy, 160
Kennelly, Brendan
  Antigone (after Sophocles), 189
Kenner, Hugh, 139
Kent, Jonathan, 265
Kiberd, Declan, 37, 70, 267
Kilroy, Thomas, 2, 5, 8, 9, 121–2, 130–157, 179, 220, 242, 243
  The Big Chapel, 130
  Blake, 131, 154
  The Death and Resurrection of Mr. Roche, 8, 130, 133–6, 138
  Double Cross, 8, 130, 144–7, 180, 211
  Ghosts (after Ibsen), 131
  The Madame MacAdam Travelling Theatre, 8, 131, 147–50, 154, 180
  The O'Neill, 8, 130, 133, 137–8
  The Seagull (after Chekhov), 8, 130
  The Secret Fall of Constance Wilde, 8, 131, 150–4
  The Shape of Metal, 8, 131, 154–7, 273
  Talbot's Box, 8, 130, 139, 141–4
  Tea and Sex and Shakespeare, 8, 130, 131, 139–40, 154
  That Man Bracken, 147
King, Martin Luther, 190
The Koran, 207
Kristeva, Julia, 170, 173–4

Lacan, Jacques, 127
Lally, Mick, 238
Lambert, Mark, 228
Lefevre, Robin, 59
Leonard, Hugh, 2, 3, 5, 45
  The Au Pair Man, 2
  Da, 164, 199
  Out After Dark, 15
  The Patrick Pearse Motel, 2
  Stephen D (after Joyce), 2
The Listener, 191
Littlewood, Joan, 17, 21, 36, 37, 69, 263
Liverpool Playhouse, 174
Llewellyn-Jones, Margaret, 261
Loane, Tim, 9, 211
Logue, Christopher, 18
Lojek, Helen, 273
Lonergan, Patrick, 221, 225–6
Lynch, Martin, 178
Lyric Theatre, Belfast, 160, 168, 270

Mac Anna Tomás, 3, 87
Macintosh, Fiona, 250, 253
MacIntyre, Tom, 139
Macklin, Charles, 59
MacLiammóir, Micheál, 15, 16, 20, 36, 273
Madonna, 209
Mahon, Derek, 9, 197
Mamet, David, 238, 239
  American Buffalo, 237
Manahan, Anna, 238, 240–1
Martin, Gus, 252
Mason, James, 61
Mason, Patrick, 75, 130, 131, 139, 160, 205, 211, 252, 267, 273
Mathews, Aidan Carl
  Antigone (after Sophocles), 189
McCabe, Eugene
  Swift, 47
McCann, Donal, 61, 232
McCarthy, John, 207
McCormack, W.J., 146
McCowen, Alec, 207
McCracken, Henry Joy, 272
McDiarmid, Ian, 265

McDonagh, Martin, 10, 11, 220–1, 236–9, 275
The Beauty Queen of Leenane, 236, 238, 240–2
The Lonesome West, 240, 243
A Skull in Connemara, 237, 239, 240
McGinley, Sean, 225–6, 231
McGowan, Shane (The Pogues), 236
McGrath, F.C., 265
McGuinness, Frank, 4, 5, 9, 61, 62, 158, 159, 179, 251, 253
The Bird Sanctuary, 273
The Bread Man, 206
Carthaginians, 35, 206
Dolly West's Kitchen, 273
Gates of Gold, 273
Innocence, 206
Mary and Lizzie, 206
Mutabilitie, 273
Observe the Sons of the Ulster Marching Towards the Somme, 9, 197–206, 211, 234
Someone Who'll Watch Over Me, 9, 207–11
Speaking With Magpies, 273
There Came a Gypsy Riding, 273
McHugh, Roger, 16
McKenna, Siobhán, 7
McMaster, Anew, 36
McPherson, Conor, 10, 179, 220–4, 274
Come On Over, 227
The Seafarer, 224
St. Nicholas, 224
The Weir, 10, 110, 223–31
This Lime-Tree Bower, 222
Mercier, Vivian, 34
Merriman, Brian
The Midnight Court/Cúirt an Mheán Oídhche, 19
Merriman, Vic, 221, 275
Methven, Eleanor, 178
Miller, Arthur, 13
Milligan, Spike, 13
Mitchell, Gary, 159, 274
As the Beast Sleeps, 9, 214–20, 228
In a Little World of Our Own, 9, 211–14, 218
Independent Voice, 211
Mooney, Ria, 46
Moore, Carol Scanlan, 178
Moore, Julianne, 245
Morash, Chris, 11
Morrison, Kristin, 114
Mullan, Marie, 238
Murphy, Tom, 2, 3, 5, 7–8, 61, 84–129, 220, 243, 266
Alice Trilogy, 8, 10, 88, 124–9, 268
Bailegangaire, 7, 84, 87, 109–22, 124, 127, 154, 167, 237, 267
The Blue Macushla, 99
Conversations on a Homcoming, 85, 96, 97, 125
A Crucial Week in the Life of a Grocer's Assistant, 87, 102
Famine, 3, 87
The Gigli Concert, 7, 84, 85, 88, 96–109
The House, 7, 123–4
The Informer (after O'Flaherty), 99
The Morning After Optimism, 3, 7, 85, 87, 96, 98
Too Late for Logic, 7, 122
On the Outside, 5, 85–6, 87
The Sanctuary Lamp, 7, 84, 85, 98, 99, 125
The Seduction of Morality, 7, 122
A Thief of a Christmas, 115, 267
The Wake, 7, 122–3
A Whistle in the Dark, 2, 7, 87, 88–96, 109, 110, 125, 212, 213
The White House, 3
Murray, Christopher, 82–3, 266
My Left Foot, 209

National Theatre, London, 42, 179, 221, 273
The New Yorker, 43, 44
Norton, Jim, 231
Ntshona, Winston
The Island, 194

O'Briain, Sean, 14
O'Brien, Conor Cruise, 190–4
Salome and the Wild Man, 193
O'Brien, Eugene
Eden, 223
O'Brien, Flann, 112
O'Brien, George, 44
The Observer, 191

O'Casey, Sean, 33, 37, 42, 60, 142, 182, 274
The End of the Beginning, 215
Juno and the Paycock, 34, 134
The Plough and the Stars, 1, 16, 175, 205
Red Roses For Me, 142–3
The Silver Tassie, 205
O'Connell, Daniel, 60, 183
O'Connor, Frank, 15, 44
O'Connor, Ulick, 17, 262
O'Donoghue, Noel, 5, 84, 86, 87
O'Faolain, Sean, 18, 44, 131
O'Flaherty, Liam
The Informer, 99
O'Hara, Joan, 233
Olivier, Sir Laurence, 46, 72
Olympia Theatre, Dublin, 2, 15, 21, 160
O'Neill, Eugene, 97
Operating Theatre Company, Dublin, 260
O'Reilly, Genevieve, 225–6
O'Rowe, Mark, 10
Howie the Rookie, 223
Osborne, John
Look Back in Anger, 13
The Entertainer, 72
O'Toole, Fintan, 84, 87, 192, 234, 238, 267
Out of Joint Theatre Company, London, 232

Pacino, Al, 237
Paisley, Reverend Ian, 192
Parker, Lynne, 160
Parker, Stewart, 8, 197
Catchpenny Twist, 161
Heavenly Bodies, 160
Lost Belongings, 159
Northern Star, 75, 160, 161, 211, 272
Pentecost, 8, 9, 159–68, 169, 180, 206
Spokesong, 160
Paulin, Tom, 9, 179, 189–90, 197, 272–3
The Riot Act (after Sophocles), 9, 180, 189–97
Peacock Theatre, Dublin, 3, 16, 130, 206, 211, 232, 246, 250, 252, 259
Pearse, Patrick, 205
Pearson, Noel, 75, 22
Peckinpah, Sam, 209
Pierrepoint, Albert, 26
Pike Theatre, Dublin, 3, 13–41
Pine, Richard, 1, 43, 59, 150, 261, 272
Pinget, Robert
La Manivelle/The Old Tune, 33
Pinter, Harold, 10, 36, 110, 114, 222, 229, 237, 238, 239, 265
The Homecoming, 212, 230
Pope John Paul II, 128

Queen's Theatre, Dublin, 16, 46
Quintero, José, 61

Radio Telefís Éireann, 237
Rank Otto, 105
Rea, Stephen, 5, 75, 77, 130, 160, 161, 179, 207, 211
Red Kettle Theatre Company, Waterford, 131
Redmond, John, 234
Reid, Christina, 159, 161
Tea in a China Cup, 8, 168–74
Reid, J. Graham, 161
Richardson, Sir Ralph, 46
Rickson, Ian, 228, 231
Riverdance, 78
Robinson, Mary, 6
Roche, Billy, 179
Roddy, Lalor, 9, 211
Rogers, Ginger, 79
Rough Magic Theatre Company, Dublin, 131, 139, 160
Royal Court Theatre, London, 130, 144, 169, 174, 223, 224, 228, 232, 238, 259
Royal Shakespeare Company, Stratford, 206, 273

San Francisco Actors' Workshop, 30
Schopenhauer, Arthur, 122
Shaffer, Peter
Equus, 103

Shakespeare, William, 15, 16, 273
Hamlet, 47, 57, 96, 257–8
Henry IV, Part One, 180–2, 184–5, 187–9
Henry V, 182
King Lear, 109, 233
Macbeth, 73, 89, 258
Othello, 140
The Tempest, 140. 193
The Winter's Tale, 140
Shaw, Fiona, 259
Shaw, George Bernard, 1, 131
John Bull's Other Island, 105–6, 184
Pygmalion, 49–50
Shepard, Sam, 110
Sheridan, Richard Brinsley, 1
Sihra, Melissa, 270, 276
Simon, Neil
The Odd Couple, 237
Simpson, Alan, 3, 13–41, 262
Sophocles
Antigone, 180, 189–97, 252
Oedipus the King, 52
Spenser, Edmund, 273
Stafford, Maeliosa, 238
Stafford-Clark, Max. 232
Steiner, George, 180, 189–97, 272
Stevenson, Juliet, 124
Stewart Parker Trust/Award, 160, 211
Stoppard, Tom
Rosencrantz and Guildenstern Are Dead, 181
Stove, Betty, 210
Strachan, Kathy, 248
Stuart, Francis, 146
Studies, 130
Swift, Carolyn, 3, 13–41, 262
Synge, J.M., 8, 14, 34, 47, 71, 72, 112, 159, 236, 238, 239, 242, 251
The Aran Islands, 113–14, 267
Deirdre of the Sorrows, 70
The Playboy of the Western World, 1, 12, 24–5, 51, 70, 96, 114, 133, 237, 252
Riders to the Sea, 35, 111–12
Shadow of the Glen, 246
The Well of the Saints
When the Moon Has Set, 71

Taibhdhearc Theatre, Galway
Tarkovsky, André, 38
Thatcher, Margaret, 207
Théatre de Babylone, Paris, 18
Thompson, Sam, 161, 270
Over the Bridge, 270
Times Literary Supplement, 191
Tinderbox Theatre Company, Belfast, 9, 211
Travers, Sally, 20
Tricycle Theatre, London, 211
Turgenev, Ivan, 59

Vanek, Joe, 75
Virgil
The Aeneid, 188

Wade, Virginia, 210
Walcott, Derek
Viva Detroit, 34
Waters, Les, 174
Watt, Stephen, 263
Whatever Happened to Baby Jane?, 237
Williams, Tennessee, 13, 16, 17
The Rose Tatoo, 41
Wilcox, Angela, 204
Wilde, Oscar, 1, 8, 131, 145, 150–3
An Ideal Husband, 151
The Importance of Being Earnest, 140, 145
Salomé, 260
Wyler, William, 15

Yeats, W.B., 34, 60, 69, 71, 193, 196, 255
Cathleen ni Houlihan, 25, 111
The Dreaming of the Bones, 73
On Baile's Strand, 33
Purgatory, 71
Young Vic, London, 168